AF838016

Misbuilding Brains

Misbuilding Brains

Biological Causes of Neurodevelopmental Divergence and Disorders

David J. Price
University of Edinburgh
United Kingdom

WILEY

This edition first published 2027

Registered Offices
John Wiley & Sons, Inc., 111 River Street, Hoboken, NJ 07030, USA
John Wiley & Sons Ltd, New Era House, 8 Oldlands Way, Bognor Regis, West Sussex, PO22 9NQ, UK

For details of our global editorial offices, customer services, and more information about Wiley products visit us at www.wiley.com.

The manufacturer's authorized representative according to the EU General Product Safety Regulation is Wiley-VCH GmbH, Boschstr. 12, 69469 Weinheim, Germany, e-mail: Product_Safety@wiley.com.

Wiley also publishes its books in a variety of electronic formats and by print-on-demand. Some content that appears in standard print versions of this book may not be available in other formats.

Library of Congress Cataloging-in-Publication Data
Names: Price, David J., 1957 December 20– author
Title: Misbuilding brains : biological causes of neurodevelopmental divergence and disorders / David J. Price.
Description: Hoboken, NJ : Wiley, 2027. | Includes bibliographical references and index.
Identifiers: LCCN 2026003550 (print) | LCCN 2026003551 (ebook) | ISBN 9781394164738 paperback | ISBN 9781394164745 adobe pdf | ISBN 9781394164752 epub
Subjects: LCSH: Developmental neurobiology | Developmental disabilities–Etiology | Nervous system–Abnormalities | Nervous system–Variation | Brain–Abnormalities | Brain–Variation
Classification: LCC QP363.5 .P75 2026 (print) | LCC QP363.5 (ebook)
LC record available at https://lccn.loc.gov/2026003550
LC ebook record available at https://lccn.loc.gov/2026003551

Cover Design: Wiley
Cover Image: © Modified versions of a graphical representation of data from developing human brain cells, generated by Dr. Wai KitChan, University of Edinburgh, UK

Set in 9.5/12.5pt STIXTwoText by Straive, Pondicherry, India
Printed and bound by CPI Group (UK) Ltd, Croydon, CR0 4YY

C9781394164738_080826

Contents

Preface

In a previous book, *Building Brains; An Introduction to Neural Development*, Andrew Jarman, John Mason, Peter Kind and I outlined our current understanding of brain development based on studies in a wide range of organisms, from worms and flies to mammals including humans. It seemed only natural to follow that account with one focused specifically on our current understanding of *human* brain development and to include a discussion of the causes of neurodivergent conditions and neurodevelopmental disorders.

I have tried to make *Misbuilding Brains* accessible to non-specialists. For those readers with little background in biology, I have explained the relevant principles of how genes and cells work and minimized my use of specialist terms and abbreviations, which I have defined and limited to those I think are useful or essential. If you fall into this category of reader, I hope you will not be put off by finding that some elements of the book contain a lot of detail, which is relegated mainly to the figures and tables. These are intended for more expert readers who might want to dig deeper into specific topics. You should be able to gloss over these details and still follow the key points that I make in the text. For those who already have a grounding in topics discussed in the book, I hope you will find the breadth and detail of coverage offer valuable new information and perspectives or suggest new avenues to explore.

On a personal level, it has been a fulfilling experience to draw together many of the elements of my own training and experience that have fascinated and inspired me over many years. These range from the management and treatment of patients to research into the ways in which genes work and combine with environmental factors to control brain development. I started my career in clinical medicine before moving into basic research on the neurophysiology and molecular genetics of brain development. I am privileged to have had many outstanding and inspiring scientists and clinicians around the world as my teachers, mentors and collaborators and to have worked with many wonderful research students and postdoctoral scientists in my own laboratory. I am grateful to you all; I have learned so much from each and every one of you. A very special heartfelt thanks goes to Professor Norah Spears, my most effective, patient and constructive critic, for spending hours reading and discussing the manuscript with me.

I hope you will enjoy reading the book as much as I have enjoyed writing it. I apologize in advance for all the inevitable errors, inconsistencies and unclear explanations. Please let me know about them.

September 2025

David J. Price
Edinburgh, UK

Glossary

Terms used throughout the book that might be unfamiliar, especially for non-experts, are listed here. In every chapter in which a term in this list is used, it is shown in bold on first appearance in either the text or in the figure captions. Terms shown in bold in the explanations that follow have their own entry in the list.

Action potential Electrical impulse travelling along an axon, allowing nerve cell communication.
Adenosine triphosphate (ATP) Primary energy-carrying molecule in cells; it can have other functions, for instance as a **neurotransmitter**.
Agenesis Absence or incomplete development of an organ or tissue.
Allele A variant form of the nucleotide sequence at a specific location on a DNA molecule.
AMPA (α-amino-3-hydroxy-5-methyl-4-isoxazolepropionic acid) receptor **Ionotropic receptor** for **glutamate** involved in fast synaptic transmission in the central nervous system.
Anencephaly Severe birth defect where a major part of the brain and skull is missing.
Apoptosis A process by which cells kill themselves by triggering an internal systematic programme of self-destruction.
Astrocyte A type of glial cell in the brain and spinal cord shaped a bit like a star.
Ataxia Lack of muscle coordination affecting voluntary movements.
Atrophy Wasting away or reduction in size of a body part or tissue.
Autoimmunity When the immune system mistakenly attacks the body's own cells.
Autophagy Cellular process of degrading and recycling damaged organelles or proteins.
Autosome Any chromosome that is not a **sex chromosome**.
Basal ganglia Groups of cells at the base of the brain involved in movement control and other functions.
Cell cycle Ordered series of events for cell division and replication.
Centrioles Structures involved in cell division that form **centrosomes**.
Centromere Chromosome region linking **sister chromatids** during cell division.
Centrosome Organelle serving as the main microtubule-organizing centre in cells.
Cerebellar ataxia Disorder of movement coordination caused by cerebellar malfunction.
Chorea Involuntary, rapid and irregular movements.

Chromatin Complex of DNA and protein.

Codon Sequence of three nucleotides in mRNA coding for an amino acid; different codons code for different amino acids.

Copy number variation (CNV) A difference in the number of copies of a particular DNA segment between the two chromosomes in a pair or between the chromosomes of one person and another.

Corpus callosum A nerve fibre bundle connecting the two brain hemispheres.

Cytoskeleton A network of molecules, many of which are long polymers, providing structural support for cells and facilitating their movement.

De novo mutation Genetic alteration appearing for the first time as a new mutation.

Dendrite Branching extensions of neurons that receive signals from other neurons.

Depolarization A reduction in the potential difference across the cell membrane, crucial for **action potential** initiation.

Diencephalon Brain region containing the **thalamus**, **hypothalamus** and other structures, involved in sensory and other functions.

Domain (protein) A distinct functional and structural unit within a protein.

Dominant (allele) An **allele** that expresses its phenotype, overshadowing a **recessive** allele in **heterozygotes**.

Ectoderm Outer embryonic layer forming the nervous system and skin.

Enhancer DNA sequence increasing gene **transcription** levels.

Epigenetics The study of how gene expression is changed without altering DNA sequences.

Excitatory (neuron, synapse, **neurotransmitter** and so on**)** Neuron or other object increasing the likelihood of generating an **action potential** in other neurons.

Exencephaly Severe **neural tube** defect where the brain is formed outside the skull.

Exome The parts of the genome composed of exons, coding for proteins.

Extracellular matrix Network of proteins and polysaccharides around cells supporting their adhesion, intercellular signalling and other functions.

Forebrain Most anterior part of the brain, comprising the cerebral cortex, **basal ganglia**, **thalamus** and other structures.

GABA (γ-aminobutyric acid) Major **inhibitory neurotransmitter** in the central nervous system.

Ganglionic eminences Transitory **forebrain** structures adjacent to the embryonic cerebral cortex in the region that will become the **basal ganglia.**

Genome wide association study (GWAS) An epidemiological study that identifies links between genetic variants and traits.

Germ cell Cells, including sperm and eggs, that contain one of each pair of chromosomes.

Glutamate Major **excitatory neurotransmitter** in the brain.

Growth cone A dynamic, motile structure at neuronal extensions guiding axon growth.

Heritability The proportion of phenotypic variation in a population attributable to genetic variation.

Heterotopia Collections of displaced cells.

Heterozygous Having two different **alleles** at a particular location on each of a pair of chromosomes.

Hippocampus Part of the **forebrain** adjacent to the cerebral neocortex involved in memory and other functions.
Histone Protein around which DNA winds, forming **chromatin**.
Holoprosencephaly Developmental disorder causing improper **forebrain** division into hemispheres.
Homozygous Having identical **alleles** at a particular location on each of a pair of chromosomes.
Hydrocephalus Condition involving fluid accumulation in the brain's **ventricles**, causing pressure.
Hypoplasia Underdevelopment or incomplete development of a tissue or organ.
Hypothalamus Brain region linking nervous and endocrine systems, involved in involuntary control of many body parameters, such as temperature, heart rate and blood pressure.
Hypotonia Decreased muscle tone leading to floppy limbs and reduced strength.
Inhibitory (neuron, synapse, **neurotransmitter** and so on**)** Neuron or other object reducing the likelihood of an **action potential** being generated in other neurons.
Ion channel Protein allowing specific ions to flow across a cell's membrane, thereby changing the potential difference across it.
Ion pump Protein actively transporting ions across membranes, maintaining concentration gradients.
Ionotropic receptor A receptor that is also an **ion channel**, which opens when its **neurotransmitter** binds.
Karyotyping Method of identifying the number and structure of an individual's chromosomes.
Leukodystrophies A group of disorders affecting the brain's white matter.
Lineage (cell) Sequence of cell divisions tracing a cell's ancestry.
Linkage (genetic) Tendency of genes close together on a chromosome to be inherited together.
Lissencephaly Brain malformation characterized by smooth brain surface.
Megalencephaly An abnormally large brain.
Membrane potential Difference in electrical charge across a cell membrane.
Metabotropic receptor Receptor linked to a signalling pathway, affecting cell metabolism.
Micrencephaly An abnormally small brain.
Microglia Resident immune cells in the central nervous system.
Mitotic spindle Structure separating chromosomes during cell division.
Monozygotic twins Twins from a single fertilized egg.
Myelin Insulating layers around axons that speed up **action potential** transmission.
Neural induction The process by which neural tissue is specified during early embryogenesis.
Neural plate Embryonic structure that develops into the central nervous system (made of **neuroectoderm**).
Neural tube Embryonic precursor of the central nervous system, formed by folding of the **neural plate**.
Neuroectoderm Embryonic tissue giving rise to the central nervous system.

Neurotransmitter Chemical messenger transmitting signals across synapses between neurons.

NMDA (N-methyl-D-aspartate) receptor Subtype of **glutamate** receptor crucial for **synaptic plasticity**.

Non-coding RNA (ncRNA) RNA molecule whose functions do not involve **translation** into protein.

Nucleosomes Basic units of **chromatin**, comprising DNA segments wrapped around **histones**.

Oligodendrocyte Glial cell forming **myelin** sheaths around axons in the central nervous system.

Pachygyria Condition with abnormally broad and flat cerebral cortical folds.

Patterning Spatial and temporal regulation of gene expression during development, shaping cell fate.

Penetrance The proportion of individuals carrying a particular genotype who express the associated phenotype.

Polymorphism (genetic) Genetic variant present in more than 1% of a population (can be a **single nucleotide polymorphism** or larger).

Porencephaly Condition in which the brain develops cystic cavities.

Postsynaptic Refers to the receiving side of a synapse.

Precursor cell Undifferentiated cell capable of differentiating into a specific cell type without dividing again.

Presynaptic Refers to the transmitting side of a synapse.

Progenitor cell A cell that will divide at least once more.

Promoter DNA region that binds specific proteins to initiate gene **transcription**.

Prosencephalon Another word for **forebrain**.

Proteostasis Maintenance of protein balance in cells through synthesis, folding and degradation.

Radial glial progenitor A **progenitor** that also helps guide neuronal migration during brain development; not a glial cell, despite the name.

Reactive oxygen species Chemically reactive molecules containing oxygen that have the potential to damage cells.

Recessive (allele) **Allele** whose effects on phenotypes are only seen when both chromosomes carry the same allele and are, therefore, not overshadowed by the presence of a **dominant** allele.

Regulatory element DNA sequence controlling gene **transcription** levels.

Schizencephaly Rare brain malformation causing clefts or splits in the brain.

Second messenger Molecules mediating a cell's internal responses to **signalling molecules** arriving at the cell surface; each second messenger cascades its responses through chains of target molecules within the cell.

Sex chromosome X and Y chromosomes in humans; primary determinants of biological sex.

Signal transduction The conversion of one type of signal to another, such as the process by which a signal initiated by **signalling molecules** outside the cell is converted into a response inside.

Signalling molecules Chemicals released by cells to influence others.

Single nucleotide polymorphism (SNP) Genetic variation at a single nucleotide position in the genome present in more than 1% of the population.
Sister chromatids Two copies of a chromosome connected at the **centromere** during cell division.
Somatic cell Any cell in the body except the **germ cells**.
Splicing (of RNA) Process removing introns from pre-mRNA and joining exons, often in multiple different configurations, to make mature mRNAs (messenger RNAs); a single gene can usually make multiple mRNAs that differ in which exons are included (splice variants).
Structural protein Protein providing support and shape to cells and tissues.
Synaptic plasticity Ability of synapses to strengthen or weaken over time and with experience.
Telencephalon A part of the **forebrain** (or **prosencephalon**) that develops into the cerebral cortex and **basal ganglia**.
Thalamus Brain structure situated deep inside the mature brain through which much neural information passes.
Transcription factor A protein that regulates gene **transcription** by binding to relatively specific DNA sequences.
Transcription The process of copying a DNA gene sequence into messenger RNA (mRNA).
Translation The process of synthesizing proteins from messenger RNA (mRNA) sequences.
Trinucleotide repeat expansion Increase in the number of times a specific nucleotide triplet is repeated sequentially at a genomic locus.
Trisomy Conditions in which there is an extra chromosome.
Ventricles Fluid-filled cavities in the brain through which cerebrospinal fluid circulates.
Ventriculomegaly Enlargement of brain **ventricles**, not necessarily due to increased pressure inside them.
X-linked Traits that are influenced by genes on the X chromosome.
Y-linked Traits that are influenced by genes on the Y chromosome.
Zygote The cell that results from the union of an egg with a sperm.

Introduction

All brains are misbuilt in the sense that they contain errors of some sort. Most of these errors do not matter to the individual carrying them and will languish undetected during their lifetime. If errors pass to subsequent generations, they might become important and, in the long term, they might contribute to gradual evolutionary change. In the short term, some errors might actually bring benefits to an individual by enhancing the functions of their brain in some way, although in most cases this would be almost impossible to prove. Only a small minority of errors affect people in a detectably and incontrovertibly negative way during their lifetime. For understandable reasons, however, it is to them that we pay most attention. In rare cases, they cause severe problems with fundamental aspects of brain function, hindering a person's ability to walk, talk, eat and so on, but more often they have milder effects that might be difficult to identify, describe and define as 'normal' or not. There is no gold-standard error-free brain against which to compare all other brains and yet the way we discuss our brains is often based on some ill-defined notion of normality. Even taking normal to mean a range, it can be very hard – or currently even impossible – to define its limits, especially when it comes to complex aspects of our perception, thinking and behaviour.

A more neutral and perhaps more useful way to look at all of this is to say that all our brains *differ*, with some differing more than others. From the moment the first brain cells make their appearance in the young embryo, we can find differences from embryo to embryo if we look hard enough. As development progresses before and then after birth, the differences become increasingly obvious. The exact causes of most of them are virtually impossible to nail down. In part, this is because many arise simply from random variation in the developmental processes that make our brains. Variation in our brain cells' genetic material and environment also contribute, but exactly how they do so is often hard to tell. This state of affairs is likely to persist for some time to come because the sheer complexity of the mechanisms that create the human brain makes them challenging to study and ethical considerations severely limit the research that we can do on ourselves.

Despite these challenges, we have now reached a point in our understanding where we can explain how some of the largest and most obvious differences among human brains arise and how they affect the ways in which individuals think and act. In this book, I will outline what we know about human brain development and the origins of its diversity, how genetic and non-genetic factors control and influence developmental processes and how some brains become so different that we categorize them as either divergent or disordered. I will explain how we have obtained our knowledge and how we are trying to find out more. I will expose the current gaps in our understanding and discuss how we might fill them in the future.

All Brains Differ

We can safely assume that all human brains *must* be different to some extent. Why do I say this? After all, the brains of most adult humans look much the same, weigh about the same (around one and a half kilograms) and their surfaces fold with similar patterns. Moreover, we can subdivide brains into regions responsible for functions such as sensory perception, the control of movement and so on, that occupy consistent locations and that connect and interact with each other in consistent ways. For example, the brain's grey matter contains nerve cells arranged in characteristic patterns interconnected via stereotypically organized bundles of nerve fibres; damage to the same part of the brains of any two people would cause them similar problems. But, notwithstanding these and many other properties that our brains share, a deep dive into the fine structure of our brains and the workings of their microscopic components would surely identify many important differences between them all.

You might ask whether we can absolutely rule out the possibility that identical brains exist, or have ever existed. We cannot. We do not have the technology to do this. We have no way of analysing every detail of a brain's microscopic structure and how all its elements work. Our brains contain somewhere in the region of 100 billion nerve cells interlinked in a network of about 100 trillion interconnections. We cannot even count the cells and their interconnections precisely, let alone analyse the fine details of them all. Nevertheless, it seems safe to assume that the possibility that identical brains exist, or have ever existed, is so vanishingly small that we can ignore it for all practical purposes.

It might be worth explaining in a bit more detail why I am so confident this is the case. First, the development of any organ or organism involves extensive interactions between our genetic material (in the form of deoxyribonucleic acid, or DNA) and its environment, neither of which are entirely reproducible from one developing individual to the next. But even if they were identical, it is extremely unlikely that they would create identical brains because of the nature of developmental processes and their control. The mechanisms that build our brains are *probabilistic* in nature. By this, I mean that they have evolved to regulate the probability – or the likelihood if you prefer – that the outcomes at every step in the process fall within a range. They do not provide exact certainty of outcome.

The events that conception triggers amount, essentially, to an enormous biochemical chain reaction that is controlled tightly enough to produce similar numbers of cells of consistent types at similar times and locations and doing similar things at every step of the way from the fertilized egg onwards. At any given time, the reactions are constrained within limits by factors such as the availability of molecules encoded by the genome, energy levels within the cells and the nature of the cells' environment. These factors provide the conditions in which the reactions take place, affecting their types, rates and efficiencies. The DNA's molecular environment is constantly updating as development proceeds, and this feeds back to the DNA to alter which bits of it are active, which in turn affects the molecular environment, and so on around and around it goes. If all goes well, and the chain reaction unfolds uninterrupted and with sufficient stability, it will lead inexorably to the reproducible generation of living structures, including brains, with similar features. But those features can never be identical, especially for structures as complex as brains, because at each step of the way all the vast numbers of biochemical reactions, and hence the cells and structures they build, will show at least slight, often random, variation.

These variations will increase enormously as the new individual increases their ability to interact with the world around them, both before but particularly after birth. It is impossible to imagine how evolution could have produced mechanisms capable of providing absolute certainty of outcome at all stages of development and hence identical products. While manufacturers might be able to generate computers that are effectively identical in high-tech, tightly controlled environments, the likelihood of nature evolving the ability to produce identical brains is essentially zero. So, for anyone intent on using biological cloning to recreate themselves or a loved one, I suggest you forget it!

The consequence of developmental variation is to give each of us our unique mental abilities and personalities. When we consider our brains at this holistic, functional level, it becomes easier to appreciate that the physical nature of our brains must differ in significant ways between us all. Our personalities and mental abilities are manifestations of the vast, highly organized network of electrically and biochemically active cells and connections in our brains. It follows, therefore, that variation in our personalities and mental abilities must be due to variation in the exact molecular, cellular and structural nature of each of our brains. Although this much is clear, exactly how variation in our brains' molecular, cellular and structural nature affects our mental abilities and personalities is much less so.

Brains Have Key Functions in Common

The mechanisms constraining the likely outcomes of the developmental processes that create our brains lead to similarities in the ways in which most of our brains work to create our perceptions, thought processes and behaviours. Some of those similarities are what makes us human: for example, our brains differ from those of all other extant species by their universal potential to develop complex language skills, dextrous movements and complex tools. Humans across all cultures share many cognitive and behavioural traits, which psychologists often refer to as psychological universals or core mental attributes. These are often described under broad headings, such as cultural universals, which include the creation or endorsement of commonly held beliefs (sometimes called myths), social norms and enforced rules; linguistic universals such as grammar; social universals such as ethnocentricity (which means judging the attributes of ethnic groups by the standards of others, usually one's own). It is because most of us share similar ways of thinking and behaving that we have been able to create large societies held together by shared beliefs and conventions, agreed rules and laws and so on. Indeed, our evolution has depended on our ability to work collectively to compensate for our disadvantages as individuals in attributes such as strength and speed. This leads societies to develop and accept norms against which they can, and often do, compare the thinking and behaviour of individuals.

One common feature of our brains, which might also distinguish us from many other highly evolved species with complex brains, is our tendency to suffer a high incidence of mental health problems. It is hard to say whether this is a true difference or one we perceive because we detect mental health issues more easily in ourselves than in other species, with whom communication is limited at best. It is possible that our brain's tendency to manifest mental health problems arises as a side effect of the way that it has evolved to give us our cognitive potential. Researchers have noticed that brain structures that are particularly highly evolved in humans, such as the frontal regions for example, are those often implicated in mental health problems. Our brain has evolved relatively very rapidly, over just a few million years, and yet our genomes are remarkably similar to those of many other mammalian species with much smaller and simpler brains. It is possible that the essential changes that allowed our brains to evolve did so by loosening developmental constraints, thereby increasing the ranges of possible outcomes that developmental mechanisms could produce. This expansion of possibilities might have opened the way to more rapid change across the population by allowing natural selection to operate on a more variable set of phenotypes, but might have come at the cost of increasing an individual's risk of developing as neurodivergent or disordered. The very same genetic changes that allowed our brains to evolve so rapidly might have had the unfortunate effect of making them more liable to experience mental health problems.

The Language We Use to Discuss Brain Diversity Matters

Many brains develop in ways that create varying degrees of difficulty, or in extreme cases make it impossible, for their owner to think and behave in ways that are similar to most others in either their specific social groups or the human population in general. It is very important to think about the terms that we use to discuss this issue, since words are often ambiguous and can carry unintended meaning. You will see all kinds of words used by people from clinicians and scientists to members of the public. Broadly speaking, my preference in this book will be to use 'disorder' (or 'abnormality' or occasionally 'disease', where this is included in the commonly accepted name of certain specific disorders) to describe more extreme situations that are relatively easily defined and usually have a negative impact on the individual. I will use the terms 'divergence' or 'neurodivergence' for situations that are more subtle, harder to define and have a more variable impact on an individual's high-level functions. Sometimes I will use the general term 'condition' in a neutral way to describe phenotypes that might or might not be definable as disorders or as types of neurodivergence. I will use the term 'neurodiversity' to mean all the variation that we see among all our brains, including those we might label as disordered or neurodivergent.

There is little difficulty applying the terms disordered or abnormal to those relatively rare individuals whose brain development suffers such severe derailment as to threaten their very survival and gives them no chance of behaving like most others. For example, such severely affected individuals might be unable to move and/ or communicate in any meaningful way and/ or eat and drink unaided. This can happen for many reasons, including the inheritance of faulty genes or exposure of the fetus to harmful substances. In such cases, we can usually detect features of the brain's anatomy and function that we can legitimately describe as being disordered or abnormal because we have a clear understanding of what ordered or normal would be. In other cases, we can describe brains as disordered based mainly on measurable differences from the majority in how they carry out fundamental perceptual, cognitive, movement and behavioural functions. Examples include some types of neurological disorders such as epilepsy or major psychiatric conditions such as schizophrenia, whose symptoms usually include abnormalities in a person's perception. We can define the symptoms of such conditions with relative precision, measure them and show that they deviate from normality. For example, the prevalence of hallucinations (that is, perceptions without a stimulus outside the body) is much higher – and can be shown to be significantly so in a statistical sense – in people with schizophrenia than in the general population.[1]

Things start to get more complicated when we consider how best to describe people who have come to the attention of clinicians because they exhibit unusual behaviours that are relatively hard to measure precisely and whose causes are uncertain. Many people have brains that act in ways that appear to diverge from those of most others. Examples would be autistic people, dyslexic people or people with a learning difficulty. The nature of the atypical behaviours can be highly variable and we usually think of them as being on a spectrum, one end of which overlaps the ways the majority of people typically think and behave. Following common practice, I will refer to such people as neurodivergent. I will avoid describing them by terms such as disordered or abnormal, in part because, when it comes to measuring complex aspects of our behaviours, our ways of thinking and how we communicate with each other, it is usually difficult to define with any statistical certainty what normal is.

The value of the term neurodivergent for individuals is that it implies that the behaviour in question differs in some recognizable way from the behaviours of most others without prejudging whether the differences are a good or a bad thing. The terms disordered or abnormal have negative connotations and increase the risks of a person becoming stigmatized. Neurodivergent individuals can come to the attention of medical professionals for many reasons. For example, their behaviours might not conform to those expected of them in particular contexts, such as in social groups at home or at school. A critical point here is that the evaluation of such an individual's ways of thinking and behaving are in the context of what it is they are trying to do. What might be acceptable and unremarkable in one societal context might be problematic in another. Although a neurodivergent individual's atypical reactions might be unhelpful in some scenarios, they might prove very successful in others. For example, many highly creative individuals went through childhood education with difficulty due to their neurodivergent ways of thinking and behaving. It is easy to see how actors, artists, scientists, writers and comedians, for instance, can reap the benefits of an ability to focus deeply, to bring unique perspectives and to think and behave in ways that would not conform comfortably with many classroom regimes and styles of teaching.

As you can imagine, attempting to generate scientifically robust definitions of neurodivergence is fraught with difficulty. Nevertheless, clinical research has devised categories (or diagnoses) of neurodivergent conditions, although there is much overlap between them, meaning that many neurodivergent individuals fall into more than one. Categorizing types of neurodivergence under discrete headings is a pragmatic approach in the clinical or socioeconomic management of individuals, but can be risky in research. It might bias scientists to search for discrete explanations for different diagnoses, whereas in reality the causes of different diagnoses might overlap significantly. Much modern research on neurodivergence tries to avoid such traps or is busy climbing out of them.

What I Mean by Brain Development

Much of the brain's spectacular complexity forms before birth. Construction and refinement continue for many years after. Even in adulthood, the brain's connections are not fixed but are dynamic and malleable, capable of substantial biochemical and structural modification in response to physiological changes in the brain's activity. This ability is important for allowing the creation of new memories throughout life. Whether one describes all of this as development is a matter of choice; one could argue that the brain never stops developing until it dies. Development is not a precisely defined term and it means different things to different people. In its broadest sense, it means a change with time, and it is undeniable that brains continue to change with time, either in positive or in negative ways, throughout life. In everyday conversation we might say that a particular situation is 'developing' – for example in the world's economy or in international relations – no matter whether it is constructive or destructive. In contrast, most of the time we think of brain development as the constructive processes that get us to some ill-defined peak of cognitive ability, after which aging brings changes that we usually describe as degenerative.

In this book, I have included some common clinical conditions that are usually first diagnosed in adolescence or in adulthood, including old age. The reason I have done this is because researchers now think they are probably caused or predisposed to by events that occur earlier in brain development. Examples include psychiatric disorders such as schizophrenia and bipolar disorder, and even some neurodegenerative conditions that usually appear in the elderly. I will describe evidence suggesting that such conditions might have been initiated many years or decades before diagnoses are made, perhaps by reducing the brain's resilience to future events that might trigger the onset of detectable symptoms.[2]

Ultimate and Proximate Causes of Brain Disorder and Divergence

I will discuss the causes of brain diversity under two headings: ultimate causes and proximate causes.

What I mean by ultimate causes are the genetic or non-genetic factor(s) that we can incriminate as the root cause(s) of a particular phenotype. There are several types: factors arising from within our genetic material, from the brain's environment or most likely from both, as well as the random variation (or noise) inherent in developmental processes. Genetic ultimate causes arise from variations in regions of the DNA that affect the developing brain, either directly or indirectly. Environmental ultimate causes might come from inside the body (the internal environment) or from outside it (the external environment).

For example, the ultimate cause of Down syndrome is the inheritance of too much genetic material (specifically, chromosome 21); the ultimate cause of fetal alcohol syndrome (also known as fetal alcohol spectrum disorder) is maternal consumption of alcohol. In other words, the ultimate causes are the factors that trigger the disorders in the first place. Our knowledge of ultimate causes has increased massively in recent decades as revolutionary discoveries in genetics combined with advances in computer technologies allowed us to analyse large volumes of data on large numbers of people. Advances have been particularly rapid where there is a discrete ultimate cause, such as a mutation in a single gene, a large chromosomal abnormality, or something specific identified in the external environment. Working out the ultimate causes of some disorders and most types of neurodivergence is more challenging, since the clinical conditions themselves are hard to define and the ultimate causes likely involve combinations of many genetic and environmental factors. Researchers are now able to devise ways of rising to these challenges, and this is starting to provide valuable insights.

Proximate causes are those that link ultimate causes to their observed effects: for example, they would address the question of *how* an extra chromosome 21 causes the symptoms of Down syndrome, or *how* ingestion of alcohol causes fetal alcohol syndrome. They are the events that are closest to, or immediately responsible for, the observed effects. For most cases of neurodevelopmental divergence and disorder, our understanding of the proximate events that link ultimate causes to their effects is at best incomplete, even where the ultimate causes are clearly defined, as in the examples of Down and fetal alcohol syndromes. In part, this is due to the sheer number and complexity of the developmental processes among which we are trying to find the proximate causes.

Another complication is that obtaining evidence for proximate causes ideally requires experimental manipulations in which developing brain is subjected to a factor that has been identified as an ultimate cause so that we can follow all the steps that then give rise to the phenotype. Clearly, such manipulations would be unethical in humans: we cannot deliberately add an extra chromosome 21 to a person, or fill a pregnant mother with alcohol, and see what the cells of the brain do. There are some things we can do to learn from ourselves, such as analysing post-mortem samples or living human cells in culture in the laboratory, but these have limitations. Post-mortem samples are usually obtained long after the ultimate causes have had their effects. Culture systems are always to some extent artificial, which introduces a degree of uncertainty as to whether what we learn from them applies faithfully in real life. We end up extrapolating much of our understanding of human brain development, its diversity and disorder, from knowledge of how non-human organisms develop their brains. This approach makes the reasonable assumption that there are many similarities in human brain development and the development of the brains of other species, particularly other mammals, but we have found that there are many differences too. In the end, there is no single way to discover how neurodevelopmental divergence and disorder arise; collectively, scientists need to use many approaches in parallel to build up the evidence.

The Layout of This Book

I have divided the book into three main parts.

Part I will be largely descriptive. I will describe the way in which the major structures of a typical human brain develop in the embryo and fetus (Chapter 1), how a typical human brain develops its major functions (Chapter 2) and the types of disordered and divergent phenotypes that we find in the human population (Chapter 3).

In Part II, I will focus on the ultimate causes of neurodevelopmental disorder and neurodivergence. I will discuss types of genetic (Chapter 4) and environmental (Chapter 5) variation that can be causative, before describing what we know about the ultimate causes of specific clinically recognized conditions (Chapter 6).

In Part III, I will attempt to link information in Parts I and II by addressing proximate causes of disordered and divergent phenotypes that arise before birth (Chapter 7), in the early postnatal years (Chapter 8) and later in life (Chapter 9).

In an Afterword, I will consider where major breakthroughs might come in the years ahead.

Part I

Our Growing Brains

1

The Development of Brain Structure

The Spanish scientific superstar Santiago Ramón y Cajal (1852–1934) opened our eyes to the fact that nerve cells, or neurons, are the brain's fundamental functional units. His meticulous microscopic studies and beautiful anatomical drawings of these cells laid the foundations of modern neuroscience. In this chapter, I will focus on how neurons and other equally important cell types are made and how they combine to create the brain's main structures. I will touch on some of the mechanisms that control these events, but I will discuss most in Part III in the context of our current understanding of how disorder and divergence occur. I will describe how the brain and its cells and structures develop their functions in Chapter 2.

A Framework for Understanding Prenatal Brain Development

Ways of Discovering How Brain Cells Are Created

In a fantasy world, we would have the technology to watch and film continuously the development of each of the vast number of cells that human brains contain, without interfering with them or harming them in any way. Goodness knows how we would do it! The human brain is far too complex for this to be achievable in the foreseeable future. At the moment, scientists are struggling to analyse the microscopic structural details of just a single cubic millimetre of post-mortem human brain tissue. Even a piece of brain this small contains many thousands of neurons linked by over a hundred million connections. Watching such complexity unfold in a living human brain is effectively impossible.[3] In the real world, we can only do such things in simple, transparent and rapidly growing organisms like some types of worms – most famously the nematode *Caenorhabditis elegans* – whose several hundred nerve cells we can watch and film individually as they develop under a microscope.

In the real world, we get useful knowledge about the creation of the cells that make up our brains by studying them as best we can at a series of different ages. It is like looking at stills from a movie and working out what happened in between. We can obtain our stills in a number of ways. For example, we can collect the brains of early human embryos donated by mothers who have elected to terminate their pregnancy, up to legally defined ages that vary between countries, or brain tissue of any age donated under other circumstances, for example after death or brain surgery. We can use such material to study anatomical structures, zoom in on individual cells, test their biochemistry and sometimes, if we can obtain samples that are still alive, their activity. These approaches have taught us a lot about how our brain cells are created and mature, although they have not provided the complete or detailed understanding we would like. We need to derive additional knowledge by assuming that the development of our own brain cells is much like the development of brain cells in other species that we are better able to study, particularly those we are most closely related to. While this assumption is often reasonable, it does have caveats. For example, the cerebral cortex of humans is not only much larger than that of most other mammals but has several unique features of organization and cell type diversity that arise through processes that do not take place in other commonly studied species.

An Overview of the Development and Diversification of Brain Cells

We start life as a fertilized egg, which biologists call a **zygote**. This single cell divides and then the two new cells each divide, then the four new cells divide, and so on. By the time the embryo is about 5 days old, it contains in the region of 70–100 cells. At this stage, it attaches to the uterus and implants into it over several more days. The brain starts to form a couple of weeks later. As with any organ, it is very hard to define exactly when the brain first appears: in reality, it involves the transition of cells from unspecialized states to organized, specialized states, which is a gradual process.

Various practical issues make it particularly difficult to study the onset of human brain formation and its timing. Women who elect to terminate their pregnancy and donate embryos for research are unlikely to do so at such early stages, when many will not yet even know they are pregnant. This makes it hard to obtain many naturally conceived human embryos at the stages when the brain first emerges. If human embryos are made by *in vitro* fertilization, they are not usually allowed to grow beyond about two weeks, which is before the brain starts to form (this limit is the subject of ongoing international debate). Steps towards countering this difficulty have been made relatively recently by research showing that biological structures resembling specific organs (called organoids), including the brain, or very young human embryos in their entirety (called gastruloids), can be made in the laboratory from cultured human stem cells (Figure 1.1) or from fetal brain tissue.[4–8] Despite the promise that these embryo-mimics hold, they still need validation, ideally by comparing them to real human embryos, which brings us back to the same problem. As David Turner and Alfonso Martinez Arias have pointed out, 'Gastruloids are not a substitute for embryos but rather a complementary experimental system that allows many experiments ... that are not possible with embryos'.[9]

The relatively simple first steps that a zygote takes in life set off a cascade of events that in most cases lead eventually to the production of an adult brain containing approximately 100 billion neurons. It is very hard to count the numbers of neurons in our brains; there are so many of them. It is also hard to conceptualize such large numbers. You could try by imagining every neuron as a small coin: 100 billion coins would fill about 250 double-decker buses. Or, if you like, a hundred billion is more than 12 times the number of people alive on Earth today. The truly remarkable thing about our brains is not simply that they

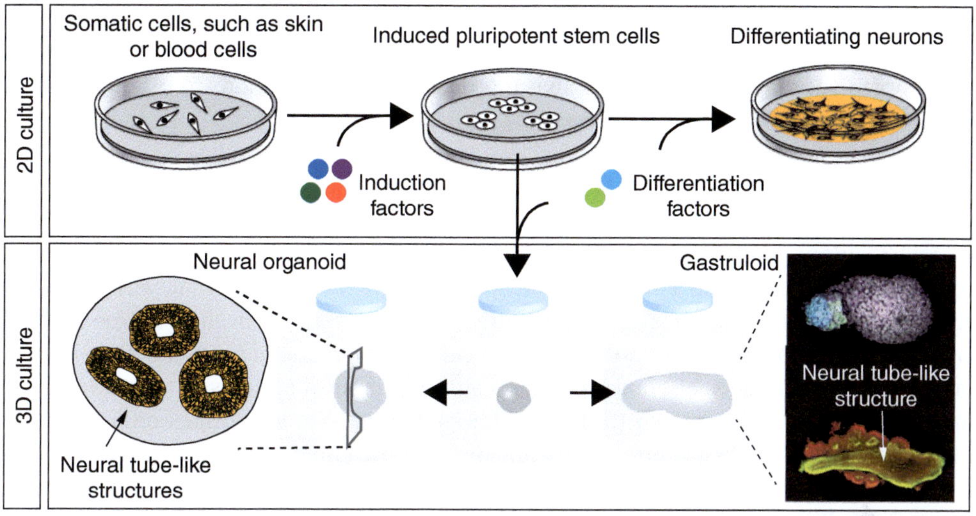

Figure 1.1 **Using human pluripotent stem cells to study human embryogenesis.** Stem cells are cells with the ability to self-renew continuously through cell division and to generate cells that turn into more specialized types; pluripotent stem cells can produce any cell in the body; induced pluripotent stem cells (iPSCs) are a type of pluripotent stem cell made in the laboratory from **somatic cells**, such as skin or blood cells, taken from a donor and grown in culture. The somatic cells are converted to iPSCs by adding induction factors (the ones originally used, known as Yamanaka factors after their discoverer, are **transcription factors**; I will explain what transcription factors are in Chapter 4). By adding further substances to the cultured iPSCs, they can be coaxed into specific cell types (we call this differentiation), such as neurons, on the two-dimensional (2D) surfaces of culture dishes. Alternatively, they can be grown floating in culture medium, in which case they can form three-dimensional (3D) tissue. Depending on how exactly this is done, the tissue might come to resemble either a particular organ such as the embryonic nervous system (or neural tube, see later in this chapter; we call them organoids for obvious reasons) or an entire embryo (we call them gastruloids, for reasons that should become obvious later in this chapter). In neural organoids, the neural tube-like structures are jumbled together like lengths of spaghetti and their profiles appear as illustrated here when they are cut across. In gastruloids, things are more orderly, as seen in the photographs: the top one is a view from outside a gastruloid and cell types such as those coloured blue or green are where they would be in an embryo; in the bottom one, which is a view inside the gastruloid, the developing neural tube (yellow) is laid out linearly surrounded by other cell types (red), as it would be in an embryo (if you look forward to Figure 1.4, you will see it looks a bit like the drawing of a Carnegie stage 8 embryo). One advantage of iPSC-based ways of studying human development is that the iPSCs can be made directly from patients with neurodevelopmental disorders that have a genetic cause to try and discover the basis of that disorder. Alternatively, researchers can relatively easily engineer the genomes of iPSCs (using methods we described in *Building Brains; An Introduction to Neural Development*[10]), for example to inactivate specific genes or to add genetic material in ways that recapitulate changes causing disorders. This potential goes a long way towards explaining why their use has become so widespread in recent years. *Photographs of gastruloids are reproduced from Moris et al.*[11] */ with permission of Springer Nature. (top) and Hamazaki et al.*[12] */ Springer Nature / CC BY 4.0. (bottom).*

contain so many cells. After all, starting with a single cell, just 36 rounds of cell duplication would almost get there, 37 rounds would overdo it. Cells are produced in vast numbers all around us all the time. For example, it is easy to have a single bacterium produce 100 billion descendants in about a day and a half, if we provide the right nutrients and temperature.

What is truly remarkable about the 100 billion neurons in our brains is that they develop a wide array of complex and highly specific functional specializations and they combine in highly organized ways via something like 100 trillion interconnections to produce a biological machine of immense cognitive power. Understanding how this is done remains one of our greatest scientific challenges.

So much for the general picture. What about the development of the differences between our brains? Every brain is unique when looked at in detail. When do our brains start to diversify? The answer: from the very beginning. In fact, even before the brain starts to form, we can already see subtle embryo-to-embryo variation in the cells that will make it. Diversity increases as embryos continue to grow, due to variations in the genome and its environment and to the random variations inherent in developmental processes. Developing brains are good examples of complex and sensitive dynamical systems in which initially imperceptible changes can lead to major differences later. We sometimes refer to these initial slight variations as butterfly effects and describe the emergence of the large variability they produce using the mathematical language of chaos theory. It is very likely that some of the diversity present at the earliest stages of brain formation in the embryo is amplified by subsequent events to give the diversity we see in adult brains, although it would be hard to provide direct evidence since we cannot follow individual embryos through to adulthood.

Ways of Describing Structures Anatomically

For those who are not aficionados of anatomy, I will give a brief explanation of some commonly used anatomical nomenclature. Anatomists use two special sets of nomenclature to describe (i) the stages of development of human embryos and (ii) the locations of cells and structures in the body from the early embryo onwards.

Embryologists often prefer to describe the relative ages of human embryos by the appearances of their characteristic features using what we call Carnegie stages (named after the Carnegie Institution for Science in the United States of America, which housed the collection of embryos whose study generated the nomenclature). One reason for preferring Carnegie stages to absolute time elapsed after conception is that it can be very hard to know exactly when fertilization took place. An alternative option of relying on overall embryo size is problematic since embryos can grow at slightly different rates, achieving different sizes at the same stage. There are 23 Carnegie stages in all, covering the first 60 days or so of human embryo development. After this, Carnegie stages are not used and people usually start calling the embryo a fetus. Carnegie stage 1 is the zygote; by Carnegie stage 23, the great majority of the structures of the body are present and identifiable. Figure 1.2 shows some Carnegie stages that are important milestones in brain development and how they approximate to days after conception.

To describe relative locations in the body, we tend to avoid using words like top, bottom, front, back and sides because they can be ambiguous. For example, you would find a head at the top of an adult person but at the front of a crawling baby or animals that walk on all fours, although the head is clearly an equivalent structure in the same locations relative to the rest of the body in all of them. Instead, we prefer the words anterior (or sometimes rostral), posterior (or sometimes caudal), ventral, dorsal, medial and lateral. I have

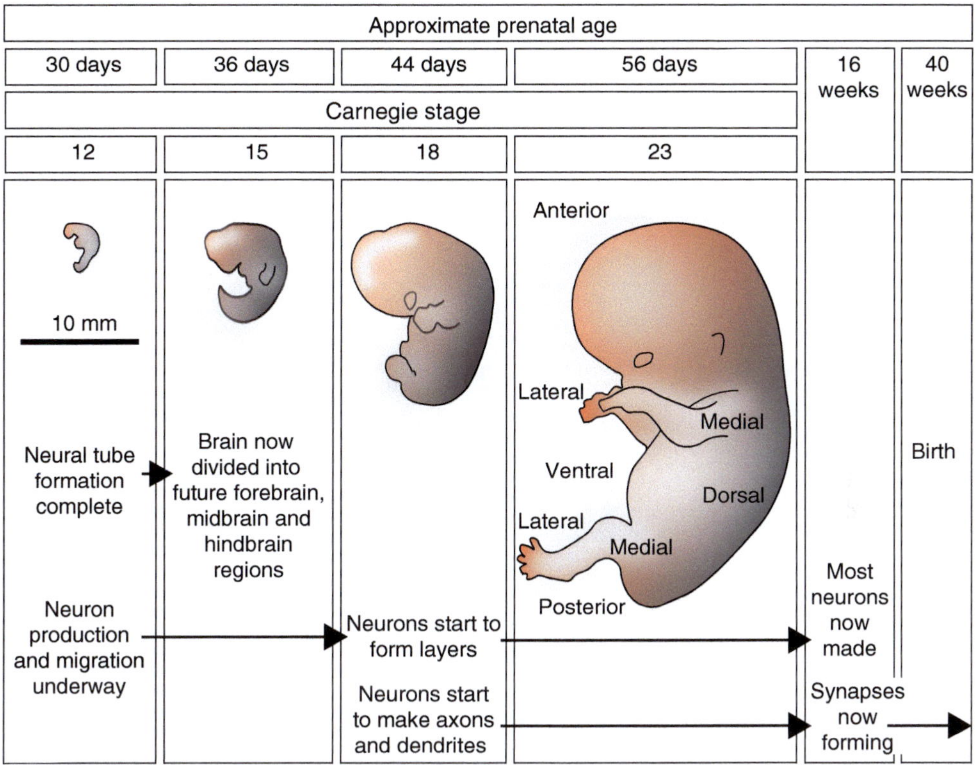

Figure 1.2 Some key stages of human prenatal development. By Carnegie stage 12, which is reached around the end of the first prenatal month, the formation of the nervous system is well underway. Formation of the neural tube is complete, and it is starting to produce large numbers of neurons that migrate to the locations in which they will mature. Over the next week or so, by Carnegie stage 15, major brain regions become visible. By Carnegie stage 18, around midway through the first trimester, neurons that have migrated are organizing themselves into layers. They are starting to become increasingly complex by generating the elements (axons and dendrites) they need to make electrical connections with each other. Most neurons are made and the specialized structures that allow them to communicate (synapses) are forming at around 16 weeks. The formation of electrical connections between neurons continues after birth. The scale bar at Carnegie stage 12 applies to all drawings.

illustrated their meaning in Figures 1.2 and 1.3. The brain forms at the anterior end of an embryo (or rostral end, from *rostrum* meaning a beak in Latin). If we move progressively from the anterior-most part of the brain to the end of the spinal cord, we say that we are progressing in a posterior direction (or a caudal direction, from *cauda* meaning a tail in Latin). This delineates the anterior–posterior axis (or the rostrocaudal axis means the same thing). Perpendicular to this axis is the dorsoventral axis, which runs from ventral (from *ventrum*, meaning belly in Latin) to dorsal (from *dorsum*, meaning back in Latin). The third axis (mediolateral) runs perpendicular to the other two from medial (which means towards the midline, an imaginary line running along the middle of the anterior–posterior axis) to lateral (away from the midline).

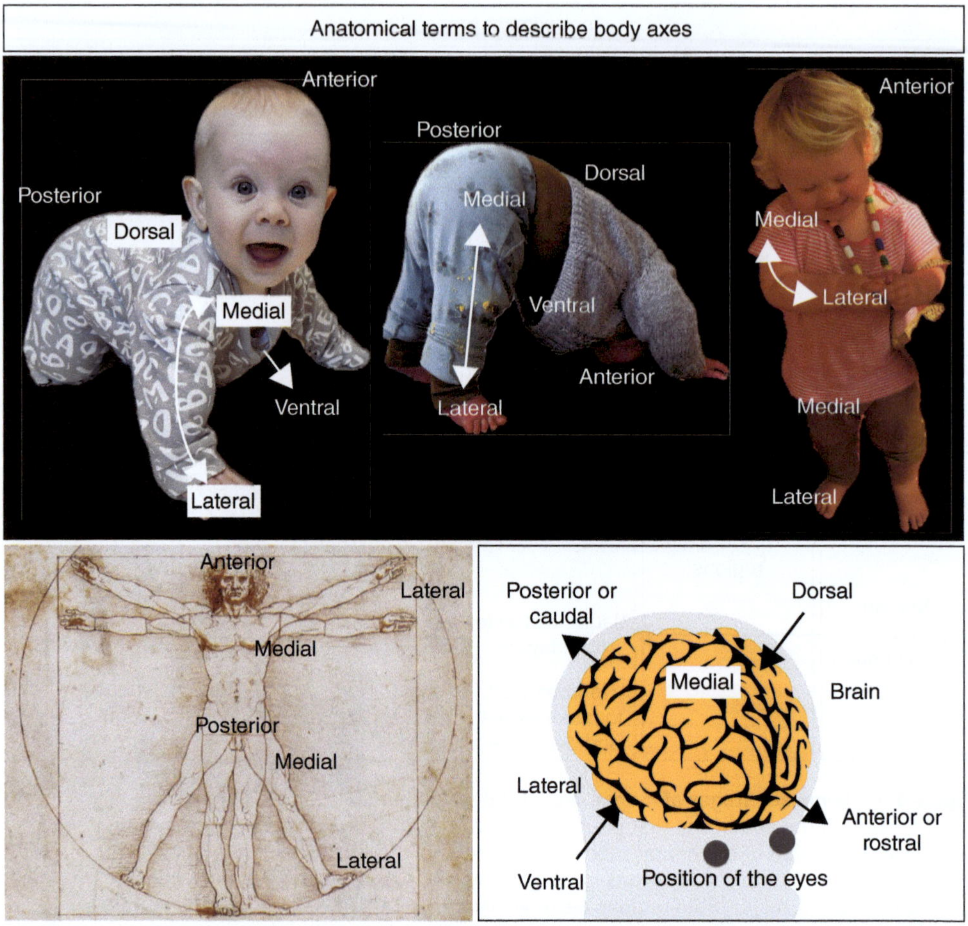

Figure 1.3 **Anatomical words used to define unambiguously the axes of the body and its parts.** The words describe the relative positions of body parts including those in the brain, no matter whether a person is crawling, adopting the downward dog yoga pose or standing (the terms are also used in Figure 1.2). *Leonardo da Vinci / https://commons.wikimedia.org/wiki/File: Homme-vitruve.jpg / Public domain.*

The Early Stages of Brain Growth

Gastrulation and the First Signs of Brain Tissue

The developmental biologist Lewis Wolpert (1929–2021) is quoted as saying: 'it is not birth, marriage or death but gastrulation that is truly the most important time in your life'. Gastrulation means formation of the gut, and when it forms, our basic body plan is created: gut inside, skin outside and bones and muscles in between. Starting between Carnegie stages 5 and 6, about two to three weeks after fertilization, it involves the movement of cells from the outer surface of the embryo to its inside, where they will eventually make the gut (Figure 1.4).[13,14] This movement creates two of the embryo's three germ layers: the endoderm on the inside (future gut and associated organs) and the **ectoderm** on the outside (future skin and nervous system). The third germ layer, the mesoderm (future bones and muscle), develops between the endoderm and ectoderm. This organizational tour de force ensures that the major types of tissue develop in the correct relative positions.

As gastrulation gets underway, ectodermal cells lying along the future dorsal surface of the embryo become more specialized **neuroectodermal** cells. Because the ectoderm generates the body's covering, which we call epithelium, we sometimes refer to neuroectoderm as neuroepithelium. I will not go into too many details of how this neuroectodermal specialization is brought about because we do not know much about whether, and if so how, its disruption might be a source of disorder or neurodivergence. Briefly, we think it involves a process we call **neural induction** in which intercellular signalling triggers neuroectodermal development. The overall principle of neural induction is that ectodermal cells become neuroectodermal cells because they receive inductive signals from cells around them. These signals come in the form of **signalling molecules*** of specific types. You can find more information on mechanisms of neural induction, drawing on evidence from other species, in our book *Building Brains; An Introduction to Neural Development.*[10]

At the earliest stages of neural induction, neuroectodermal cells viewed under the microscope would be hard to tell apart from other surrounding cells in the embryo, but, invisibly, their biochemistry is already diverging from that of their neighbours. Most importantly, different types of cells make different – although partially overlapping – sets of proteins. Indeed, it is *because* of this invisible diversity that cells develop to become specific types with specialized functions. Researchers often say that these differences give cells their molecular identities that specify their future developmental trajectory towards different fates.

From Neural Plate to Neural Tube

As neuroectodermal cells proliferate, the neuroectoderm becomes noticeably thicker than surrounding ectoderm and starts to bulge on either side of a central groove that runs along the midline in an anterior–posterior direction. We refer to this structure as the **neural plate** (Figure 1.4). The bulges on either side of the neural plate's midline rise up to form what we call neural folds, which bend towards each other and fuse along the dorsal midline to form the **neural tube**: effectively, the neural plate rolls itself up. The process is complete by Carnegie stage 12, which is one to two weeks after neural induction.

The transformation of the neural plate into the neural tube comes about in ways that we do not fully understand. I will return to this topic in more detail when I discuss neural tube defects in Chapter 7. For now, I will just say that the mechanisms that fold the neural plate to form the neural tube probably include an elevated rate of cell production and changes in the shapes of cells along the centre of the neural plate's long axis. These would

* Developmental biologists often refer to signalling molecules that guide or regulate developmental processes (such as neural induction) as morphogens. You will not find a strict definition allowing you to say whether any given molecule is or is not a morphogen. This is because many molecules have multiple roles, and if they mainly do things that are not guiding or regulating development, they are usually not thought of as morphogens even if they can act as such in some circumstances.

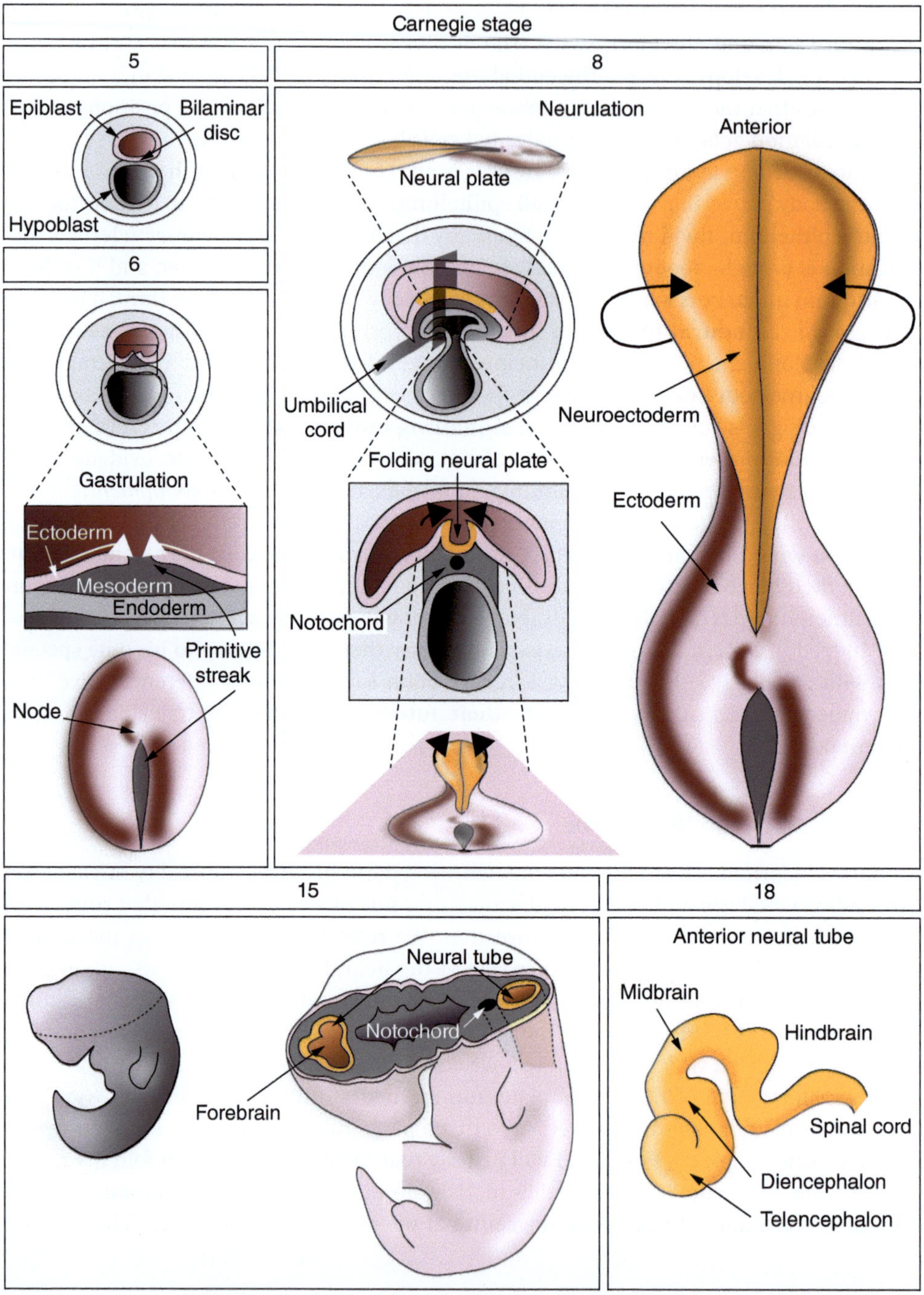

Figure 1.4 Key elements of human nervous system development up to Carnegie stage 18. At Carnegie stage 5, two main tissues are in contact. These are called the epiblast (the primitive ectoderm; future skin and nervous system) and the hypoblast (the primitive endoderm; future gut). They make contact at a region called the bilaminar disc, from which the embryo itself develops

Figure 1.4 (Continued) (cells in other regions make other structures including the membranes that surround and protect the embryo and fetus). By the way, you will see the prefix or suffix 'blast' used in many words applied to early embryos or immature cells; it comes from the Greek word *blastos* meaning sprout or shoot and implies that, whatever the thing is, it is at an early stage of development. Gastrulation is underway by Carnegie stage 6 and is marked by the formation of the primitive streak, with the node in front of it (the node is an organizer of gastrulation; you can see the node in the drawing of the gastrulating bilaminar disc viewed from above). Epiblast cells move into the primitive streak (white arrows) to form the mesoderm (dark grey; future muscles and bones). At Carnegie stage 8, the formation of the nervous system has begun (we call this neurulation): some ectodermal cells have become neuroectodermal (orange). These cells initially form a sheet called the neural plate but quickly fold to form a tube (the neural tube). Here you can see a partial zippering together of the two edges of the folding neural plate, viewed from various angles: along its length (top left), in cross section (middle left), angled away from you (bottom left) and from above (right: the arrows highlight the movement of the two edges towards each other). By Carnegie stage 15, the neural tube is closed and has sunk beneath the surface ectoderm (which becomes skin). It is now bending along its length as its major regions start to form. By Carnegie stage 18, major forebrain, midbrain and hindbrain components are readily visible: the forebrain is subdivided into the telencephalon and the diencephalon.

make it buckle along this axis, bringing its lateral edges progressively closer together dorsally. As these edges touch, they fuse together, closing like a zipper to complete the tube in a process called neural tube closure (Figure 1.4).

Once the neural tube has formed, the embryo has completed the basic layout of its central nervous system with future brain located anteriorly and spinal cord posteriorly. It has become a tube, and it will remain so for the rest of its life. Maybe it is not immediately obvious that the adult central nervous system is a tube, but it is, albeit a distorted one. It has a narrow cavity filled with cerebrospinal fluid running through it, called the central canal in the spinal cord and the **ventricles** in the brain. The reasons the tube is less obvious in adults are that the thickness of its wall far exceeds the diameter of its narrow central space (or lumen, to use the biological term) and the shape of the lumen is greatly distorted in the brain.

By Carnegie stage 18, the major subdivisions of the brain are visible (Figure 1.4). In the next sections, I will explain how these subdivisions develop.

Patterning: Specifying the Subdivisions of the Early Brain

To explain how the subdivisions of the embryonic brain that we can see at Carnegie stage 18 come about, I need to go back to the neural plate stage. The cells in the anterior part of the neural plate are *already* specified (or instructed, if you like) not only to become the brain but also to become particular parts of the brain. The anterior neural plate is already subdivided into regions that will, broadly speaking, translate through development into the major components of the adult brain. If you were to just look down the microscope at the neural plate, you would not see much evidence of these subdivisions. This is because the main differences are in the molecules that the cells in each subdivision contain. Many of these molecules are directing how the cells in each region should develop in the future, including what shapes they should adopt, what biochemical processes they should initiate, whether they should move or stay put, and so on. Through their actions, these instructive molecules will guide cells to become particular specialized cell types in appropriate locations (which we often describe as their fates). We refer to the early molecular subdividing of structures such as the neural plate and neural tube as **patterning**. Patterning is like a gameplan for what will come next. As the neural plate becomes the neural tube and continues its growth, you would see the game itself unfold as the structural differences that are specified by the earlier molecular patterning become progressively clearer. I will describe the anatomy of these structural differences in the next section. First, some comments on the principles by which the cells in each region are specified during the process of patterning.

Cell specification tends to occur collectively to entire populations of cells located together in the same region. By specifying each contiguous population differently, early brain tissue is patterned into regions with different molecular identities (or containing different gameplans, to continue my analogy from above). As was the case for neural induction, it is driven by specific signalling molecules in each cell's immediate environment. Across the anterior neural plate, the types and levels of signalling molecules vary from region to region, specifying cells similarly within a region but differently between regions. Some of these signalling molecules might be the same or similar to those involved earlier in neural induction, but they are now being deployed for a different purpose. In fact, it is common to find the same signalling molecules being reused in multiple processes throughout development. The signals that impinge on cells specify what they should do by affecting what molecules they make, and those molecules then instruct the development of each cell towards its eventual fate, as I outlined earlier. Figure 1.5 gives an overview of ways in which developmental biologists think about the processes of patterning and fate specification.

Early patterning creates the framework for the development of discrete brain regions. As a rule, patterning occurs progressively, with initially broadly defined regions subdividing further to build more and more complexity. This is the general picture, but, as you might imagine, things are more complicated when you dig deeper. For example, patterns of signalling molecules are dynamic. They change with developmental stage, which can help explain why the same region can make predominantly different cell types at different ages. A striking illustration of this is the so-called neurogenic-to-gliogenic switch in the developing cerebral cortex, during which production of neurons ceases and production of glial cells begins (I will return to this topic in a later section).

Subsequent Growth of the Neural Tube

The brain forms through the expansion of the anterior neural tube into three major swellings: the most anterior one makes the **forebrain** (also called the **prosencephalon**), the middle one makes the midbrain (or mesencephalon) and the most posterior one makes the hindbrain (or rhombencephalon), posterior to which is the spinal cord (Figures 1.4 and 1.6). The forebrain subdivides into the **telencephalon** anteriorly and the **diencephalon** posteriorly. The telencephalon makes structures that include the cerebral cortex (dorsally) and **basal ganglia** (ventrally), and the diencephalon makes structures such as the **thalamus** (posteriorly). The cerebral cortex outgrows other structures, including the basal ganglia and thalamus, and ends up engulfing them. As it does so, it subdivides further into the neocortex (the relatively recently evolved part that has expanded massively in us compared to other animals) surrounded by evolutionarily older regions such as the **hippocampus**.

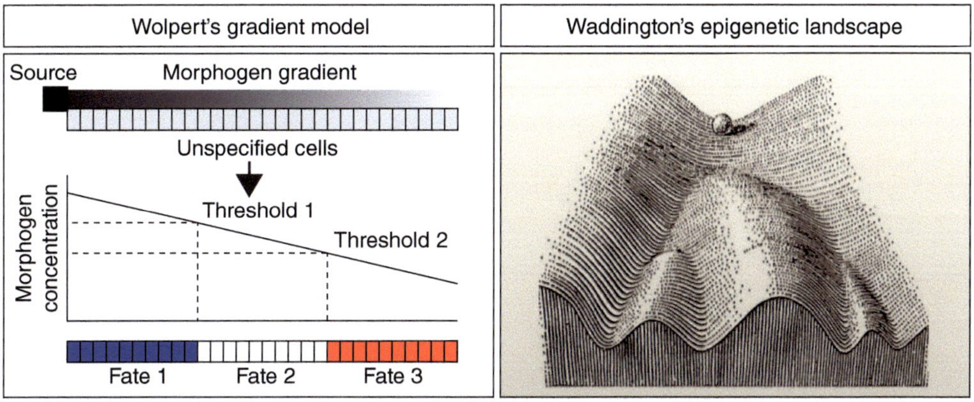

Figure 1.5 **Two iconic models that help us understand tissue patterning and cell specification.** There are many types of mechanism that might explain how tissues become patterned, with different regions adopting different fates, and there are many ways of illustrating how the fates of individual cells within those regions become specified. Two models have had enormous influence over many decades: Wolpert's gradient model to explain what he termed the French flag problem (a flag being a pattern); Waddington's **epigenetic** landscape to illustrate the process of cellular decision-making during development. Wolpert's gradient model: Lewis Wolpert devised this model in the 1960s, while working in London, England. One of its key elements is a gradient of signalling molecules, or morphogens. This might be produced by the diffusion of morphogens released from discrete groups of cells, called signalling centres, among populations of unspecified cells in their vicinity (I will provide more details on some known signalling molecules and centres in Chapter 7). The other key element is that the unspecified cells react differently if morphogen levels are above or below threshold values: here, morphogen concentrations above Threshold 1 specify Fate 1, those between the two thresholds specify Fate 2 and those below Threshold 2 specify Fate 3. This creates discrete regions of cells with similar identities (illustrated here as blue, white and red: the French flag). Waddington's epigenetic landscape: In the 1950s, Conrad Waddington, working at the University of Edinburgh in Scotland, introduced this model to help us think about the complex process by which genes dictate cell types. Whereas Wolpert's model addresses patterning across populations of cells, this model focuses on the specification of individual cells. The way a cell develops is imagined to be like the path taken by a ball rolling downhill through valleys whose shapes are created by the actions of that cell's genes. The ridges in Waddington's model represent barriers preventing a cell (or ball) that is going down one route switching to another. In the real world, barriers might arise because the actions of some of the genes used by the cell as it moves along its chosen developmental pathway block the activity of other genes whose actions would allow it to take the alternative route (on the other side of the ridge). A valley represents a route that the cell's genes allow it to take. As the cell (or ball) progresses, it encounters regions at the heads of valleys where alternative routes are available. Cells could, at these points, develop in a variety of ways. The two models can be brought together. The route the ball takes at decision points in the epigenetic landscape might depend largely on factors such as morphogen levels in the surrounding environment. The actions of the morphogens can be thought of as tilting each cell's landscape to ease the ball down a particular path. Thus, a high level of a morphogen might tilt the landscape so as to coax the ball down a route towards a blue fate; a low level might tilt it the opposite way, leading the cell towards a red fate.

The cerebellum forms in the hindbrain. If you are not familiar with all of these structures, do not worry, I will say more about them and what they do later in this chapter and in Chapter 2.

As brain cells proliferate and start producing an increasingly large array of different cell types specialized to carry out particular functions in specific locations, they interact with

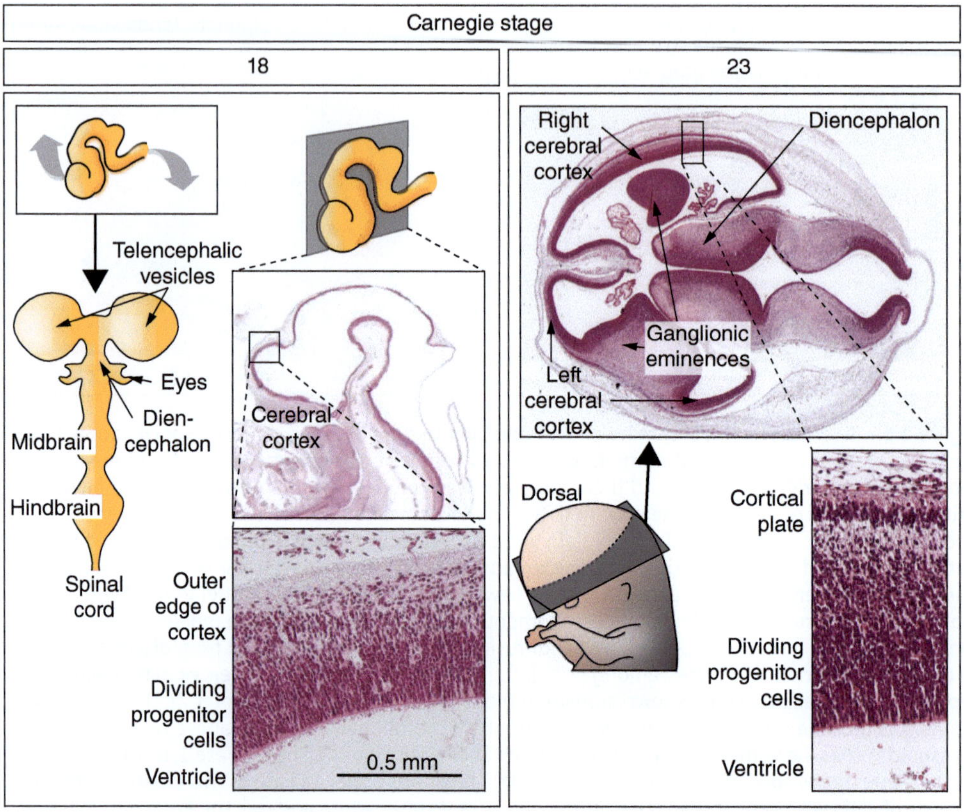

Figure 1.6 Development of the brain at late Carnegie stages. If we were to stretch out the embryonic brain at Carnegie stage 18, it would show its major components more clearly. The telencephalon consists of two swellings that will become the right and left cerebral hemispheres (these swellings are often referred to as telencephalic vesicles since they have a thin wall and much of their interior is fluid at this stage). The diencephalon is posterior to the telencephalon but will become engulfed by it later, which is why its derivatives (such as the thalamus) end up deep inside the adult brain. If we cut the unstretched Carnegie stage 18 embryo along its length, you can see how thin its wall is. The cerebral cortex is only about half a millimetre thick, but is organizing itself into a layer of dividing cells next to the ventricle and differentiating cells beneath its outer edge. By Carnegie stage 23, a horizontal cut in the plane shown reveals more advanced development of many of the brain's structures. The 3-dimensional anatomy gets a bit complicated now: the ganglionic eminences (which will become the basal ganglia in the mature brain) are mostly ventral to the cerebral cortex. If you were to look at sections more dorsal to the one shown here you would see the two bits of the left cerebral cortex join together as the left ganglionic eminence disappears. You see this on the right: the cut is at a slight angle such that the right is a bit more dorsal, and part of the right ganglionic eminence is just clipped and appears as a dense mass in the middle of the right ventricle (the frond-like structures in the ventricles are choroid plexus, which makes the fluid in the ventricles). By Carnegie stage 23, more differentiating cerebral cortical neurons have migrated away from the dividing progenitor layer in which they were generated to form an obvious layer (which we call the cortical plate at this stage; see Figure 1.7 for further development of the cerebral cortex). *HDBR* / https://hdbratlas.org / *Last accessed by 22 Oct 2025.*

each other and become interconnected in ever more elaborate ways. They form the neuronal networks that allow neurons to communicate with each other via their axons (the biological wires of the brain) and synapses (the points at which electrical impulses travelling along one neuron's axon are passed on to another neuron). In the following sections, I will outline how brain cells proliferate, migrate and mature to produce the appropriate numbers of cells of the required types in the correct places and how brain cells develop their interconnections.

The Development of Brain Cell Numbers, Locations and Types

Brain Cell Proliferation

Mechanisms have evolved to ensure the reproducible production of the vast numbers of cells that populate our brains. They control which cells proliferate, where they proliferate, how long it takes them to divide and when they should stop doing so. We refer to cells that are actively proliferating as **progenitors,** and we say that they are going through the **cell cycle** (I will say more on its details in Chapter 7). We use the term **precursors** to describe cells that have stopped dividing but have not yet acquired their specific mature properties (or not yet fully differentiated, which means much the same thing). Since cells in the brain divide by mitosis, we often say that cells that have stopped dividing are postmitotic. Our brains create most of their postmitotic neurons before birth.

Cells proliferate very rapidly in the embryonic and fetal human brain. To give you an idea of the scale of growth, it is likely that as many as 15 million new cells per hour are added at some stages. Do not take this number too seriously, it is just a rough estimate, but is probably in the right ballpark. The rate at which cells are added to the brain is controlled in a variety of ways. First, the numbers of progenitors (which we often refer to as the sizes of the progenitor pools) that generate each part of the brain are regulated throughout development. This involves mechanisms that determine, at each developmental age, the proportions of cells that continue to divide as opposed to the proportion that stop dividing to differentiate, which we call terminal differentiation. The general trend is for the sizes of progenitor pools to drop with age as more and more cells differentiate. Second, the speeds with which dividing cells progress through the cell cycle are regulated. Third, some cells in the developing brain, including progenitors, are removed by naturally occurring cell death; I will say more about this type of cell death at the end of this chapter.

Researchers sometimes classify progenitor divisions as either symmetric or asymmetric. A symmetric division produces two cells of the same type; they might both be progenitors of the same type or precursors of the same type. An asymmetric division is one that produces two cells of different types; for example, one daughter might become a progenitor and the other a precursor, they might both become progenitors but of different types or they might both become precursors of different types. Although this classification can be useful, not everyone uses it, and, as we learn more, it is likely to become increasingly problematic. This is because the definition of cell types depends on the degree of resolution used to study them and, as I will explain later (Chapter 7), modern technologies allow us to

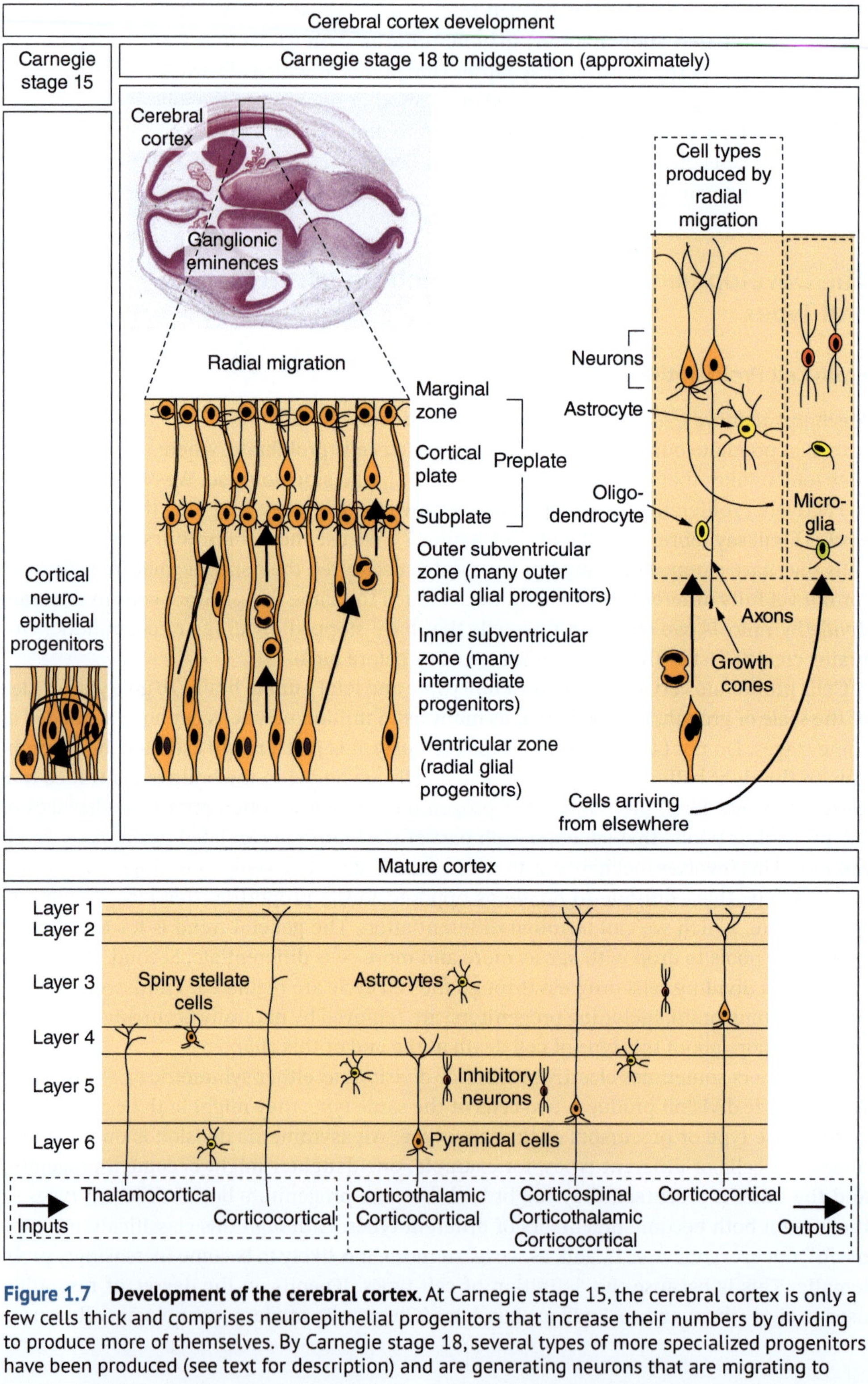

Figure 1.7 Development of the cerebral cortex. At Carnegie stage 15, the cerebral cortex is only a few cells thick and comprises neuroepithelial progenitors that increase their numbers by dividing to produce more of themselves. By Carnegie stage 18, several types of more specialized progenitors have been produced (see text for description) and are generating neurons that are migrating to

Figure 1.7 (Continued) form layers beneath the outer surface of the brain. These processes continue until approximately midgestation. By this time, most neurons derived from cerebral cortical progenitors have been made and these progenitors are now starting to make glia (astrocytes and oligodendrocytes). Inhibitory neurons, some oligodendrocytes and microglia have arrived or are arriving from outside the cortex. Neurons are making axons and their growth cones are seeking targets to synapse with. The mature postnatal cortex develops from the cortical plate and marginal zone. It is made of six main layers, from 1 (most superficial) to 6 (deepest), with deeper layers having been created before more superficial layers during prenatal development. Major types of cortical cells (orange: excitatory neurons; red: inhibitory neurons; yellow: glia) and major inputs and outputs are shown (these are among the most functionally important, but there are others): thalamocortical axons arise from the thalamus (in the diencephalon) and bring information into the cortex from the sense organs of the body; corticocortical axons mediate communication between different parts of the cerebral cortex; corticospinal axons send signals to the spinal cord to control movement; corticothalamic axons allow the cortex to modulate the function of the thalamus. Chapter 2 contains more information on these connections.

get ever more fine-grained information on cells. For example, two cells that are considered to be the same type at low resolution might be different at a higher one.

Divisions that produce two new progenitors have the potential to increase the production of cells more rapidly than any other mode of division because they expand the size of the progenitor pool exponentially. Divisions producing one new progenitor and one precursor will produce at best a steady stream of precursors from a progenitor pool of constant size. Divisions producing two precursors will shrink the progenitor pool. Broadly speaking, brain cell proliferation involves a progression from divisions of the first type to those of the second and then the third types.

Many researchers are now working out the details of this process in humans. Most studies have focused on the developing cerebral cortex (Figure 1.7), but their conclusions also appear to apply to many other brain regions. When the neural tube first forms, its wall is only a few cells thick and comprises relatively unspecialized progenitors that we refer to as neuroepithelial progenitors. As these cells continue to proliferate, they turn into progenitors that are more specialized and that we usually call **radial glial progenitors** (or radial glial cells).[15] The name is somewhat unfortunate because they do not function as glial cells, which generally provide support and protection to the brain's neurons. Scientists including Camillo Golgi and Santiago Ramón y Cajal described radial glial cells about 150 years ago and named them based on their glial-like appearances and their long processes that often span the full depth of the wall of the neural tube (Figure 1.7), which can be analogized to the radial spokes of a wheel. We now know that they are, in fact, multipotent[†] progenitors capable of producing a range of both neuronal and glial cell types (neurogenesis is the process of creating neurons and gliogenesis is the process of creating glia). Throughout the developing brain, many radial glial progenitors lie in a zone adjacent to the ventricles that we call the ventricular zone.

In the developing cerebral cortex, some radial glial progenitors lie in additional layers of proliferating cells called subventricular zones that form beneath the ventricular zone (beneath, or 'sub', if you look at it from the ventricle): we call them outer radial glial progenitors (Figure 1.7). There are two subventricular zones, one called the inner

† In Figure 1.1, I explained that the word pluripotent means a cell that is capable of making all cell types in the body (so far as we know, that is). A multipotent cell can make many different cell types, but not all.

subventricular zone and the other called the outer subventricular zone; the outer subventricular zone is closer to the outside of the brain. Most outer radial glial progenitors lie in the outer subventricular zone. A note in passing: you might sometimes see radial glial progenitors in the ventricular zone described as apical radial glial progenitors and those in the subventricular zones described as basal radial glial progenitors. Recent studies are now subdividing these sets of radial glial progenitors further.

The radial glial progenitors in the ventricular and subventricular zones produce another type of progenitor that amplifies the eventual output of neurons and glia, the so-called intermediate progenitor cells (Figure 1.7). These cells are most numerous in the inner subventricular zone, where they undergo multiple rounds of cell division contributing massively to the growth of our cerebral cortex. We sometimes describe them as examples of transit amplifying cells, a general term that also applies to progenitors that make other types of cells in other parts of the body.

Things are a little different when it comes to the developing cerebellum, which develops from the anterior part of the hindbrain (Figure 1.8). Some of its cells come from progenitors in this region's ventricular zone while others come from a nearby dorsally located structure called the rhombic lip. These structures contain radial glial progenitors and intermediate progenitors arranged into ventricular and subventricular zones, although not obviously into inner and outer ones. The main difference in cerebellar development is that the rhombic lip progenitors go on to make another set of progenitors called granule cell progenitors, which form a proliferative layer called the external granule layer around the outside of the developing cerebellum, rather than along the inside, as occurs in the cerebral cortex.

As the embryonic brain's progenitor cells undergo their divisions, they or the neuronal and glial precursor cells they produce migrate away from the locations in which they were created, in directions that depend on where they are and what cell types they are turning into. In the next section, I will describe in more detail the routes taken by these cells.

Migration: Getting the Right Types of Cells to the Right Places

In the previous section, I described the production and positioning of specific types of *progenitor* cells. In this section, I will explain how these progenitors go on to make neuronal and glial precursors and eventually differentiated cells of the right types in the right locations.

Much of what we first learned about the migration of early brain cells came from studies in non-human species. For decades, these studies used various methods to label dividing cells and their descendants, so they could be followed microscopically either directly or indirectly. Direct methods involved watching the movements of cells in pieces of brain tissue kept alive under the microscope (we say these are studies in real time). Indirect methods inferred movement from snapshots of the changing locations of identified cells in different brains spanning a range of ages (we described ways of doing this in *Building Brains; An Introduction to Neural Development*[10]). More recently, samples of living human embryonic brain obtained from elective terminations of pregnancy have also been used to watch the movement of selected cells in real time. Combining conclusions from these various approaches has given us a good understanding of the steps by which the brain gets specific cell types to their correct locations.

Initially, mostly neuronal and later mostly glial precursor cells migrate away from the ventricular and subventricular zones where they were made, continuing to differentiate into mature neurons and glia as they go. Many neuronal precursors move along the radial processes of radial glial progenitors until they reach the outer surface of the neural tube where they accumulate, forming future grey matter. We call this radial migration. The first neurons to be made come to lie directly outside the progenitors that made them, where they form a new layer (in the cerebral cortex, this is called the preplate: Figure 1.7).

In the developing cerebral cortex, more and more cortical neurons arrive in the middle of the preplate, forming a rapidly thickening layer called the cortical plate, which splits the preplate into an outer layer called the marginal zone and an inner layer called the subplate (Figure 1.7). The marginal zone forms cortical layer 1, the cortical plate forms cortical layers 2–6 and the subplate is a transient structure that, after growing much thicker than the marginal zone, largely disappears after birth. The subplate is an elaborate structure that is thought to act as a biological scaffold guiding the development of persistent cells and their connections. In the developing cerebral cortex, as too in other parts of the developing brain, two of the main types of glial cells – the **astrocytes** and the **oligodendrocytes** – are generated by progenitors after they have finished producing neurons (this is the neurogenic-to-gliogenic switch I mentioned earlier). Astrocytes and many oligodendrocytes migrate radially (Figure 1.7).

Similarly, in the developing cerebellum, Golgi cells, Purkinje cells, Bergmann glia, basket cells and stellate cells generated in the ventricular zone migrate radially towards the outer surface of the developing cerebellum (Figure 1.8). Granule and unipolar brush cells also migrate radially, but in the opposite direction, starting from their birthplace in the external granule layer, which is around the *outer* edge of the cerebellum. They pass through the Purkinje cells beneath to form the internal granule layer, guided on their inward migration by the processes of Bergmann glial cells. As they migrate, granule cells trail behind themselves long thin structures that subsequently form synapses with the **dendrites** of the Purkinje cells, creating a critical component of the circuitry of the cerebellum (note: dendrites are branched synapse-covered extensions that sprout in a wide variety of shapes and sizes – which I will illustrate with examples in the next section – from nerve cell bodies and carry the impulses from the synapses to the cell body).

Not all migration is radial. Many neuronal and glial precursors move roughly parallel to the surface of the brain for at least a significant part of their journey. We call this tangential migration; we have already seen an example of it in the case of progenitors, specifically the granule cell progenitors, in the cerebellum. It is a fascinating aspect of brain development that many major neuronal and glial cell types tend to originate and acquire their identities in different regions and then mix later. For example, one of the two major types of neurons that contribute to the cerebral cortex, the **excitatory** neurons, arrive by radial migration from the ventricular zones immediately beneath them, whereas most of the other major type, the **inhibitory** neurons, migrate tangentially into the cortex from the **ganglionic eminences** (Figures 1.6 and 1.7) (incidentally, the ganglionic eminences are so-called because they turn into the basal *ganglia*). Another example of this is in the developing cerebellum, where the excitatory neurons originate from the rhombic lip, whereas inhibitory neurons come from the cerebellar ventricular zone (Figure 1.8) (I will say more about the differences between excitatory and inhibitory neurons in Chapter 2; essentially,

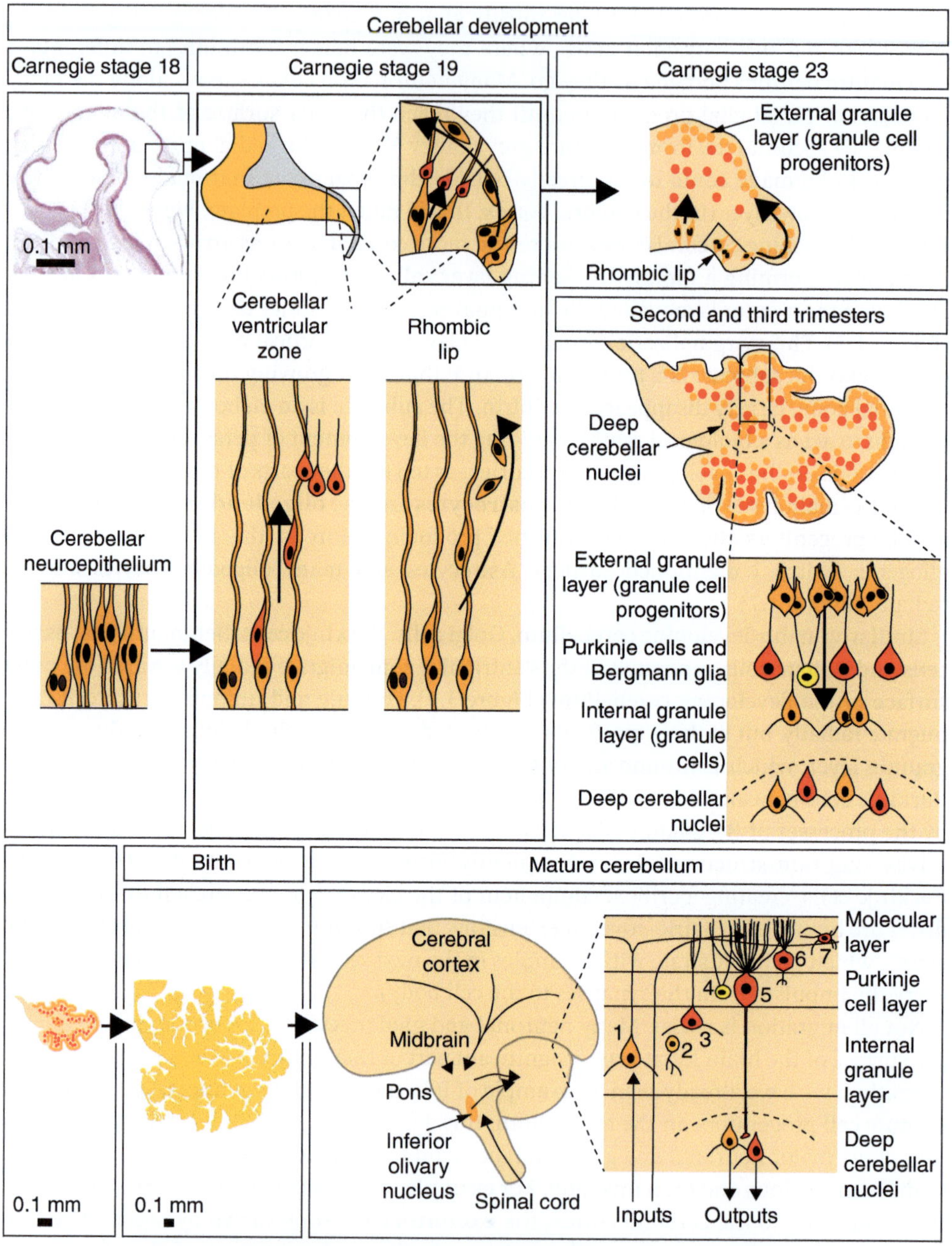

Figure 1.8 Development of the cerebellum. The cerebellum develops from anterior hindbrain neuroepithelial cells, starting around Carnegie stages 18–19. Some cerebellar cells come from progenitors in this region's ventricular zone; others come from the rhombic lip. These structures contain similar types of progenitors to those in developing cerebral cortex, arranged into ventricular and subventricular zones. The ventricular zone gives rise to inhibitory neurons of various types (red) and Bergmann glia (yellow). The rhombic lip progenitors make a second set of progenitors called granule cell progenitors, which migrate to form a proliferative layer called the

Figure 1.8 (Continued) external granule layer around the outside of the developing cerebellum before generating excitatory neurons of various types (orange). Neurons of the deep cerebellar nuclei come from ventricular zone and granule cell progenitors. The mature cerebellum comprises deep cerebellar nuclei surrounded by three highly convoluted layers containing: 1. granule cells and 2. unipolar brush cells, both of which come from granule cell progenitors; 3. Golgi cells, 4. Bergmann glia, 5. Purkinje cells, 6. basket cells and 7. stellate cells, all of which come from ventricular zone progenitors (orange: excitatory neurons; red: inhibitory neurons; yellow: glia). Axonal inputs converge on the cerebellum from many parts of the brain and spinal cord; outputs are returned to many regions of the brain, including the thalamus which in turn passes outputs to parts of the cerebral cortex that control movement.

excitatory neurons increase brain cell activity; inhibitory neurons do the opposite). Another example of tangential migration is that, according to current evidence, a first wave of oligodendrocytes migrates tangentially from the ganglionic eminences into the cortex.

Some migration cannot be described as radial or tangential. For example, **microglia**, a third major type of glial cells in the brain, do not originate in the brain at all. They come from the yolk sac of the embryo and migrate into the brain during embryogenesis. Once there, they proliferate and disperse.

Long tangential migratory routes are still being discovered, and some might be unique to human brain development. For example, researchers have found evidence, specifically in human brains, of tangential neuronal migration from the ganglionic eminences to the diencephalon and into particular regions of the cerebral cortex, with the latter continuing for a year or so after birth.[16]

The Structural Maturation of Brain Cells of Various Types

Newly created neurons and glia change their shapes and structures spectacularly during development, mainly by growing long extensions of their cell bodies (anatomists often call them cellular processes, which is a different meaning of the word process to that describing an action such as cell division or migration). Examples are shown in Figure 1.9. These extensions comprise cytoplasm surrounded by cell membrane. Most are long and thin and branched. Exceptions are the extensions that come from oligodendrocytes, which flatten into sheets that wrap themselves around axons to create their **myelin** sheaths (we call this myelination; it plays a crucially important function by increasing the speed at which electrical signals travel in the brain and I will have more to say about it in later chapters). Most myelination takes place after birth, increasing through childhood and adolescence (Figure 1.10).

Neurons produce two types of long thin extension: dendrites and axons. The main functional difference between them is that dendrites integrate incoming signals from other neurons, whereas axons carry signals away from the cell body (I will explain how this happens in Chapter 2). We call the dendrites that extend from a neuron's cell body a dendritic tree because of all the branches they make. These branches allow individual neurons to receive many contacts from other neurons. The dendrites of many neurons become covered with many thousands of tiny protrusions (dendritic spines), which are the sites of synaptic contact with incoming axons. Dendritic trees develop a large array of shapes, aspects of which are characteristic of different types of neurons in different parts of the brain (Figure 1.9

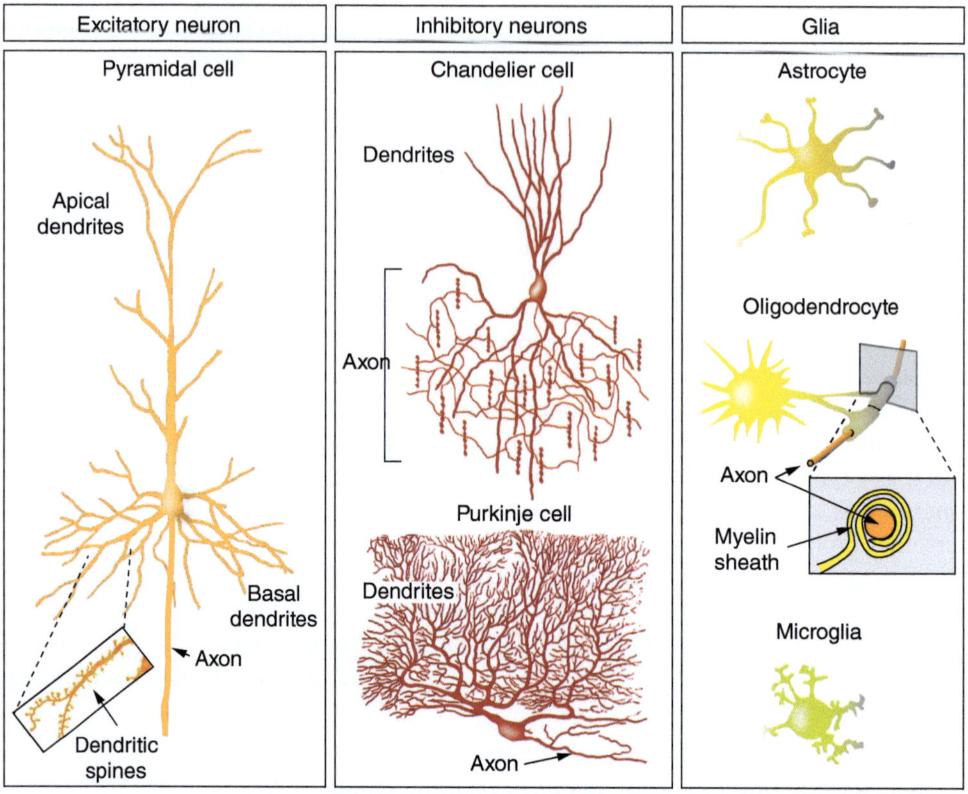

Figure 1.9 A more detailed view of a few of the numerous cell types in the brain. Pyramidal cells are the main type of excitatory neuron in the cerebral cortex (they are not the only one, for example there are the spiny stellate cells, Figure 1.7). Their apical dendrites stretch in the direction of the brain's outer edge while their axons exit in the opposite direction (I showed simplified versions in Figure 1.7). Inhibitory neurons have many different shapes. Chandelier cells are found in many regions of the brain, including the cerebral cortex, and their name reflects the chandelier-like shape of their complex array of axons ending in synaptic arrays called cartridges. They can contact many other neurons in their vicinity. Purkinje cells, which we encountered earlier (see Figure 1.8) have massive dendritic trees allowing them to receive vast numbers of synaptic inputs. The glial pictures illustrate the key differences between the three major types of glial cells found in the brain. Oligodendrocytes make the myelin sheaths that surround many axons.

gives only a few examples). They can be very extensive: for example, each Purkinje cell in the cerebellum has a huge dendritic tree that receives an estimated 100,000 synapses. Axons tend to be longer than individual dendrites, some very much longer. While some neurons in the brain make connections with other neurons lying close by, many connect with neurons as far away as the other side of the brain or far down the spinal cord. Synapses start to form before birth, with many more added during early childhood (Figure 1.10). The explanation of a subsequent decline in synaptic density through later childhood and adolescence is unclear: it might be due to a change in the numbers of synapses in the brain, the enlargement of brain volume or a combination of the two.

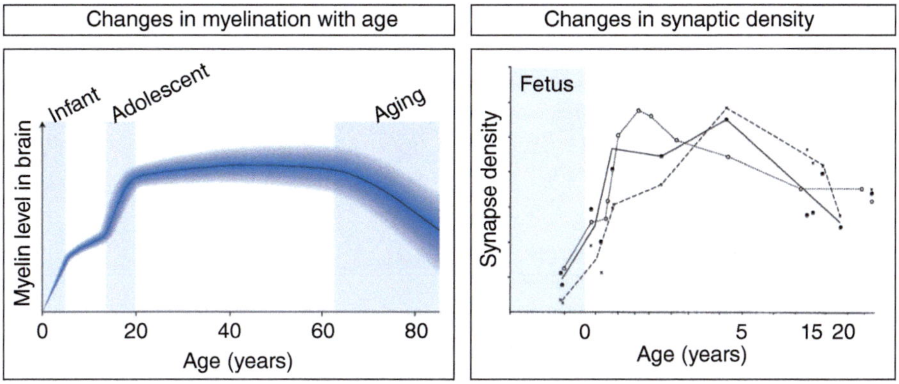

Figure 1.10 Changes in myelination and synaptic density with age. Left: a schematic graph of forebrain myelin changes during the human life span. Right: a graph showing data on mean synaptic density in three regions of the cerebral cortex (open circles and dotted line, visual cortex; filled circles and solid line, auditory cortex; crosses and dashed line, prefrontal cortex). At birth, the density of synapses increases rapidly and this continues for up to several more years afterwards. *Graphs are reproduced from Faria et al.*[17] */ with permission of Springer Nature (left) and Huttenlocher et al.*[18] */ with permission of John Wiley & Sons. (right).*

Astrocytes (from *astron*, the Greek for star, which they look a bit like: Figure 1.9) are the most abundant cell type in the brain and play diverse supportive and protective roles.[19] They elaborate increasingly branched processes that insinuate themselves throughout the brain with relatively little overlap between the territories covered by each cell. This type of arrangement, referred to as tiling, ensures uniform coverage of the brain. Tiling is lost if astrocytes become reactive, meaning that they undergo morphological, molecular and functional changes in response to injury, disease or infection of the CNS. As astrocytic processes mature, specialized structures called perisynaptic astrocytic processes appear at their ends. They are positioned near synapses, allowing them to play an active role in synaptic function. As I mentioned earlier, oligodendrocytes (derived from Greek words *oligo*, meaning few, and *dendron*, meaning branches) create the myelin sheaths around axons.

Microglia (which are small, hence *micro-*) protect the brain by scavenging particles or molecular aggregates that might endanger it. They are a central component of the brain's immune system. They populate the brain as small cells with long thin extensions that constantly survey the tissue around them. Collectively, they produce a grid-like system that guards the brain. They are usually immobile, but if they detect a harmful event or agent, they move rapidly towards the site of harm and proliferate and release protective molecules (called cytokines). After resolution, the activated microglia return to their previous state. There is evidence that they help maintain the integrity of the boundaries between developing brain regions during the early stages of their formation.[20]

The brain also contains other cell types, such as those that make up its blood vessels, which include endothelial cells and surrounding pericytes. Blood is supplied to the human brain via an extensive and complex network of vessels including about 100 billion capillaries with an arrestingly high total length of about 650 km. This network is made during embryonic and postnatal development by a process called vasculogenesis (meaning the

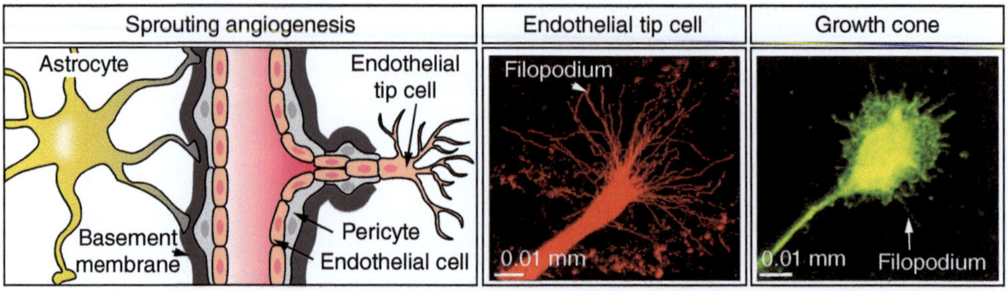

Figure 1.11 **The growth of blood vessels and axons share similarities.** Endothelial tip cells and growth cones are at the front, respectively, of blood vessels that are produced by sprouting angiogenesis and of growing axons. Both tip cells and growth cones are specialized, hand-like structures that sense environmental cues using finger-like filopodia, allowing them to find their way to their targets. The blood–brain barrier is created by endothelial cells, pericytes, astrocytes and neurons (the last of which are not included in the illustration here). Basement membranes are thin layers of a specialized matrix that form outside cells to give support to epithelial and endothelial cells. *Photographs are reproduced from Wälchli et al.*[21] */ with permission of Elsevier.*

formation of blood vessels). At the early embryonic stages of this process, endothelial progenitor cells generated in the mesoderm coalesce to start making the developing brain's main arteries and veins. These progenitor cells are specified to produce the endothelial cells that form the single-cell thick lining of all blood vessels. Once the major blood vessels have developed, sprouting angiogenesis leads to the formation of new blood vessels from pre-existing ones (Figure 1.11). A critical function of the brain's vascular endothelial cells is to work with other cell types, including pericytes, astrocytes and neurons, to create the blood–brain barrier during embryogenesis and early postnatal life. A key function of this highly selective semipermeable barrier is to protect the brain from harmful molecules and pathogens in the blood, while allowing the entry of substances required by the brain.

The process by which the tips of capillaries grow during sprouting angiogenesis is remarkably similar in principle to the way in which the tips of axons grow (Figure 1.11); I will discuss the latter in the next section. In fact, the similarity is more than skin-deep since both processes involve the same molecules, which move the growing tips forward and allow them to navigate in response to substances in their surroundings; I will describe these molecules in Chapter 7. This means that if the functions of these molecules are disrupted early in development, disorders of both axon growth and angiogenesis are likely to result.

The Development of Neural Circuits

Axon Development

In the last section, we saw how neurons of different types, such as excitatory or inhibitory ones, migrate over often long distances to mix together in appropriate proportions in each part of the developing brain. The development of the axonal connections that link them into functioning networks starts in the early embryo. By the time this process gets underway, some regions that must eventually become interconnected are already several millimetres apart. This means that axons need to lengthen and navigate to find their targets. The tips of growing axons – known as their **growth cones** – do the navigating (Figure 1.11). They are only about a hundredth of a millimetre in diameter and must hit specific target cells, or groups of target cells, over distances that are up to about a thousand times their diameter. Their task is made even more complicated because, as they grow, their targets are likely to be getting further and further away as the brain gets bigger and because the shape and composition of the tissue that they are navigating through is changing. How growing axons are guided has fascinated developmental biologists for well over a century. It involves the use of various molecular guidance cues called chemotropic factors that are distributed in the environment of the navigating growth cones and either attract them (a process called chemoattraction) or repel them (chemorepulsion). Much of what we know comes from research on non-human, often non-mammalian, species and you can read more about it in *Building Brains; An Introduction to Neural Development.*[10] Once growth cones have reached their targets, they form synapses in a process known as synaptogenesis.

Circuit Formation

Axonal growth and synaptogenesis create huge bundles of axons linking the major regions of the fetal brain. They connect the two sides of the brain, give the cerebral cortex its inputs from and outputs to other parts of the brain and the spinal cord, and so on. I will describe some of these major interconnecting bundles, which we often call axonal tracts, in Chapter 2 (you can look forward to Figure 2.1 if you like). You could think of them as being like the networks of undersea fibre-optic cables used for communication between countries or continents. Just as these large bundles of connections transmit vast numbers of signals that are then distributed for analysis within the receiving countries, so axonal tracts send information between major brain regions that process the inputs they receive using their internal, local circuitry. Brain regions then output information through other axonal tracts.

The local circuits within each region of the brain are organized in fairly standard ways that are characteristic for each part. For instance, in many parts of the cerebral cortex, inputs to the cerebral cortex from the thalamus synapse on neurons in layer 4 (as illustrated earlier in Figure 1.7). These layer 4 neurons pass information to neurons in layers above and below them. In turn, neurons in these layers output information to other excitatory and inhibitory neurons either in their immediate vicinity, in more distant cortical areas or outside the cortex. Inhibitory neurons play a modulatory role, refining and controlling circuit activity (I will say more about this in Chapter 2). In the cerebellum, which we saw earlier is formed of three layers – an internal granular, a molecular and a Purkinje layer (Figure 1.8) – inputs arriving from other parts of the brain and spinal cord synapse on excitatory granule cells in the granular layer or on the dendrites of Purkinje cells in the molecular layer (the latter inputs are called climbing fibres). Granule cells make axons that run through the molecular layer (they are called parallel fibres) where they too synapse on the dendrites of Purkinje cells. Purkinje cells are inhibitory and their axons provide the only output from the cerebellum (Figure 1.8).

We are a long way from understanding how these local circuits form. Similar guidance mechanisms to those that create major axonal tracts might be employed, but evidence is much harder to find since the local circuits are more intricate than large axonal tracts. It is likely that additional mechanisms come into play, such as those that shape or refine local circuitry based on the nature of the neural activity that flows through them. The creation of local circuits continues postnatally, when their development can be influenced by increasingly complex patterns of neural activity triggered by sensory stimulation from the outside world. There is evidence that some aspects of the development of neural circuitry involves an initial overproduction of connections with the subsequent pruning of some, most probably the least active ones. In the next chapter, I will explain how neural activity can shape the connections between neurons.

A Role for Cell Death Throughout Brain Development

You might think it odd that cell death plays a physiological role in the developing brain. In fact, we have known for many decades that cell death is widespread during the normal development of both neural and non-neural tissues. It is important not just for removing damaged cells but also for removing healthy ones. The removal of some healthy cells helps sculpt developing brain tissues to give them their mature shapes and sizes. It also removes cells that have completed their transient functions and so are no longer needed. This type of cell death is often referred to as naturally occurring cell death.

A role for naturally occurring cell death in sculpting early brain tissue is exemplified by its higher incidence along the dorsal midline of the early forebrain, keeping the growth of this region lower than that of the ballooning right and left hemispheres on either side of it. Another example is that, later in development, cell death can provide a mechanism for fine-tuning the numbers of neurons in some brain regions to match the numbers in regions they connect to. Scientists have observed this happening in many circumstances. They have often seen instances of the brain initially producing more than enough neurons and then removing some later, once neural circuits have formed, leaving only the numbers required for those circuits to work efficiently. In some cases, neurons carry out specific functions before being removed. The subplate, which I mentioned earlier (see Figure 1.7), is an example of a transient neuronal structure providing temporary circuitry that acts as a scaffold for persistent circuitry.

Naturally occurring cell death is accomplished through a well-defined biochemical process that we call **apoptosis**, in which cells effectively commit suicide (you can read more about this process in *Building Brains; An Introduction to Neural Development*[10]). They activate an intracellular death program, which is sometimes referred to as programmed cell death. For development, survival and well-being, control over such a potentially lethal intracellular program is essential and its breakdown is likely to be a factor in many disorders.

Summary

Figure 1.12 summarizes the main cellular events occurring during human brain development.

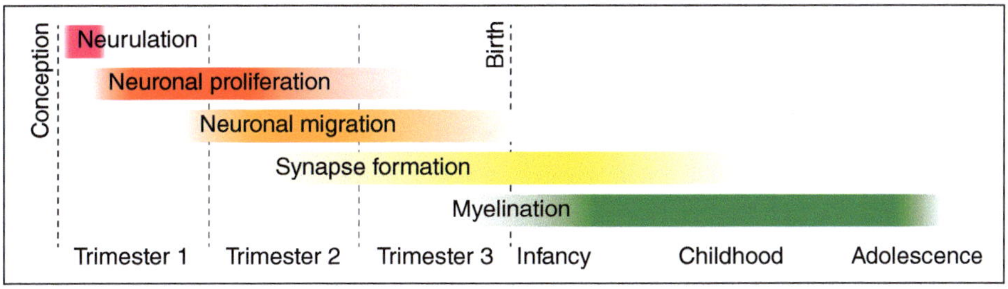

Figure 1.12 **A summary of major cellular events in brain development.**

Major take-home messages from this chapter are as follows:

- Neurons are the brain's fundamental functional units, although many other cell types are essential for the brain's structure and function.
- Research on human brain development presents many challenges, necessitating the use of multiple scientific methods involving, for example, the use of non-human species or human cells grown in culture.
- Our brains start to diversify from the moment they first emerge in the developing embryo, which occurs during the first few weeks after conception.
- During the first trimester, signalling molecules allow cells to communicate with each other in the rapidly growing brain and this process guides the formation of specific brain regions and structures.
- Prenatal brain growth involves the production of cells in specialized proliferative zones followed by their subsequent migration to the sites where they will develop their mature cellular structures and functions.
- The process by which neurons become interconnected in complex neural circuits, which involves the growth of axons and the creation of synapses, begins prenatally but much remains to be done by birth.
- Paradoxically, naturally occurring cell death is critical for brain development because it sculpts brain tissues, regulates cell numbers and removes unwanted cells.

2

Brain Function and Its Development

It is only in the last 200 years or so that we have reached broad consensus on the brain's main functions. Although Egyptians wrote the first known descriptions of the brain and its anatomy about 3,700 years ago, it was another 1,200 years before Greek philosophers identified the brain as the organ responsible for our cognitive functions. Around 400 BCE, for example, Hippocrates (460–375 BCE) declared: '... from the brain, and from the brain only, arise our pleasures, joy, laughter and jests, as well as our sorrows, pains, griefs and tears'. However, not everyone agreed. For example, Aristotle (384–322 BCE) believed that the heart was responsible for sensation, intelligence and mental processes.

Over the next 2,500 years, the work of scholars including Galen of Bergama (129–216), Ibn al-Haytham (965–1040), Leonardo da Vinci (1452–1519) and Rene Descartes (1596–1650) improved our understanding until, by the start of the 19th century, the brain's importance as the organ of perception and higher mental function was beyond serious doubt. Much remained to be done, however, to figure out how it works. The first stabs in the late 18th and early 19th centuries were severely misguided, mainly because they were built on little or no real scientific evidence. The popular pseudoscience of phrenology, for example, proposed that each of the numerous aspects of our personalities, such as our tendencies to love, hope, fight, create, exercise caution, act benevolently and so on, are each controlled by a specific part of the brain. As the 1800s wore on, these erroneous ideas were quickly overtaken by others that have survived the test of time.

Through the late 19th and early 20th centuries, scientists including Pierre Paul Broca (1824–1880), Carl Wernicke (1848–1905) and Wilder Penfield (1891–1976) found credible scientific evidence supporting the subdivision of the brain into discrete areas with different specific functions. The functions they studied were not the elaborate, high-level types envisaged by the phrenologists, but more fundamental ones such as the ability to see, hear, touch, control movement and so on. Their work was based on studies showing that patients with brain damage limited to anatomically distinct regions have difficulty with specific functions and that stimulating specific parts of the brain cause movements in specific parts of the body.

Exactly how our complex perceptual, cognitive and behavioural abilities are created by the maturation and interaction of the brain's component parts is still shrouded in mystery. We can grasp the principles, but cannot follow the detail. We are certainly unable to explain how our brains create *specific* perceptions, thoughts or behaviours in a given individual in particular circumstances. This contrast with analyses of simple organisms that contain only a few neurons and a small repertoire of behaviours, such as hydra, a tiny creature related to jellyfish. In such creatures, we can study each neuron simultaneously in real time and follow how the activation of simple networks of neurons can produce specific responses to particular stimuli in an individual organism. In the human brain, we cannot do this; everything is so many orders of magnitude more complicated. It is beyond our reach to follow, step by step, how specific activation of our brains' neural networks generate their behavioural consequences. Despite all we have learned, when it comes to explaining how the components of our brains produce our individual personalities, we are not much further forward than the phrenologists, although we are at least heading in the right general direction. This lack of detailed knowledge limits our ability to explain how variation in the nature and interactions of the brain's major components can cause neurodevelopmental diversity and disorder.

In Chapter 1, I explained how the brain's main regions emerge during prenatal development and how they get connected up. I will start this chapter by explaining what these main regions do. I will then dive down to consider how they do it by discussing the microscopic workings of their individual cell types and how their properties allow the brain's experience and patterns of activity (which become increasingly elaborate postnatally) to contribute to their development. I will then resurface to provide a holistic view of the brain's functional development. Linking the broad principles of what goes on at the different levels is possible, but cellular explanations for most specific brain functions are highly incomplete. We have limited understanding of how the activities of individual cells and their networks within and across the brain's major regions generate the complex ways in which our brains function.

The Functions of the Brain's Major Regions

The Brain's Major Anatomical Features

In Chapter 1, I described where many of the brain's major regions came from during prenatal development. I will briefly review the anatomy of the mature brain here, before moving on to discuss the functions of its main components.

If you held a post-mortem human brain in your hands, the main thing you would see is the cerebral cortex, which lies just beneath the skull and spreads across the brain's two hemispheres. It looks like a giant walnut (Figure 2.1). If you were to cut into the brain, carefully dissecting its visible parts by naked eye, you would see it is made up of different shaped regions distinguished by subtly different colours and textures (for example, the brain's grey and white matter, which differ in appearance because the former contains many cells and the latter contains many myelinated axons). These regions surround fluid-filled spaces (the **ventricles**, which arose from the lumen of the **neural tube**, see Chapter 1). These regions would appear more or less the same in virtually all of us. We call this the gross anatomy of the human brain; you do not need a microscope to see it.

The walnut-like cerebral cortex that is visible without cutting into the brain is actually just a part, albeit a big part, of the cortex. It is the part we call the neocortex (*neo* means new

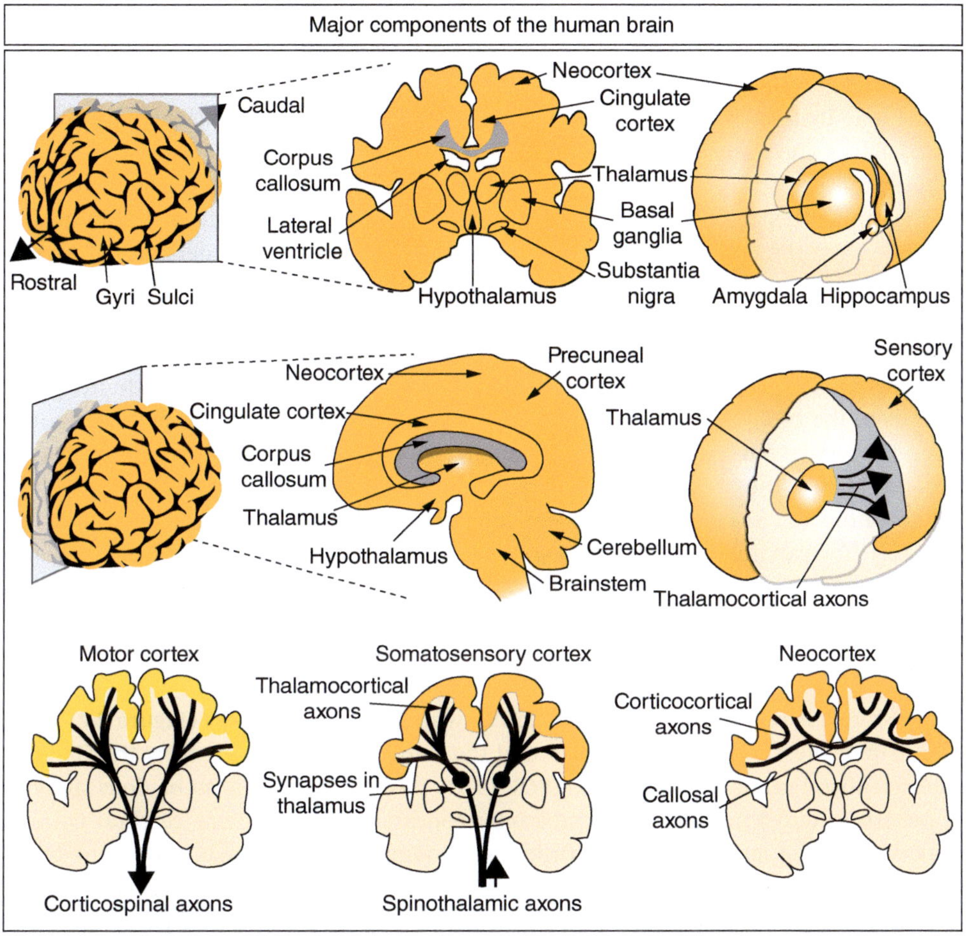

Figure 2.1 **Some of the mature brain's main regions and axon tracts.** In the top and middle rows, sections cut through the brain (centre) or the removal of the front of the left cerebral cortex expose some of its main internal structures. The bottom row shows some of the brain's major axonal tracts, descending from the motor cortex, ascending to the somatosensory cortex or linking different parts of the cortex together.

in Greek and refers to the fact that this is the part of the cortex that has expanded massively in mammals, and particularly in humans compared to other species). Neocortex is sometimes called isocortex or neopallium, but this is relatively uncommon. Its ridges are called gyri (singular, gyrus) and the valleys between them are called sulci (singular, sulcus), and if you were to stretch it out, you would see that it has a large surface area roughly equal to a square with sides half a metre long. The neocortex has six major layers of cells; you would need to stain slices of it to see this under a microscope. The allocortex (*allos* means other in Greek) is the much smaller but no less important part of the cortex tucked away from immediate view around the edges of the neocortex. It has less than six major cell layers, most parts of it have only three. Its best-known component is the **hippocampus**, which is essential for memory (the left hippocampus is drawn in Figure 2.1). The cerebellum is partially visible under the posterior part of the neocortex. It is much smaller than the cerebral cortex in volume, but its very highly convoluted surface, if stretched out, is only about 20% smaller than that of the neocortex. It is important for a range of functions including movement, balance and cognitive activities.

Slicing through the brain would reveal its internal structures. Neurons are packed into the grey matter beneath the surface of the cerebral cortex and cerebellum and in deeper regions such as the **basal ganglia**, with white matter containing high densities of **myelinated** axons in between. The two hemispheres of the brain are connected along their anterior-to-posterior length by a massive bundle of axons that we call the **corpus callosum**. Underneath are two major structures: the **thalamus**, a dense collection of neurons that transmit signals arriving via axons from sensory cells and organs outside the brain onwards to the cerebral cortex; the **hypothalamus**, which regulates hormone secretion. Around these structures cluster the basal ganglia, which are large groups of neurons whose functions include the control of movements among other things (Figure 2.1).

Some of the largest axonal tracts that link the major parts of the brain are illustrated in Figure 2.1. You can link these drawings with the finer details of connections in the cerebral cortex shown in Figure 1.7 in Chapter 1.

The Functions of Human Neocortex

The neocortex is regionalized into many areas with different functions (Figure 2.2).[22] An early map of these areas was drawn around the start of the 20th century by the German neuroscientist, Korbinian Brodmann (1868–1918). It was based on his meticulous study of characteristic differences in the microscopic appearances of different cerebral cortical areas. With some subsequent modifications, Brodmann's map remains relevant today because the areas he defined turned out to correspond well with functionally distinct regions of the cerebral cortex, as defined by evidence including the effects of localized brain damage and regional activity recorded from live brains using modern imaging methods (which I will describe later in the chapter). These functional differences are due in large part to regional differences in patterns of incoming and outgoing axonal connections (which we call afferent and efferent connections respectively). These link each region in specific ways with cells in other parts of the brain and throughout the body. The bottom line is that the cerebral cortex of humans can be subdivided into more than 50 distinct areas based on features including their microscopic appearances, their patterns of connections with other brain regions and their functions.

For example, Brodmann's Areas 1–3 (Figure 2.2), which are rather centrally positioned in each hemisphere, are innervated by afferent axons carrying electrical signals that will generate an awareness of parameters such as touch, temperature, body position and sensations of pain originating from sensory cells throughout the opposite side of the body (note that it is a general rule that each side of the brain represents or controls mainly the opposite side of the body). We call these inputs somatosensory (*soma* means body in Greek), and we call Brodmann's Areas 1–3 somatosensory cortex. Brodmann's Areas 17–19, which are in the posterior-most cortex, are major components of the visual cortex; Areas 41 and 42 are auditory cortex; Areas 44 and 45 correspond to Broca's area, which is important for producing

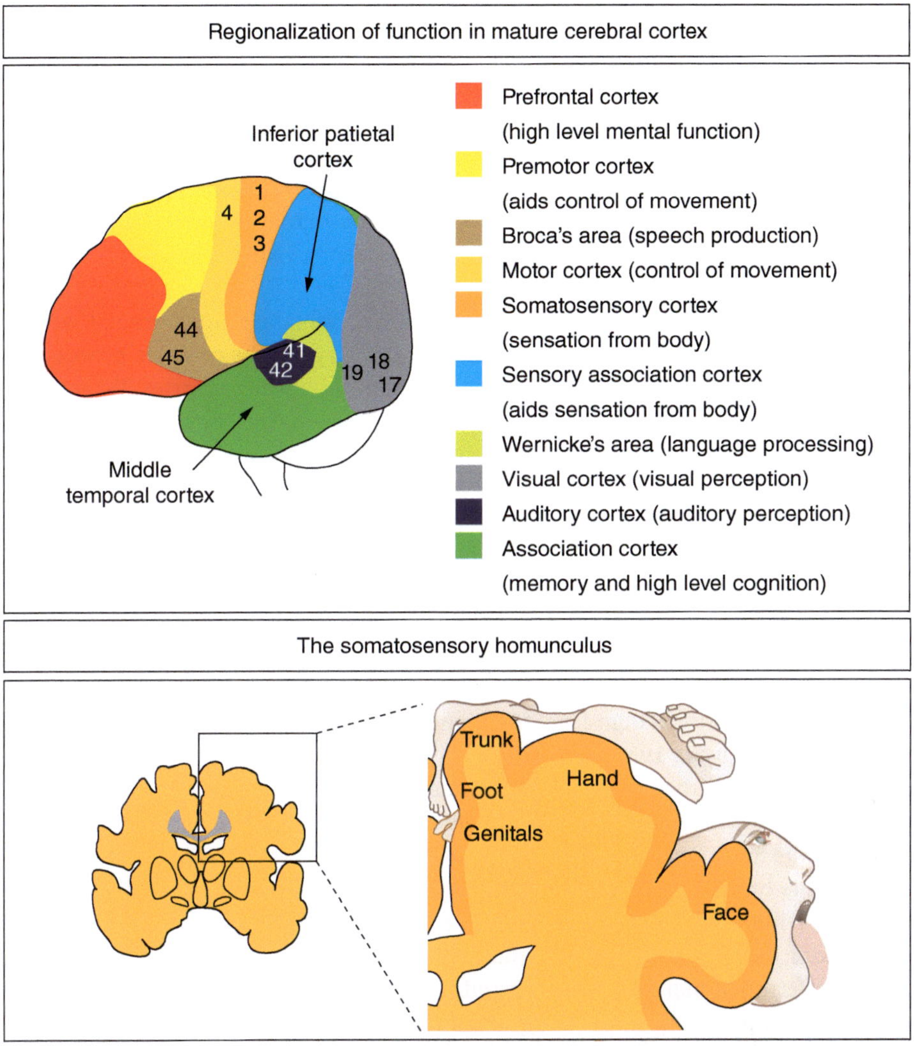

Figure 2.2 **The organization of the cerebral neocortex.** The regionalization of function is illustrated, with the numbers marking the positions of some of Brodmann's areas (see text for explanation). The somatosensory homunculus, in which the parts of the body are mapped onto the neocortex in an orderly manner, shows how some parts – the ones that are most sensitive, such as the face and hands – occupy more space in the brain than others. Essentially, this illustrates the general principle that our brains have evolved to care more about some things than others.

speech; Area 4 is motor cortex, which signals through axons descending to the spinal cord to control muscles on the opposite side of the body (corticospinal axons, see Figure 2.1).

The functions of some areas of the cerebral cortex, including those discussed in the previous paragraph, are understood better than others. We tend to know much more about those areas that contain representations of the world around us (encoded in the form of electrical impulses) or that generate movements of specific parts of the body. This is because we can study them with a variety of well-developed and precise methods: to study sensory areas, we can use methods such as live brain imaging or electrical recording devices to figure out how their cells respond to precisely defined stimulation of their sensory inputs; to study motor areas, we can electrically stimulate cells to discover what movements ensue. Much of what we know about these areas comes from work on non-human species in which their structures and functions are highly conserved. Neuroscientists sometimes describe these types of cortical areas as unimodal, in that their cells respond mainly to electrical signals representing just one type of sensation (such as somatosensory, visual or auditory) or their main function is to activate movement of the body.

There are, however, many parts of the brain, and particularly the human brain, that are much harder to study even in non-human species. They carry out the complex processing that occurs between input arrival in sensory cortex and output via motor cortex. We refer to these as areas of association cortex.[23] They are located in prefrontal cortex, cingulate cortex, inferior parietal cortex, precuneal cortex and middle temporal cortex (you will find these areas marked in either Figure 2.1 or Figure 2.2). Association cortex tends to be multimodal; for example, its cells often respond to and integrate electrical signals representing multiple types of sensation. The evolution of some of these areas in the human brain has so outstripped that of equivalent parts of other species' brains that it is questionable how much we can learn about them without focusing on human brains themselves. The most obvious part of the brain where this applies is the prefrontal cortex, which is the region of our brains that lies anterior to the motor cortex and which has expanded massively in us.

Some insight into the critical high-level function of the prefrontal cortex came from now infamous attempts, from the 1930s to the 1950s, at surgical treatment of individuals who were psychologically disturbed or whose behaviours were difficult to manage. Many thousands of such individuals around the world were given prefrontal lobotomies (another name for them was leucotomies), in which the connections between the prefrontal cortex and the rest of the brain were severed in a variety of ways. This had profound effects on individuals' behaviour and personality, inducing a state that might best be summarized as apathy and risking damage to their intellectual abilities. Since then, the use of non-surgical techniques such as functional brain imaging and electroencephalography (which I will discuss later in this chapter) has convincingly demonstrated the importance of human prefrontal cortex for our brains' most complex cognitive functions. Much emphasis has been placed on its roles in the flexible planning and execution of actions aimed at achieving specific, often intellectually challenging, goals in ways that draw on prior knowledge and experience, current context and imagined possible consequences. Prefrontal cortex is also critical in other cognitive attributes such as the generation and regulation of emotions.[24]

The Functions of the Hippocampus, Cerebellum and Basal Ganglia

Known publicly for decades only by his initials H.M., Henry Molaison (1926–2008) is one of the most famous patients in the history of neuroscience. In 1953, a surgeon removed parts of his right and left hippocampi and some adjacent brain regions, including the part of the neocortex next to the hippocampus (the parahippocampal gyrus) and the amygdala (Figure 2.1), as a treatment for severe debilitating epilepsy. While this did control his epilepsy, it also interfered profoundly with his memory.[25] The effect was quite specific; other aspects of his perception and thinking were unaffected. His case, and other similar ones, established the hippocampus and surrounding regions as areas with major functions in the formation of memories.[26] This work also laid a foundation for current theories that distinguish different types of memory, including short-term memory, episodic memory and procedural memory. Short-term memory refers to the retention of information for time periods most appropriately measured in seconds. Episodic memory is what most people would intuitively think of as memory: the ability to remember specific events or episodes in the past. Procedural memory refers to the skills learned gradually through repetitive practice, for example riding a bicycle. Whereas a person's episodic memory relies heavily on activity in their hippocampus, procedural memory involves additional structures such as the amygdala, the neocortex, the cerebellum and the basal ganglia.

The cerebellum's best-known and probably main function is in controlling and refining movements that are initiated elsewhere. Its damage causes problems with motor control and posture, for example **cerebellar ataxia** (see Chapter 3). The cerebellum makes movements more precise by modifying the signals that descend from the cerebral cortex. We do not fully understand how it does this, but the end result is that it coordinates the timing and force of action of the different muscle groups required to produce fluid movements. One of its coordination roles is achieved by integrating sensory information from multiple sources throughout the body, allowing it to initiate actions that compensate for shifts in body position and thereby maintain balance. The cerebellum is important for motor learning and higher cognitive functions, including memory and language.[27–30]

The basal ganglia are large clusters of neurons that lie beneath the cerebral cortex. They are involved in generating precision of movement through their interconnections with the motor cortex and cerebellum. Their disruption causes movement disorders. The basal ganglia are also involved in other processes such as learning motor skills, decision-making and emotions. Their connections with the prefrontal cortex probably mediate a significant part of their contribution to such high-level functions. The basal ganglia are often thought of as gatekeepers, choosing which actions to allow and which to prevent.[31]

The Organization of Function Within Brain Regions

The activities of neurons in most if not all of the regions of the brain that I have discussed above are organized into some type of map. In daily life, most of us will use the term map in its geographical sense, meaning a representation of features on the surface of Earth, such as roads, mountains and so on. In these maps, adjacent points on the planet are recorded as adjacent marks on a suitable surface, such as paper or a computer screen. In the broader sense of the word, however, a map is any spatially ordered representation of a given area or feature of an object. The axonal wiring of the brain is such that adjacent points in each region of the brain tend to interact with (meaning they either respond to or activate) adjacent points either in other brain regions, or in the rest of the body or in the outside world.

Probably the most recognizable examples of maps in the human brain are the homunculi in the cerebral neocortex, which are representations of the human body (Figure 2.2). For example, the somatosensory homunculus maps the sensory input from adjacent body parts to adjacent groups of cells across the somatosensory cortex; the motor homunculus maps adjacent groups of cells in the motor cortex to adjacent muscles in the body. These maps were described in the 1930s by Wilder Penfield, a neurosurgeon working in Canada, while he was conducting surgery under local anaesthesia to treat epilepsy and brain tumours in awake patients. Penfield stimulated either the sensory cortex, asking the patient to report the locations of the sensations, or the motor cortex, asking them to report the locations of muscle movements. By systematically moving the stimulating electrodes, Penfield drew the first maps of the body as they were represented in the brain.

Over the decades, we have learned a lot about the ways that different parts of the brain are organized to generate maps. On the sensory side, these maps model various features of the world around us, such as the positions, shapes and colours of objects and the sources and frequencies of sound waves and so on. On the motor side, they model orderly sequences of muscle contractions that would lead to useful movements. You can read more about the details of all of these various types of maps in *Building Brains; An Introduction to Neural Development.*[10]

Neuronal Function and Its Development

Now I will dive down to the level of individual neurons and discuss how their ways of working generate the kinds of high-level functions I have been outlining, starting with how nerve cells convey information through the brain's circuits.

How Neurons Send Electrical Impulses

Neurons are excitable cells, as are some other cell types, such as muscle cells. A neuron has the ability to respond to electrical stimulation by producing impulses known as **action potentials**. These travel all over the neuron, including all the way down its axon, no matter how long it is, without fading away. This is what we mean when we say, in shorthand, that action potentials, once initiated, propagate themselves.

Excitable cells such as neurons have this ability because particular proteins (**ion pumps** and **ion channels**) in their cell membranes give them certain unique properties. There are many different types of ion pumps and ion channels, and different types of neurons use different complements of them but, essentially, they do two main things.

1) As if charging a battery, pumps actively move specific types of ions across the cell membrane (ions are particles carrying electrical charge), creating differences in ionic concentrations and hence overall electrical charge inside and outside the cell (we call this the **membrane potential**). At rest, the inside of the membrane is negatively charged compared to the outside.
2) When the membrane is subjected to localized external stimulation (which can be electrical or chemical), ion channels change their properties. This changes the cell membrane's permeability to specific types of ions in the stimulated region of the membrane. If the change allows ions to move more freely across the membrane, then the membrane potential will decrease in the stimulated region of the membrane (this is a bit like discharging a battery): we call this **depolarization**.

If depolarization in the stimulated region of the membrane is big enough, it will stimulate adjacent membrane to change its ion channel properties and depolarize sufficiently strongly to produce a similar depolarization of the next bit of membrane, and so it goes on until a wave of depolarization has travelled all over the cell – this is a propagated action potential. Figure 2.3 illustrates and provides more details on the process.

Comprehensive studies of changes in ion channels in the maturing neurons of the developing brain are difficult because there are so many of them: for example, there are more than 1,000 different types of potassium channels alone. We do know enough, however, to say that the development of neuronal excitability is highly dynamic and involves substantial changes. For example, the first action potentials to appear during early development have a longer duration than mature action potentials because they result from the activation of calcium channels, which open and close more slowly than sodium channels (Figure 2.3). Moreover, the immature cells do not yet have the ability to repolarize rapidly, which relies on their potassium channels being fully functional.

How Neurons Communicate with Each Other

Axons end in synapses. The brain's neurons affect each other by releasing chemicals called **neurotransmitters** from their synapses. When electrical activity spreading through a neuron reaches a synapse, the synapse releases neurotransmitters that, in turn, affect the electrical activity of the cell on the other side of the synapse, which we call the **postsynaptic** cell. The cell that releases the neurotransmitter is called the **presynaptic** cell.

Many axons branch many times, and they can create synapses on hundreds to thousands of other neurons. Synapses are usually located on the **dendrites** of neurons; some are on other parts of neurons, such as their cell bodies. The dendrites of each individual neuron can have hundreds to thousands of synapses ending on them. So, most neurons can send signals to many thousands of other neurons and most neurons can receive signals from many thousands of other neurons.

You will often see the presynaptic sides of synapses referred to as presynaptic boutons. They contain large numbers of small synaptic vesicles, which are tiny bags made of membrane and filled with neurotransmitter that are poised for release when an action potential comes along. The two sides of the synapse are separated by a 20–25 nm wide space, the synaptic cleft. Released neurotransmitters diffuse across the synaptic cleft and bind to neurotransmitter receptors (Figure 2.4). Some neurons make neurotransmitters that increase the chances of the postsynaptic cell becoming active: these are **excitatory** neurons. Other neurons make neurotransmitters that reduce the chances of the postsynaptic cell becoming active: these are **inhibitory** neurons. We encountered both types in Chapter 1. In mature brains, most excitatory neurons use **glutamate** as their neurotransmitter and most inhibitory neurons use **GABA** (γ-aminobutyric acid); we often refer to them as glutamatergic and GABAergic neurons. In immature brains, however, GABA can have an excitatory effect on postsynaptic cells (we explain how this comes about in Chapter 11 of *Building Brains; An Introduction to Neural Development*[10]). Other types of neurons use other neurotransmitters, including dopamine, serotonin (or 5-hydroxytryptamine, abbreviated to 5-HT), histamine, acetylcholine, noradrenaline (or norepinephrine) and **adenosine triphosphate** (**ATP**), which can all be excitatory or,

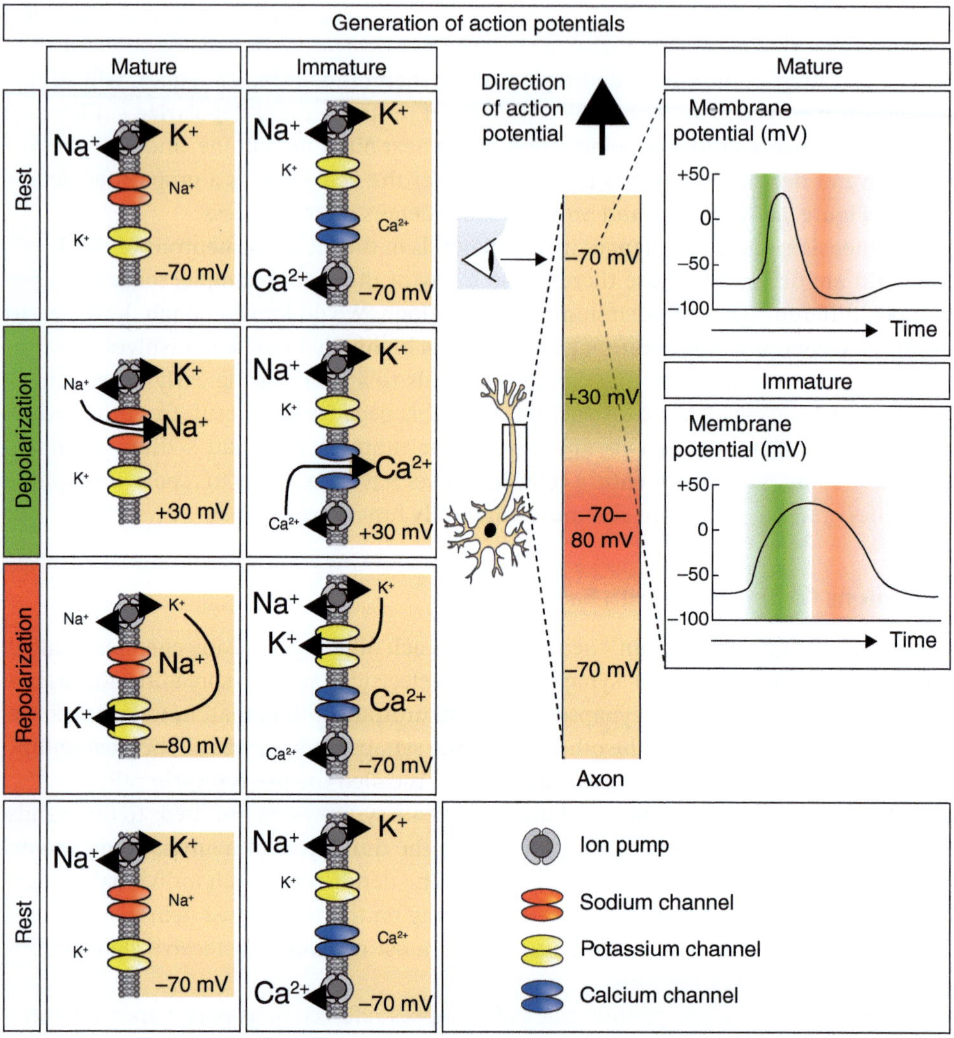

Figure 2.3 **How action potentials are propagated in mature and immature neurons.** Here I present a simplified explanation. In mature neurons in a resting state, the sodium–potassium pump exchanges three positively charged sodium ions (pumped out) for two positively charged potassium ions (pumped in), and the inside of the cell is overall negatively charged compared to the outside. It will be about −70 mV; the full explanation for this value is more complicated than the fact that more positive ions are transported out than in, relying as it does on the functions of many types of membrane pumps and channels. If a part of the cell membrane is stimulated by an external event in such a way that the inside becomes less negative, and this change is big enough (more on what 'big enough' means later, see Figure 2.5), then sodium channels in the membrane open and positively changed sodium ions rush in, making the inside positive (it will go to about +30 mV). We call this depolarization. The closure of sodium channels and the opening of potassium channels, allowing positively charged potassium ions to rush out of the cell, makes the inside negative once more (it overshoots to about −80 mV). We call this repolarization. Once the channels have all closed, the sodium–potassium pump can return the charge across the membrane to its resting state (−70 mV). In immature neurons lacking sodium channels, depolarization relies on more slowing opening calcium channels. The diagram on the right is designed to help you understand the propagating action potential. Imagine you are looking at the top of the axon (where the eye is looking), which is at rest at −70 mV. The wave of depolarization (green) followed by repolarization (red) is heading towards you as each little bit of depolarized membrane activates in turn the bit next to it. As the wave passes you, the membrane potential around you will change, as shown in the graphs.

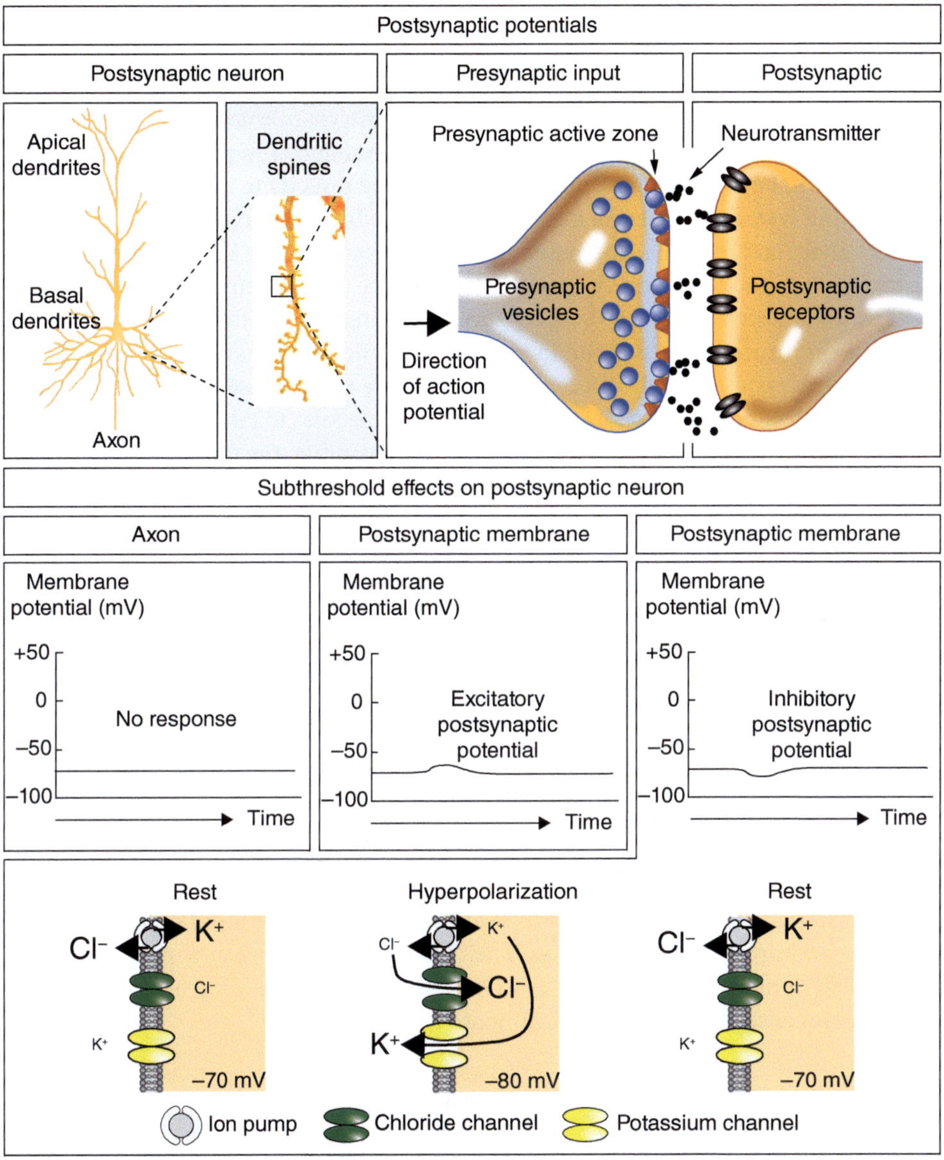

Figure 2.4 **Neurotransmission at individual synapses.** When one single lone action potential arrives at a single synapse, illustrated here as one on a dendritic spine on a cortical pyramidal cell, neurotransmitter released from presynaptic vesicles in the presynaptic endings on the input side causes small changes in the electrical potential across the postsynaptic membrane. These changes are not big enough to trigger an action potential in the postsynaptic neuron and its axon, along which any action potential would travel, shows no response (as shown in the left-hand graph, middle row). Depending whether the neurotransmitter is excitatory or inhibitory, only a small local depolarization or hyperpolarization will occur at the postsynaptic membrane (middle and right-hand graphs, middle row). The mechanism of depolarization follows the same principle as in Figure 2.3, but its size does not reach the threshold needed to trigger an action potential. Bottom row: hyperpolarization caused by the inhibitory neurotransmitter GABA is due to an exchange of chloride and potassium ions when their channels open; at rest, one of the types of ion pumps in neuronal membranes, the potassium–chloride pump, removes chloride ions from the cell in exchange for potassium.

in the case of dopamine, serotonin, acetylcholine and noradrenaline, inhibitory depending on the type of receptors they interact with on the postsynaptic cell.

When neurotransmitters bind to their receptors, the excitability of postsynaptic neurons changes. This can be because the neurotransmitters bind directly to cell membrane channels to alter the flow of ions across them (we call these **ionotropic receptors**, reflecting their direct effect on ion flow). Alternatively, neurotransmitters can bind to receptors that affect ion channels indirectly as a result of their influence on biochemical reactions within the cell (we call these **metabotropic receptors**, reflecting the fact that their direct effects are on metabolic processes, which can in turn affect ion flow). As you might imagine from this, ionotropic responses, being direct and electrical, are usually faster than metabotropic ones, which need chemical reactions involving other molecules to take place.

When postsynaptic receptors interact with a neurotransmitter, the flow of ions across the membrane through the postsynaptic channels changes the membrane potential, effectively converting the chemical signal released by the presynaptic bouton back into an electrical signal. The postsynaptic membrane potential can change in one of two possible directions. At excitatory synapses, the ionic flow causes the postsynaptic cell to depolarize (that is, the inside of the cell becomes less negative relative to the outside). We refer to this depolarization as an excitatory postsynaptic potential (EPSP) (Figure 2.4). At inhibitory synapses, the ionic flow causes the inside of the postsynaptic cell to become even more negative relative to the outside of the cell, making it less likely that the cells will generate an action potential; we say the cell becomes hyperpolarized. This hyperpolarization is called an inhibitory postsynaptic potential (IPSP). It involves not only the exit of positively charged potassium ions as their channels open but also the entry of negatively charged chloride ions, whose channels open too (Figure 2.4).

At any individual synapse, each EPSP or IPSP is too small on its own to have much effect and certainly a single EPSP will not trigger an action potential. EPSPs and IPSPs must combine to have an impact. The fact that each cell has a huge number of synapses that can generate a vast number of combinations of IPSPs and EPSPs, depending on the activity of the presynaptic cells, gives the brain its immense computing power.

How the Brain's Neurons Work Together to Make Decisions

Please do not think I am going to explain how the cells in your brain combine in specific ways to make what we call decisions in our everyday life, such as whether to read this book or not, when to have lunch and so on. That would be impossible. We can only understand the principles involved in making decisions at that level. Here, I mean decisions on whether individual neurons become active or not. The moment-to-moment everyday decisions our brains make as a whole are due to the split-second decisions made by millions or billions of their individual neurons. We cannot explain the details of how decisions by individual neurons combine to make any specific high-level decision by our brain as a whole.

The brain's neurons make decisions based on what their inputs are doing. To understand the principles by which this happens, it is essential to reiterate that the excitatory postsynaptic potentials produced at individual synapses that I described in the previous section are *not* action potentials. They are not big enough to trigger a propagated wave of depolarization all over the cell (an action potential) (Figure 2.4).

To trigger an action potential, multiple excitatory postsynaptic potentials must combine. The magnitude of excitatory postsynaptic potentials dwindles away rapidly with time and distance from the synapse, and so their combination is only effective if they occur close enough to each other and at more or less the same time. Inhibitory postsynaptic potentials counteract the effect of excitatory postsynaptic potentials. We refer to the combination of excitatory and inhibitory postsynaptic potentials at any given region on the membrane of a neurons as summation. If excitatory and inhibitory postsynaptic potentials summate to cause a big enough nett depolarization, the postsynaptic neuron will generate an action potential that will then be passed on at its synapses to affect the next set of cells and so on. In other words, it will have made a decision to activate an action potential because the simultaneous excitation from inputs at a particular place in its membrane exceed the simultaneous inhibition arriving at the same place by an amount that is big enough to trigger an action potential. Figure 2.5 illustrates how this process works.

The ways in which electrical signals ebb and flow through the complex neuronal circuitry of our brains depend on two essential aspects of that circuitry: first, its structure, principally the spatial location of all the synapses of every excitatory and inhibitory neuron; second, the activity of the synapses at any given time. Synapses that are close together have the potential to influence whether the neuron they are on sends action potentials on to the next stage of the circuitry. Whether any given neuron does or does not send action potentials to the next stage is determined by the patterns of activity in its presynaptic inputs. The processes of development outlined in Chapter 1 set up the connectivity; what every part of the brain is doing at any given moment in time determines what every other connected part does.

Now, let us explore the phrase I just used, 'every other connected part', a bit more. Really importantly, the inputs a neuron gets from 'every other connected part' are not equal in strength and their strengths are not static and unchanging. They are malleable and, what is more, they undergo long-term alterations that depend on how the brain has been activated in the past. In other words, synapses are affected by their previous activity. It follows, therefore, that the decisions our neurons – and ultimately our brains as a whole – make are affected by the activity they experienced in the past. In the next section, I will outline how activity in a brain's circuitry can modifying the strengths of synapses.

How the Brain's Activity Affects the Function of Its Synapses

Experience-dependent synaptic changes allow the brain to adapt, learn and create memories, which are essential aspects of our cognitive abilities. The evidence for this is overwhelming, and we refer to this ability of synapses as **synaptic plasticity**. On the flip side of this coin, we can speculate, with no direct evidence, that this property might provide a mechanism by which adverse experiences might put the brain at risk of generating patterns of neuronal activation that manifest as unhelpful or undesirable thoughts and behaviours.

In 1949, the Canadian psychologist Donald Hebb published his book 'The Organization of Behavior: A Neuropsychological Theory' containing a highly influential theory on how the brain learns.[32] The theory seeks to explain how connections between nerve cells change based on experience. Remember that, as far as our brains are concerned, experiences amount to previous patterns of electrical activity, which are themselves transduced

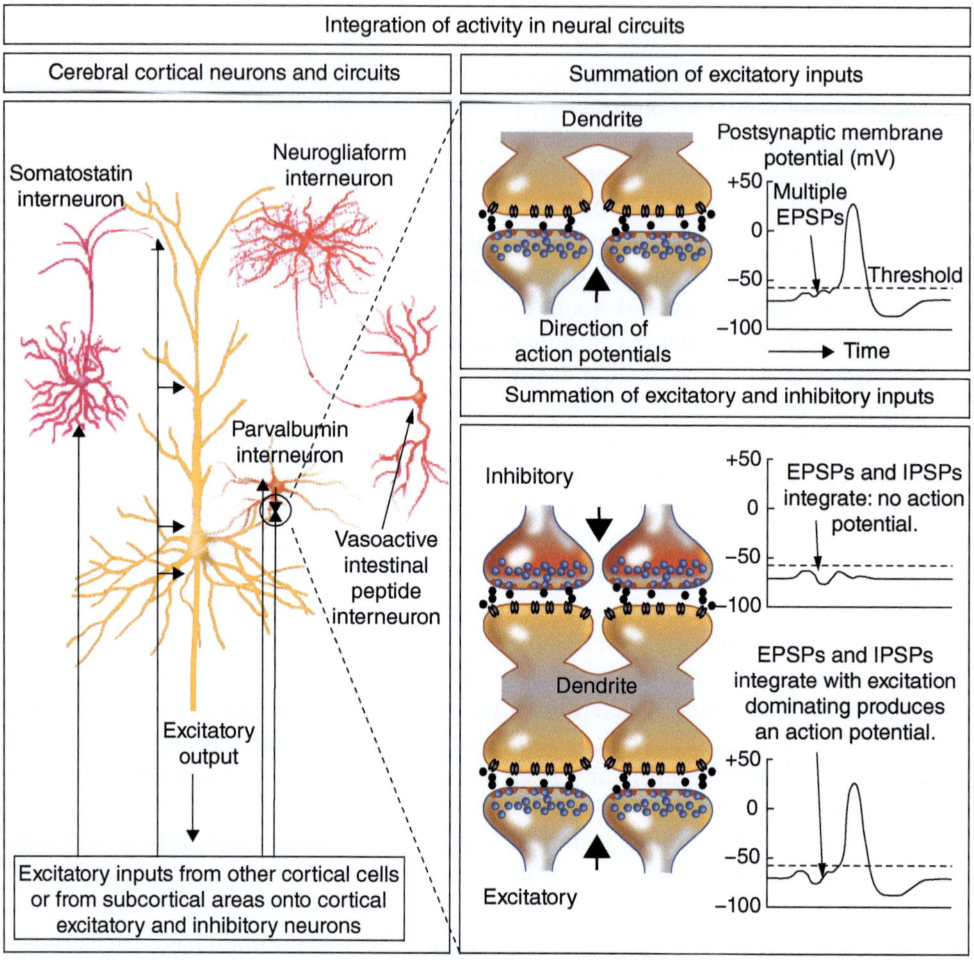

Figure 2.5 **Summation of excitatory and inhibitory inputs onto neurons.** This figure illustrates the principle that, in any postsynaptic neuron such as the cortical pyramidal neuron shown here in orange, which receives both excitatory input and inhibitory input, excitatory and inhibitory postsynaptic potentials (EPSPs and IPSPs; Figure 2.4) are integrated. If their summed value is above a certain threshold, the neuron will initiate an action potential. In this example, the cerebral cortical pyramidal cell is surrounded by four inhibitory neurons of different types, shown in red. They use GABA to produce inhibitory postsynaptic potentials in the pyramidal neuron. Cortical inhibitory neurons are commonly categorized according to several criteria including their shape (giving us some evocative names such as chandelier or basket cells, see Chapter 1, Figures 1.8 and 1.9) or, in the case of neurogliaform cells illustrated here, because they are neurons that look like glia, and/or whether or not they contain specific molecules such as somatostatin, vasoactive intestinal peptide or parvalbumin. Note that these widely used categories are valid in general but, when it comes to detail, there is no universally agreed infallible classification. Also note: you will often see these inhibitory neurons described as inhibitory *inter*neurons because they make only local connections, but do not use or interpret the term interneuron as synonymous with inhibitory because interneurons can be excitatory, as in the case, for example, of cerebral cortical spiny stellate cells (Chapter 1, Figure 1.7). In this greatly simplified circuit, excitatory inputs target both the excitatory and inhibitory cells. The diagrams on the right zoom in on an area where inputs onto the pyramidal neuron from other excitatory neurons and the parvalbumin inhibitory interneuron converge. If only excitatory inputs are active, summation is more likely to generate an action potential in the pyramidal neuron; if inhibitory inputs are similar in activity to excitatory, no action potential is likely; if excitatory inputs are more active than inhibitory ones, an action potential will be produced if the threshold for activation is exceeded.

representations of previous events experienced by the person. Hebb's theory is that connections that are used more frequently become stronger and faster. His idea, encapsulated by the following quote from his book, has formed the basis of our understanding of how the brain's development can be affected by its activity and how we store memories: 'When an axon of cell A is near enough to excite a cell B and repeatedly or persistently takes part in firing it, some growth process or metabolic change takes place in one or both cells such that A's efficiency, as one of the cells firing B, is increased'.

Hebb's postulate was strongly supported in the 1970s by the discovery that the strength of brain synapses can increase enduringly if they are activated frequently, a process known as long-term potentiation (LTP). The opposite can occur: if synaptic activation is too weak, the synapse can lose its strength enduringly, which we call long-term depression (LTD) (Figure 2.6). It is thought that this provides a mechanism for erasing a record (or memory) of previous activity at a synapse so that new ones can form.

How are these long-term changes brought about? The synapses that have been most strongly associated with synaptic plasticity use glutamate as their neurotransmitter (we call them glutamatergic synapses). We know most about how LTP occurs at these synapses.[33] It goes something like this (Figure 2.7).

There are two main types of glutamatergic receptors. Some respond by opening an ion channel and changing the membrane potential (ionotropic receptors). Others directly activate biochemical pathways in the cell (metabotropic receptors). We know enough about these receptors to see how they work together, at least in outline. One set of ionotropic glutamate receptors, including the **AMPA** (short for α-amino-3-hydroxy-5-methyl-4-isoxazolepropionic acid) **receptors** and kainic acid receptors, respond to glutamate by letting predominantly sodium ions (Na^{2+}) into the postsynaptic cell, thereby depolarizing it by an amount that is the product of summation (explained in the previous section and see Figure 2.5). Another set of ionotropic glutamate receptors, called the **NMDA** (*N*-methyl-D-aspartate) **receptors**, will only open if two conditions are met: (i) they are bound by glutamate *and also* (ii) the membrane around them is sufficiently depolarized to eject magnesium ions (Mg^{2+}) that otherwise block their channels (Figure 2.7). So, if the amount of summated depolarization due to the AMPA and kainic acid receptors is big enough, this will open NMDA receptors, which let mainly calcium ions (Ca^{2+}) into the postsynaptic cell. These Ca^{2+} ions, together with the actions of metabotropic receptors, activate biochemical pathways resulting in a long-term increase in the strength of the synapse.

There remains much to be discovered about how this final stage happens. One or more of several possibilities are likely. For example, there is evidence that molecular signals released from the postsynaptic cell might induce presynaptic changes. These feedback signals might cause the efficiency of glutamate release to increase, perhaps due to changes affecting molecules involved in synaptic vesicle fusion with the presynaptic cell membrane. Such presynaptic changes might occur not only in glutamatergic synapses but also in GABAergic ones, which can also undergo long-term potentiation and depression. When it comes to possible postsynaptic changes contributing to long-term potentiation, it is important to note that the postsynaptic elements of glutamatergic and GABAergic synapses are rather different. Glutamatergic synapses contain a prominent postsynaptic density, which is a densely organized meshwork of proteins underneath the postsynaptic membrane. It has many critical functions, including anchoring neurotransmitter receptors located at the

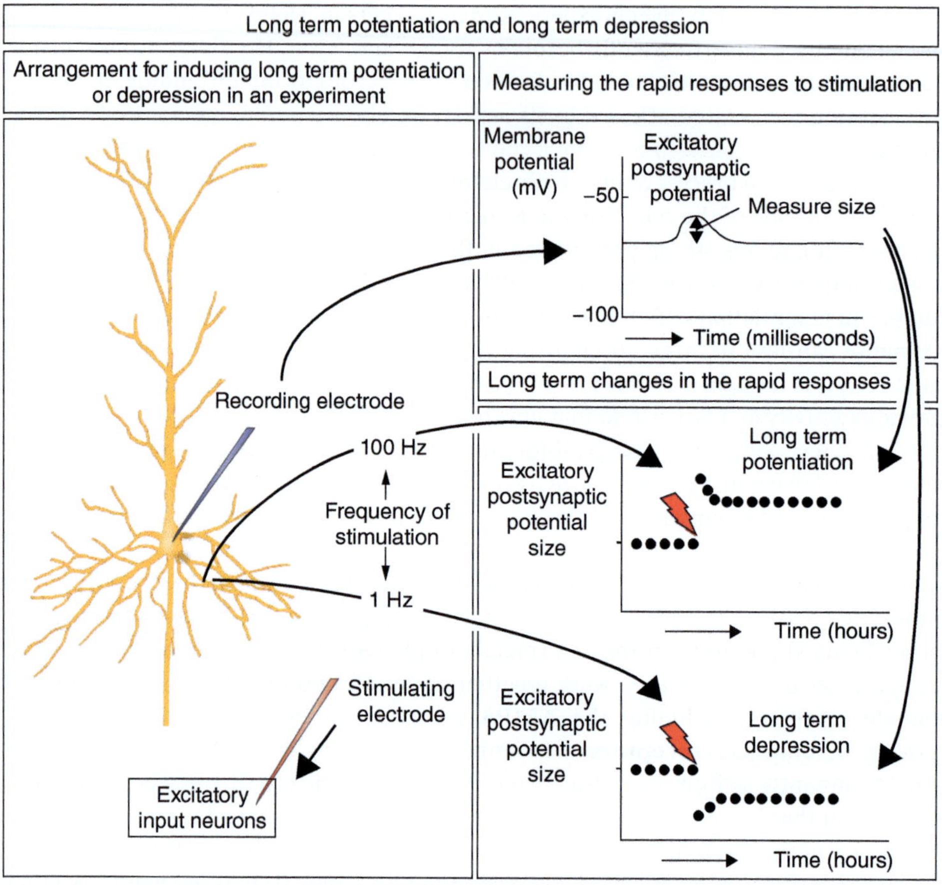

Figure 2.6 Experimental evidence for long-term potentiation and long-term depression at synapses. A typical experimental paradigm used to induce and measure LTP or LTD involves using a stimulating electrode to activate excitatory synapses ending on a neuron and a recording electrode placed into the postsynaptic neuron itself to as to discover how it responds to stimulation. The size of the excitatory postsynaptic potentials that are elicited by stimulation are measured (top right). The long-term changes, recorded at regular intervals over hours, in those sizes are shown (filled circles in the graphs below). After recording for a while, to establish a baseline, a burst of high- or low-frequency stimulation is administered. High-frequency stimulations make subsequent depolarizations of the postsynaptic cell larger when the magnitude of stimulation is kept the same as that used to establish the baseline: this is LTP. A burst of low-frequency stimulation has the opposite effect, namely a long-lasting reduction in the excitatory postsynaptic potentials elicited by the same magnitude of stimulation used to establish the baseline; this is LTD. The magnitude of the LTP and LTD varies depending on the frequency of the burst of stimulation. Most neurons are capable of undergoing either LTP or LTD.

surface of the membrane and signalling to the rest of the cell; I will discuss it further in Chapter 8. GABAergic synapses do not have a postsynaptic density like that of glutamatergic synapses. In general, how postsynaptic changes might contribute to synaptic strength is less well studied at GABAergic synapses than at glutamatergic ones. In the latter, the actions of the postsynaptic density in response to frequent activation might lead to an

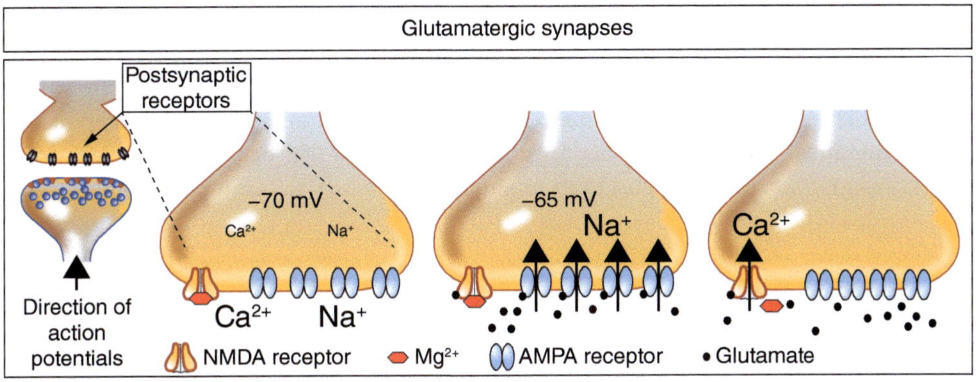

Figure 2.7 Glutamate receptors cooperate to detect different levels of depolarization. Glutamate release causes AMPA receptors (and related glutamate receptors, such as kainic acid receptors, which are not shown here) to allow sodium entry, which depolarizes the cell. If summation results in large enough depolarization – in this illustration depolarization to −65 mV is insufficient – magnesium ions blocking NMDA receptors' channels are expelled. This allows calcium ions to enter the cell to affect its biochemical processes in such a way as to strengthen the synapse.

increase in the numbers of ionotropic receptors or a modification of their efficiency, thereby increasing the strength of the synapse.

So much for long-term potentiation, what about long-term depression. This is induced when synapses are activated infrequently. If this happens at glutamatergic synapses, postsynaptic depolarization through ionotropic receptors such as the AMPA receptors will be too low to cause widespread removal of the Mg^{2+} block from the NMDA receptors, and so the rise in postsynaptic Ca^{2+} will be relatively small. While this rise is too small to activate the biochemical pathways leading to long-term potentiation, it does activate other pathways that result in the removal of ionotropic receptors from the postsynaptic membrane, which weakens the ability of the membrane to respond, resulting in long-term depression. It is likely that other changes also participate, such as feedback causing a decrease in presynaptic glutamate release (the opposite of what happens in long-term potentiation) and involving metabotropic receptors in ways that are less well understood. The mechanisms of long-term depression at GABAergic synapses are also poorly understood.

How Glial Cells Directly Assist the Functions of Neurons

I will have much to say about glial cells in Part III. Here, I will just emphasize the point that both **oligodendrocytes** and **astrocytes** *directly* affect neuronal activity in the brain. The myelin sheaths that oligodendrocytes cast around many axons in the brain allows those axons to conduct action potentials faster. This increase in speed is due a phenomenon known as saltatory conduction. Myelin is an insulator, but it does not cover the whole axon; there are gaps at regular intervals. Action potentials jump down the axons from uninsulated gap to the next uninsulated gap (saltatory means proceeding by leaping). This greatly increases the overall efficiency of the brain's activity.[34]

Among their many support functions, astrocytes often contribute to the functions of synapses, intruding via their specialized perisynaptic astrocytic processes. Together, the astrocytes, the presynaptic cell and the postsynaptic cell make what are called tripartite synapses.[35] The astrocytic processes contain many channels, transporters and receptors in their membranes, which allow them to provide stability to the local molecular environment at synapses and help support their function. I will return to this topic in Part III.

A Holistic Understanding of Brain Function and Its Development

Research on how the brain develops and works tends to focus at one of two very different levels. At one level, which I have concentrated on up to now in this book, we study how the brain's cellular components and subregions are created, how they function and how they interact with each other. This is a reductionist approach in which we try to understand the brain by analysing its parts. At the other level, which I turn to now, we take a holistic approach to look at the functions of the brain as a whole.

Ways to Study Holistically What Our Brain Does at Different Ages

How can we gain information, ideally in some quantifiable way, about the ways in which human brains function as a whole? We can learn quite a lot about the brain's functional development simply by observing an individual's behaviour. Even before birth, fetuses develop the ability to perceive and move in relatively sophisticated ways, and they can show behavioural responses to a range of external stimuli such as acoustic, tactile or chemical ones. After birth, we can find out about an individual's psychological attributes – their perceptions, thinking processes, intellectual abilities, emotions, behaviours and so on – by asking them about themselves, assuming they are old enough and able to understand and respond. We can also get information from others familiar with the individual, such as family members, friends, teachers and carers. We can do all of this by carrying out interviews, by studying a person's medical, educational or other records, or by using questionnaires (Figure 2.8). Questionnaires pose structured sets of questions that might require the respondent to select an answer from a list of options or to reply in their own words. These methods are widely used to collect information from large numbers of people including the same individuals under different circumstances or at different stages of life: for example, at my university they contribute to the analysis of the Lothian Birth Cohorts*, which is helping us to understand how our cognitive abilities change throughout life.

Another approach is to measure formally the responses of individuals under controlled conditions, for example in a laboratory or examination hall, to study any of a large range of brain functions, such as the generation, control and coordination of movement, speech, intelligence, memory, empathy, and so on. After birth, all kinds of age-appropriate assessments of an infant's or a child's perceptual, cognitive and behavioural abilities can be

* https://lothian-birth-cohorts.ed.ac.uk/.

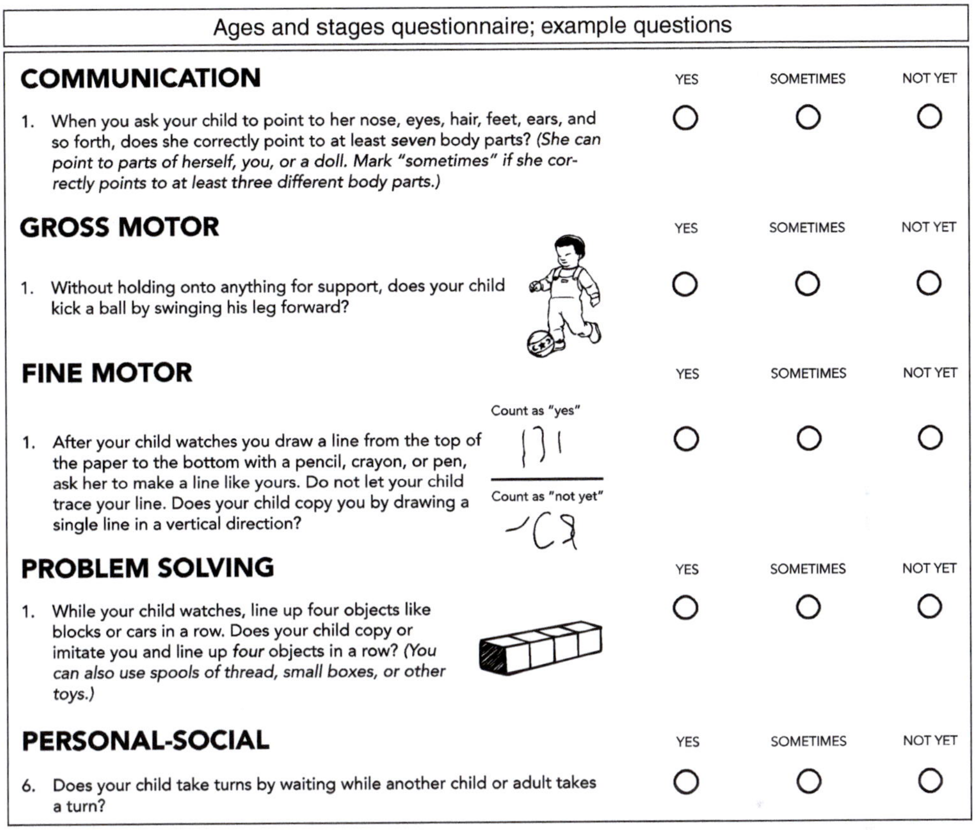

Ages and stages questionnaire; example questions

	YES	SOMETIMES	NOT YET
COMMUNICATION			
1. When you ask your child to point to her nose, eyes, hair, feet, ears, and so forth, does she correctly point to at least *seven* body parts? *(She can point to parts of herself, you, or a doll. Mark "sometimes" if she correctly points to at least three different body parts.)*	O	O	O
GROSS MOTOR			
1. Without holding onto anything for support, does your child kick a ball by swinging his leg forward?	O	O	O
FINE MOTOR			
1. After your child watches you draw a line from the top of the paper to the bottom with a pencil, crayon, or pen, ask her to make a line like yours. Do not let your child trace your line. Does your child copy you by drawing a single line in a vertical direction? (Count as "yes" / Count as "not yet")	O	O	O
PROBLEM SOLVING			
1. While your child watches, line up four objects like blocks or cars in a row. Does your child copy or imitate you and line up *four* objects in a row? *(You can also use spools of thread, small boxes, or other toys.)*	O	O	O
PERSONAL-SOCIAL			
6. Does your child take turns by waiting while another child or adult takes a turn?	O	O	O

Figure 2.8 Questionnaires for evaluating brain function. An example of a page from a questionnaire devised in the United States of America, called the Ages and Stages Questionnaire (ASQ). These questionnaires are used widely in many countries to identify developmental disorders in infants and young children. They pose age-appropriate questions aimed at detecting developmental delays between 1 month and 5.5 years of age in five sections: communication, gross motor skills, fine motor skills, problem-solving and personal and social development. An example of a question from each of these sections is shown here. The answers given are converted to numerical scores for analysis.

made (Figures 2.9 and 2.10).[36] Such tests can give valuable descriptions of behaviours, thought processes and perceptions but are unable to provide explanations for those observations. Their interpretation can be challenging, particularly in infants whose repertoire of behaviours is small and whose limited ability to communicate verbally makes their cognitive functions hard to disentangle. The exact methods used need to be appropriate for the ages and skill sets of the populations selected for study. Although this might seem an obvious point to make – for example, we can obtain behavioural but not verbal responses from infants too young to speak – it has not prevented it being overlooked or ignored, as it was, for example, in early 20th-century United States of America, when standardized intelligence tests were applied under equivalent conditions to people with different first languages and from very diverse cultural backgrounds and educational systems.[37]

Rather than evaluating the brain's functions and abilities through the responses and behaviours it produces, modern technologies allow us to record the brain's activity directly,

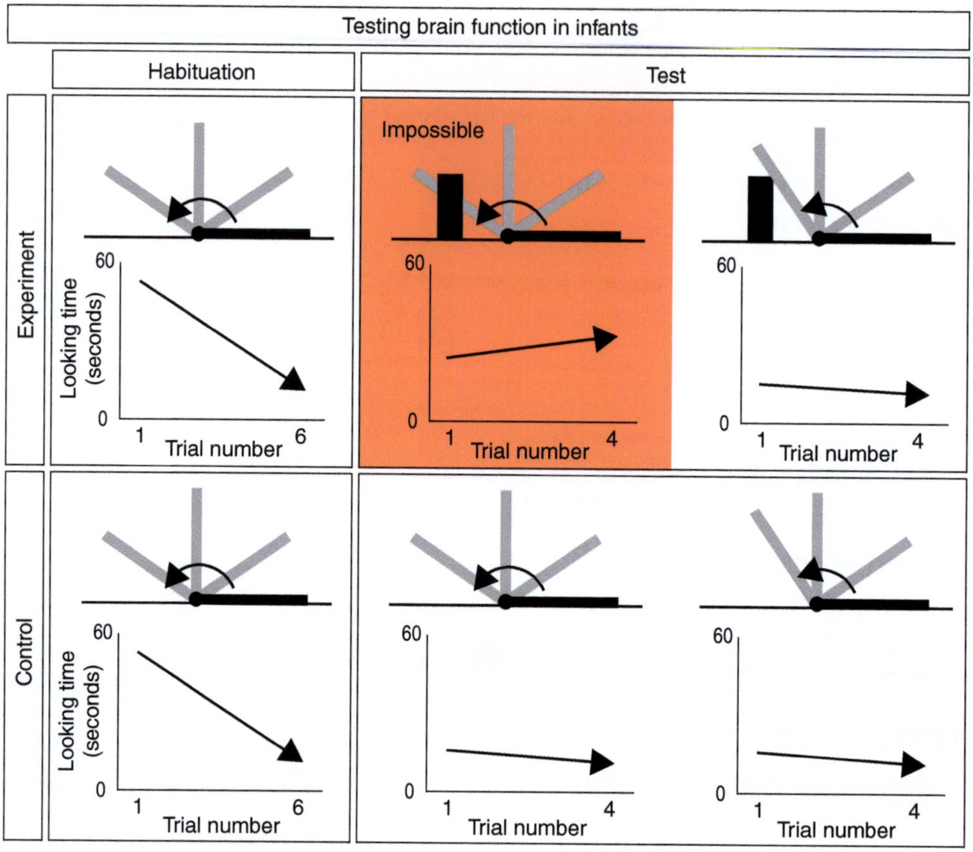

Figure 2.9 Laboratory-based tests of brain function. This shows just one of many possible examples of a laboratory test that can be applied to very young infants to study their habituation. Habituation is, essentially, their descent into boredom as they are shown the same thing in the same way over and over. The measure that is widely used to evaluate how interested or bored they are is looking time: the length of time they fixate on the stimulus (other measures, such as heart rate changes, are sometimes used). In the paradigm illustrated here, infants are repeatedly shown a sideways-on drawbridge-like motion. They rapidly loose interest over half-a-dozen or so trials, in other words they habituate. Subsequently, in the test, infants are shown one of four scenarios over repeated trials. In an impossible scenario, the drawbridge passes through a block in its path. The other three scenarios are all possible: the drawbridge stops at the block or in that same position with no block present, or it stops in the same position as in the impossible scenario, but now there is no block to impede it (this last one is just a repeat of what the infants saw during habituation). The results show that only the impossible scenario rekindled their interest, at least to some extent (this is called dishabituation), suggesting that infants' brains have expectations that are violated by this scenario.

by which I mean as directly as possible while limiting or avoiding intrusion into the body (in the jargon of the field, many of these approaches are described as non-invasive). A variety of methods developed in recent decades have enabled us to do this, in some cases even at very young stages of development (Figure 2.11), although many are currently unsuitable for studies of fetal or infant brains. Where they can be applied, they have provided useful

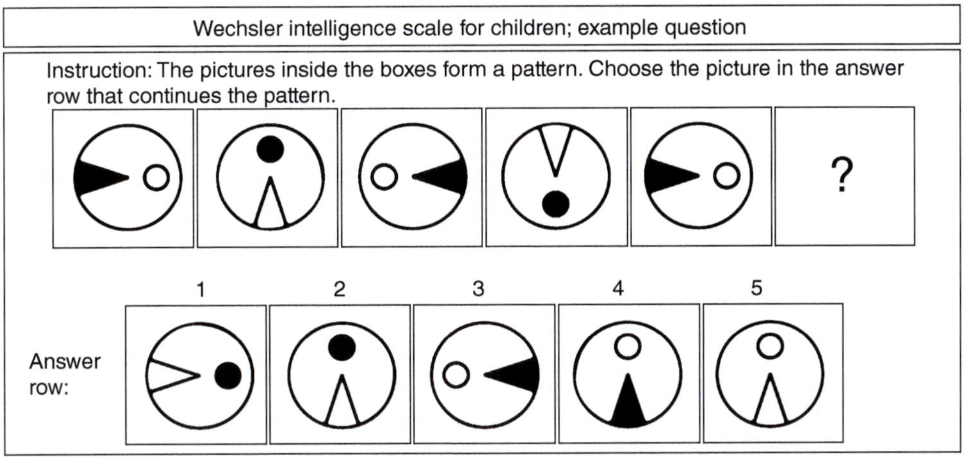

Figure 2.10 Brain function measured by standardized tests taken under controlled conditions. An example of a question from the Wechsler Intelligence Scale for Children (WISC). This intelligence test is administered individually to children aged 6–16 years. It takes about an hour and produces scores that are assumed to represent an individual's general intellectual ability (Full Scale Intelligence Quotient) as well as more specific attributes, namely: verbal comprehension; processing of visual information and spatial relationships; problem-solving using reasoning (sometimes referred to as fluid reasoning); working memory; processing speed. There is a different Wechsler Adult Intelligence Scale (WAIS) for adults. Other commonly used tests of intellectual ability include current versions of the Stanford–Binet Intelligence Scales, Woodcock–Johnson Tests of Cognitive Abilities, the Kaufman Assessment Battery for Children, the Cognitive Assessment System and the Differential Ability Scales. Whatever test is used, intelligence is a hard concept to measure and scores can be affected by factors such as nutrition, illness, socio-economic status and perinatal environment. The extent to which, and the mechanism by which, intelligence is inherited are still debated. Historically, intelligence testing has been abused to support theories of racial hierarchy now firmly rejected by modern science.

insights into cognitive activity that has not yet resulted in overt behaviour for any number of reasons, most obviously because an infant's ability to control their movements is still too immature. An advantage of directly measuring brain activity is that it might, in the future, offer a way to make earlier diagnoses of neurodevelopmental disorders and give us a better understanding of various types of neurodivergence, since these tests could be carried out before the ages at which clinicians can reliably assess a child's perceptual, cognitive and behavioural abilities and characteristics.

In general, studying brain function using advanced non-invasive functional brain imaging techniques, such as functional magnetic resonance imaging (fMRI), while subjects perform various perceptual, cognitive or motor tasks, becomes increasingly practical with age. In fact, neuroimaging in natural settings is starting to reveal important insights into the brain's ways of processing cognition, emotion and social interactions in daily life.[46] By correlating information from behavioural observations with data from brain imaging, researchers have managed to train computers, a process known as machine learning, to recognize well defined and distinct brain disorders such as epilepsy or schizophrenia.[47–50] Whether, and if so how, this approach will contribute to our understanding of other less extreme aspects of our brains and their diversity remains to be seen. One possibility is that the

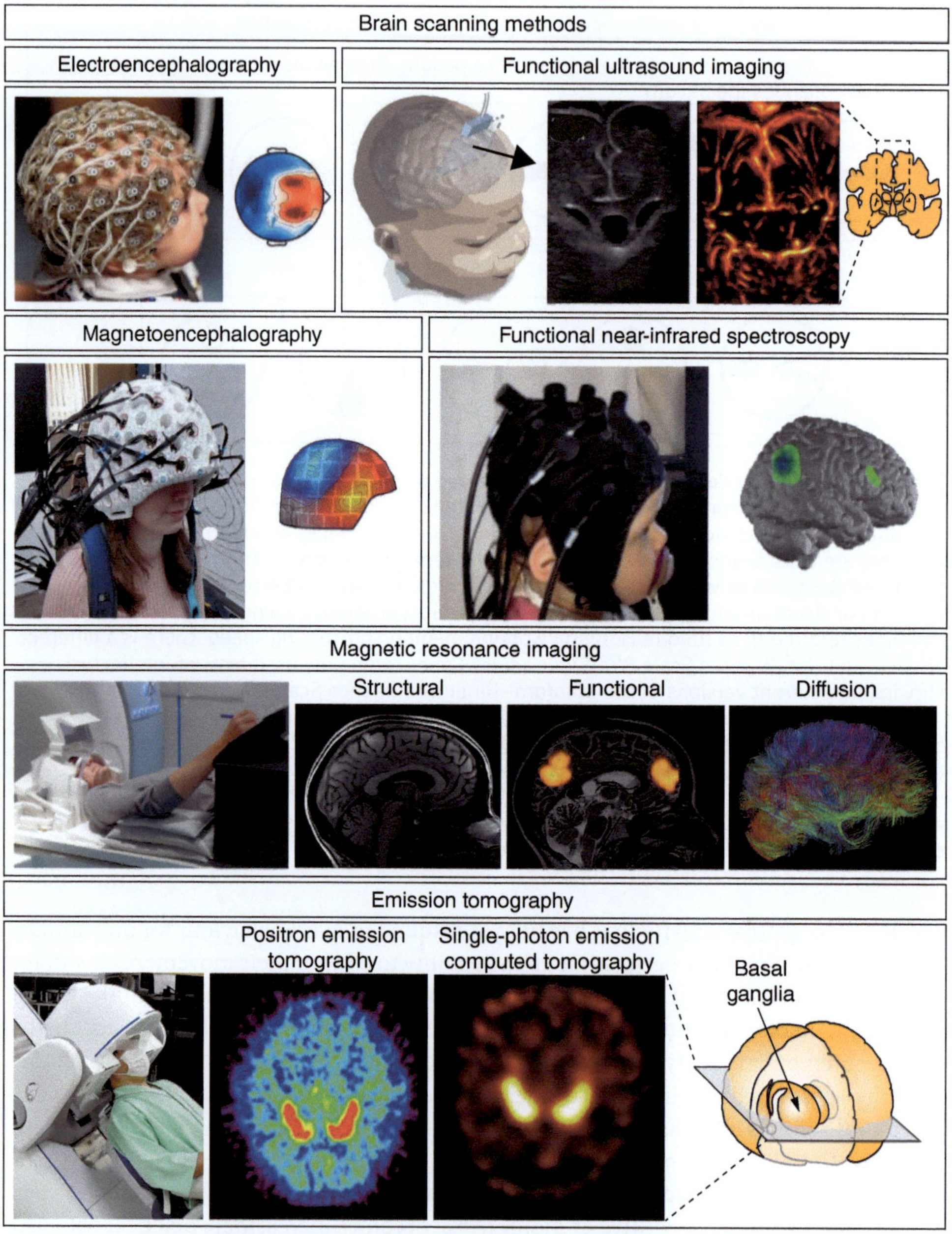

Figure 2.11 Imaging methods used to measure the brain's activity. Many non-invasive medical imaging techniques are now available for investigating brain function. Some are suitable for use in newborns and even prenatally. Each method has particular advantages and disadvantages. Electroencephalography (EEG) uses an array of electrodes placed over the scalp to detect currents at each location that arise from the electrical activity of populations of neurons at underlying locations in the brain. EEG can be used effectively even in small infants, since it requires minimal restraint to prevent artefacts due to movement. It measures the magnitude and timing of activity more precisely than its location. In the diagram here, an external stimulus (such as a word) has produced peak amplitudes (red) over the frontal lobes of the brain. Functional ultrasound imaging

imaging data will help distinguish differences in the ways in which different peoples' brains work by broadening the range of measurements we can make, taking us well beyond our current reliance on assessments of verbal or externally visible non-verbal responses to questions and other stimuli.

A Holistic View of Our Developing Brains

Health professionals refer to the approximate age ranges during which children most often acquire particular behaviours, skills or other characteristics as developmental milestones. What is included in lists of developmental milestones and how the age ranges are defined vary. The lists are not intended to give statistical precision, but rather to act as rough guides. Table 2.1 lists some examples, up to the age of six years, whose attainment involves the development of brain function (not all milestones do involve brain function, for example measures of overall body size). Milestones relevant to brain function cover gross and fine

Figure 2.11 (Continued) (fUS) uses a handheld probe to detect blood flow changes accompanying, and thereby revealing, changes in brain activity. It is useful at early ages before the bones of the skull, which are impermeable to ultrasound, fuse to close the anterior fontanel, which is ultrasound-permeable. The images show brain structure and level of blood flow in approximately the plane shown (taken from Figure 2.1). Magnetoencephalography (MEG) uses an array of sensors around the head to measure magnetic fields created by electrical activity in the brain. The diagram here shows a rapid response to a sound (red and yellow). Functional near-infrared spectroscopy (fNIRS) involves the emission of light at wavelengths suitable for measuring its absorption by blood, with changes in blood flow providing a proxy for neural activity changes, as in fUS. Altered brain activity in response to a particular task is localized to two regions (green and blue). Magnetic resonance imaging (MRI) uses powerful magnets and radio waves to affect protons in water molecules in the body in such a way that they themselves emit radio waves that a computer can use to make images of internal organs, including the brain (this is structural MRI). Functional MRI (fMRI) exploits the fact that when brain cells become more active they increase the blood flow and oxygen around themselves (this is called the blood oxygenation level-dependent, or BOLD, response; orange regions). It can localize functional signals more precisely than many other methods. It is not easy to use in infants, unless they are asleep, because it requires them to keep their heads very still for much longer than most of them can manage. More advanced MRI techniques include diffusion MRI, which examines the diffusion patterns of water and infers the existence of barriers where diffusion is found to be limited. For example, axon bundles can be detected because diffusion of water can occur along their length but is limited in other directions: this is referred to as tractography. Tractography can broadly map white matter bundles and their constituent fibres within the human brain, although its resolution is limited and its accuracy is sometimes questioned. Emission tomography involves giving radioactive substances, called tracers, to a person and then detecting where those tracers are found (note that this differs from computed tomography, or CT, which uses X-rays to analyse body structures rather than function). Positron emission tomography (PET) detects tracers that emit positrons (a type of beta radiation), whereas single-photon emission computed tomography (SPECT) detects tracers that emit gamma radiation and is less sensitive. It can be used from birth on since the levels of exposure to radiation are very low. In the images shown here, in approximately the plane shown (taken from Figure 2.1), the basal ganglia are labelled (red in PET, white and yellow in SPECT) by radioactive tracers known to localize to that specific brain region. *Images are reproduced from Turk-Browne et al.*[36] */ with permission of Elsevier (top left and 2nd row right), Baranger et al.*[38] */ Springer Nature / CC BY 4.0 (top right), Lankinen et al.*[39] *and Hill et al.* [40]*, (2nd row left), Soltanlou et al.*[41] */ Springer Nature / CC BY 4.0 (2nd row right), Lotze et al.*[42] *(3rd row left), Tymofiyeva et al.*[43] */ Frontiers Media S.A / CC BY 4.0 (3rd row right), Takahashi et al.*[44] *(bottom left), and from Verger et al.*[45] *(bottom middle).*

Table 2.1 Developmental milestones.

	PRENATAL	
WEEK	**NATURE OF FETAL ACTIVITY**	
7–8	Spontaneous fetal movements begin.	
9–27	Increasingly complex and coordinated movements, responses to various types of sensory stimulation (including touch, sound and noxious stimuli) and reflexes start.	
28–birth	Brain develops increasingly elaborate activities, such as the ability to differentiate between different tones, and gains control of breathing and body temperature.	
	POSTNATAL	
YEAR	**BEHAVIOURAL SKILLS**	**EMOTIONAL AND SOCIAL DEVELOPMENT**
Birth–1	Sits without support; pulls self to standing; rolls over without help; crawls; walks with support; drinks from a cup; uses pincer grasp; points with finger; smiles; babbles; might use first words.	From birth: shows evidence of anger, happiness or fear. 1–2 months: responds to a carer's sounds, smiles or touches. 3–12 months: starts turn-taking vocal interactions with carer and learns to manipulate immediate surroundings; distinguishes between familiar and unfamiliar; anxious with strangers; shows distress on separation; looks in the same direction as carer.
1–3	Feeds self; dresses with little help; runs; walks in a coordinated way in complex settings, for example, up and down stairs; copies a drawn line; understands simple commands; uses more words; imitates speech of others; says own name; names pictures of common objects and colours; learns to play with other children, including sharing toys and taking turns.	12–18 months: explores their environment increasingly; requests by pointing at objects of interest, coordinated with gaze to show interest; brings objects to show or give to a carer; engages in interactive play; uses gestures to wave goodbye, to communicate interests and needs; shows empathy (such as looking upset when someone else cries) and self-conscious emotions (such as pride when applauded); imitates others. 18–30 months: develops increasing autonomy; engages in pretend-play, such as talking on a toy phone; learns to moderate emotions for social acceptability (for example, saying thank you despite disliking a present); identifies themselves as 'I' or 'me', possessiveness as 'mine' and negativism as 'no'. 30–36 months: tests limits of acceptable behaviours; engages in symbolic play (for example, using an object as something different such as a block for a telephone); playing becomes more interactive and complex, with themes and storylines.
3–6	Skips and hops; has enough balance to ride a bicycle; catches a bounced ball; draws a circle and square; draws representations of people; starts to read; enjoys rhymes and word play; enjoys acting independently when possible; understands concepts of size and time.	Follows simple rules and directions; learns adult-like social skills, for example praising others and apologizing for unintentional mistakes; enjoys spending more time in peer groups and relating to a group of friends; engages in increasingly imaginative and complex play, such as dressing up and acting out fantasies.

motor skills, language acquisition, cognitive abilities, social skills and emotional development. There is considerable variation from person to person, and this becomes ever greater after about six years of age as children and adolescents develop complex motor, language, organizational and social skills. These include a deeper understanding of rules and regulations, relationships and responsibilities, the learning of more complex coping skills and the exploration of new ideas and activities with peers, who become increasingly important. By 9–10 years of age, peer and friend groups often take precedence over family, from whom children show a growing need for independence.

While transiting to an increased commitment to peer groups, adolescents often engage in activities that carry an increased risk of mental or physical harm in an attempt to avoid estrangement from their peer groups, an outcome that they consciously or unconsciously seek to avoid if they possibly can. Adolescence is typically characterized by the formation of new, sometimes complex and often long-lasting relationships, disagreements and emotional breakups. Usually, the adolescent brain learns to cope with these stresses, but it is perhaps unsurprising that it is during this particularly turbulent time that events can trigger the onset of enduring mental health problems in some, as we shall see in later chapters.

A Holistic View of Mature Brain Function

If you stand back from all the detail, our brains – in common with those of all other animals – do a very straightforward thing: they receive inputs, they process those inputs and they produce outputs. We could elaborate a little, particularly on the processing step, by saying that they gather and store information, create perceptions, abstract thought and plans, solve problems, make predictions and decisions and control our actions. Our brains get their inputs from many sources of energy in the world around them, which includes the rest of the body as well as everything outside it. For example, light and sound originating external to the body, mechanical forces either external or internal to the body or chemicals produced either by the body or outside it, which we might smell or taste, can all affect the brain's activity. Detection is achieved by specialized receptors situated either outside or inside our brains. It involves the conversion of one kind of energy (such as radiant energy in light, mechanical energy in sound or other movement, chemical energy and so on) to another in a process we call **signal transduction.**

The end result of signal transduction is the generation of new electrical activity that modulates ongoing electrical activity in the brain's neuronal circuits. It is important to appreciate that the brain does not need external inputs to be active; external inputs change the electrical activity that is going on internally all the time; they are not needed to make it.[51] Using the synaptic mechanisms of temporal and spatial summation that I outlined earlier in this chapter, the brain's neuronal networks make decisions on what to do or on how to respond to inputs. At a holistic level, these decisions appear to the outside world as anything from split-second ones allowing you, for example, to play a winning drop shot in tennis, to longer-term ones that might result in you taking up tennis in the first place.

In many ways, the most interesting part of what the brain does, and the part that probably varies the most from person to person, comes in the middle, between the input and output stages. This is where we form perceptions, learn from experience, store memories, make predictions and decisions and so on. These sorts of processes, which are sometimes referred to as our higher cognitive functions, require networks that have the potential to respond to patterns of incoming electrical activity in a selective way and to be modified by their own electrical activity. Earlier in this chapter, I explained how the properties of synapses allow modifications based on electrical activity to happen. I also explained that most of the brain's major regions are multimodal, and this is particularly the case in the brain's middle ground, between the regions that are the first port of call for sensory inputs or the last staging post for motor outputs. Few, if any, of these regions have functions that align closely with particular attributes of brain function. Even eminently definable cognitive attributes such as memory emerge from the joint activities of numerous regions distributed throughout the brain, although some brain regions – such as the hippocampus in the case of memory – play a more prominent role than others. The areas and the high-level functions of the brain become even harder to align as we consider even more complex aspects of our cognitive, intellectual and emotional abilities. It seems almost certain that the brain generates the psychological traits that define our personalities by passing signals between billions of neurons located throughout the whole brain.

Scientists consider that the brain extracts, from all the inputs it gets from the outside world, key information to create, update and check the validity of its internal representations of reality; these representations are often referred to as internal models. These models are encoded by the dynamic patterns of electrical activity across the brain's networks of neurons. The brain's patterns of electrical activity and the nature of its networks are intertwined: electrical activity influences the networks and the networks provide the substrate for electrical activity. The patterns of activity somehow give us a sense, from moment to moment, of what the world around us is like. The control of outputs from the brain is also governed by internal models that predict the patterns of activity required to have a desired effect. These patterns of electrical activity are then transduced back into other forms of energy, for example mechanical energy resulting in movement of the body.

The fact that the brain is actively creating an internal model of the outside world by extracting key information from its inputs, rather than acting as a faithful slave to their incessantly changing activity as we interact dynamically with our surroundings, implies that it has evolved to analyse primarily the constant, consistent and predictable properties of the world. Our brains have evolved to create models that predict selectively the most useful features of the world around them. A good illustration of this is the way we perceive colours. Our brains do not report faithfully on the wavelengths of light coming from an object. Instead, they predict the colour based on the wavelengths coming not only from the object itself but those around it as well (Figure 2.12). By analysing colour in this relative way, we can consistently predict the colour of an object as overall lighting conditions vary. For example, a green object will most likely appear green to us both in the full light of day and as the sun sets, although it would be reflecting a different combination of wavelengths of light under the two conditions. We call this colour constancy. If our brains did not work this way, objects would be constantly changing colour with ambient lighting conditions, which would make it much harder for us to figure out what the world around us is really like.

Another example of model-building by the brain is the famous Innsbruck Goggle Experiments that Theodor Erismann and Ivo Kohler carried out in the early 1900s. They wore spectacles that turned their view of the world upside down for days on end. Over this

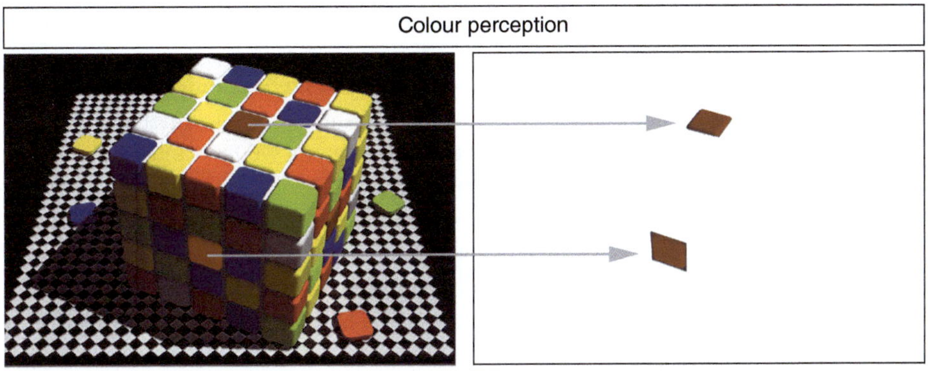

Figure 2.12 Our brains work out the colour of objects relative to the colours around them. If you look at two squares in the centre of two faces of the cube, one looks brown and other looks orange, but if you remove all the surrounding colours, you will see they are both the same brown colour. In order for the two squares to look the same brown colour in the context of the cube, the colour of the orange-looking square would need to be adjusted to a shade appropriate for a brown square in shadow, as is the case for all the other squares on that face of the cube. *Reproduced from Lotto et al.*[52] */ with permission of Springer Nature.*

period, their perception of the world reverted to the right way up, telling them that their visual perception was being created by their brains and aligned with their other sensory inputs. They were, essentially, learning to make a useful perception of the world rather than simply passively and faithfully representing within their brains the stimuli coming from their senses.

Because our brains make models that predict what is around us, they are open to being fooled; visual illusions are one example of this. What is more, by making predictions, our brains are liable to miss the unexpected. A well-known example of this is provided by the results of the 'gorillas in our midst' experiment: about 50% of observers engaged in monitoring the actions of two teams of people passing a basketball fail to notice a highly prominent event, such as a person dressed as a gorilla, walking among the players.[53] This is just one example of a phenomenon known as inattentional blindness. And, sometimes, our brains create models of things that are not there at all: we call these hallucinations and I will say more about them in Chapter 3.

The Effect of the Brain's Experiences on Its Developing Functions

After birth, the range of an individual's experiences and interactions enlarges dramatically. In the early years of life, this occurs as the numbers of synapses and myelination levels in the brain are increasing rapidly (see Figure 1.10 in Chapter 1), both of which greatly expand the brain's capacity to perceive, learn, create memories, solve problems, generate abstract thought and control movement and behaviour. The way in which this capacity is moulded to produce an individual's perceptual, cognitive and behavioural characteristics varies from person to person. This is due in part to variation in people's genetic make-up (more on this in Chapter 4), in part to random variation in the developmental processes that make the brain and in part to variation in the world each brain encounters. The last of these includes what language a person hears and learns to use, what role models they follow and what knowledge they are given or discover.

Less obviously, perhaps, experiences in early postnatal life can have long-term, even lifelong, effects on the most fundamental aspects of the brain's ability to interpret sensory information or the types of movements it induces. For example, we have known for many decades that problems with our eyes from birth, such as congenital cataracts or a squint, can irreversibly compromise the brain's ability to interpret what the eyes detect; we call this amblyopia. Evidence of the effects of unusual early experiences on movements of the body is harder to come by, although there are reports suggesting this can happen. For example, an abandoned Nigerian child who, it is thought, was reared from about six months to two years by chimpanzees in the wild was said to use their limbs in a chimpanzee-like rather than a human-like way even years after being returned to human care.

These are examples of a property of the developing brain that we often refer to as experience-dependent plasticity. I discussed the synaptic properties that underlie plasticity earlier in this chapter, although the details of how these underlying principles might translate into something as complex as limb usage are unknown. For some brain functions, such as sensory perception, we can define time windows in early postnatal development when experience has its greatest effects in shaping the brain's neuronal circuitry; we refer to these as sensitive or critical periods.[54] Their lengths are measured in years and depend on what function is being assessed and how. In general, those for more fundamental functions such as perception or the acquisition of language (no matter what the language is) are thought to end before more complex ones, such as many social skills. In reality, however, it is probably best to think of brain plasticity and critical periods during development as extreme manifestations of a property the brain retains to some extent throughout its whole life. Even the mature brain is not like an unchanging electronic circuit board; rather, its components, their circuitry and their integrative properties are capable of modification throughout life. This is, after all, how we can continue to learn and develop new skills as we age, even if we are nothing like as good at it as we were when we were young.

Language Development

Many linguists consider that the human brain is born innately prepared to acquire language.[55] Part of the reason for thinking this is that all languages, despite their obvious and often major differences, share a common framework or structure, often referred to as universal grammar, which children seem to have an intuitive ability to grasp. This is true for the many thousands of languages that are used around the world, including those that have emerged in relatively isolated communities. Moreover, the evidence suggests that children can grasp universal grammar in the absence of explicit instructions and even if their exposure to language is severely limited or absent as they grow up, as has occurred in documented cases of child abuse involving enforced isolation and deprivation. What is equally clear, however, is that this innate ability alone is not enough to equip a child with an adequate set of language skills. For this they need to experience language in their early years, probably before they become a teenager; if deprived of this experience, they will struggle to acquire any significant proficiency in language no matter how hard they try. In other words, there is a critical period for the development of language.

Most evidence on the acquisition of language does not come from extreme cases of language deprivation in childhood, but rather from studies of how children learn the languages they are exposed to. Scientists have discovered that the speech sounds a child hears influences their ability to both perceive and make those sounds. Linguists often focus on the smallest units of speech sounds that distinguish one word from another, called phonemes (for example, the phonemes 'ba' and 'pa' distinguish 'bark' from 'park'). Different languages use different sets of phonemes. Studies have shown that young infants can perceive and discriminate between all possible phonemes, but not for long. For example, Japanese infants just a few months old can discriminate reliably between the phonemes 'r' and 'l', whereas Japanese adults cannot. It seems that Japanese children lose the ability to make this distinction during the early years of their life because 'r' and 'l' are not distinct phonemes in the Japanese language that they hear around them.

The Development of High-Level Executive Function

The attainment of many high-level cognitive milestones by a developing child or adolescent, such as understanding abstract concepts and peer acceptance and recognition, requires the refinement and integration of high-level cognitive processes that psychologists refer to collectively as executive functions. This term is used because these are functions that help an individual plan and successfully execute their goals. The executive functions include control of attention, problem-solving, working memory and self-regulation. The brain's working memory refers to its ability to maintain, focus on and update information that is currently relevant. Early signs of working memory can be found in infants as young as six months, but a strong capacity for this function takes many years to develop through childhood and adolescence. Self-regulation refers to the capacity to select among competing choices of activity, such as doing work for school versus playing with friends. To make such choices, a child needs to consider the possible outcomes of the options and weigh them against their goals; for example, should they study for long-term gain or play for short-term pleasure, or is there another way to see it? Performance on psychological tests of self-regulation show steady development through to adolescence. Executive functions such as these are thought to originate, although not exclusively, in the brain's prefrontal cortex: evidence comes from neuroimaging studies that have correlated measures of brain structure and function in children and adolescents with the maturation of executive functions. We have a long way to go to understand *how* these brain regions generate these functions.

Emotional and Social Development

Most individuals' emotional and social development occurs on a roughly similar timescale. Broadly speaking, we can consider these aspects of brain function under two headings: the development of self; the development of relationships with others. By development of self, I mean the development of an individual's observable temperament – for example their level of emotionality, activity, extroversion versus introversion, adaptability, distractibility and so on – in other words, their personality as revealed by its external manifestation through their behaviour. I do not mean the emergence of an internal and unobservable sense of self, or conscious self-awareness if you like, which we assume exists in others but cannot detect directly.

Although we cannot step into the minds of others, humans have a propensity to make inferences or theories about the mental processes going on in other people's brains to explain the things they do. In the lingo often used by psychologists, we can say that when the human brain infers what is going on in the minds of others, it is generating a theory of mind. It uses this theory of mind to analyse, predict, judge and so on, other people's behaviours, all of which are critical for successful social interactions. The ability to create a theory of mind emerges in the first few years of life, but its full development continues through adolescence and possibly beyond. An important milestone, for example, is when children come to understand that others can hold different beliefs to their own and that those beliefs might be wrong. As a growing individual's theory of mind develops, their ability to recognize the intentions and desires of others, predict their actions, analyse their emotions and so on, becomes increasingly sophisticated.

Particularly strong changes in social aspects of human brain function occur during adolescence, equating approximately to the teenage years. These changes accompany the rapid development of other parts of the body, notably the reproductive, muscular, skeletal and hormonal (or endocrine) systems. Researchers have used magnetic resonance imaging (MRI) and fMRI (Figure 2.11) to identify alterations of brain structure and activity occurring during adolescence, although it is not yet clear whether and, if so, how they might account for the distinct behaviours associated with this period of life. These typically include increased emotional volatility, impulsiveness and risk-taking as adolescents start the transition to greater independence and assert their own personalities by engaging in more advanced exploration of their social environment and its boundaries, including relationships with peers.[56,57] As we will see later in this book, many relatively common mental health problems first rear their head during adolescence. Although we can rarely specify the cause of such problems, we can draw the general conclusion that the major changes that the adolescent brain undergoes are likely to make it particularly vulnerable to a range of genetic and environmental – including social – factors. I will explore this further in later chapters.

Summary

- Our understanding of how the brain works to create our perceptions, our cognitive abilities and our behaviours remains very limited despite much progress over the last 200 years or so.
- The mature brain is divided into regions, each with its specialized functions, like the cerebral cortex, hippocampus, cerebellum and basal ganglia.
- Neurons transmit electrical impulses called action potentials and communicate via synapses using neurotransmitters; each neuron's moment-to-moment activity is determined by the combination of activity it receives from its synapses at that time.
- Synapses are plastic structures whose strength can be enhanced if their activity levels are high and depressed if their activity levels are low, allowing the brain to learn from experience.

- Holistic approaches to studying human brain function and its development use methods such as behavioural observations, skills-testing under controlled conditions and non-invasive brain imaging.
- Like the synapses that they contain, our developing brains as a whole show plasticity and are influenced by their experiences, particularly during early postnatal life.
- The brain's plasticity is important for its ability to acquire language, to develop executive skills and for emotional and social development.

3

Disordered and Neurodivergent Phenotypes

So far in this book, I have described key aspects of how most brains grow and what they do. As I pointed out in the Introduction, our brains show great diversity in their structure and function. In this chapter, I will describe features of brains that lie at the extremes of this diversity. Some such brains are disordered or abnormal in a way that we can clearly define and measure. Others diverge to a degree that appears unusual or is a problem for the affected person, although often in a less clear-cut way.

In this chapter, I will be describing brain phenotypes, which are the observable characteristics of individuals' brains. They can also be called traits, meaning much the same thing. Phenotypes can range from anatomical to perceptual, cognitive or behavioural in type. Anatomical phenotypes might involve features such as brain size, structure, content and so on. Perceptual, cognitive or behavioural phenotypes might involve features such as a person's ability to create an accurate internal representation or model of the world (discussed in Chapter 2), to carry out cognitive tasks, to express themselves through movement or language and so on.

Many people, particularly clinicians, refer to the abnormal phenotypes associated with disordered brains as symptoms. (They sometimes call them signs if they are observable by others rather than being reported by a patient themselves; it is often difficult to define this distinction precisely and in the present context I do not think there is much to be gained by doing so.) The term symptom is sometimes used to describe the phenotypes of neurodivergent individuals as well as those with a clear disorder. This can be problematic since it implies that a neurodivergent individual's phenotypes are abnormal which, as I discussed in the Introduction, can be unclear and contentious. The phenotypes or symptoms that a person shows might lead to a clinical diagnosis, or a clinical categorization if you like. This involves giving a name to the individual's disorder or neurodivergence, based on a similarity between that person's phenotypes or symptoms and those of others and, if possible, some understanding that their causes might be similar. Many diagnoses are of syndromes, meaning that a particular group of symptoms (some of which might be outside the brain) co-occur in the same individual (the term syndrome comes from the Greek *sundromon*, meaning concurrence or running together). In this chapter, I will not be describing specific diagnoses and syndromes. I will leave that to Chapter 6. Here, I will concentrate on observable characteristics that allow diagnoses to be made.

Neurodevelopmental disorders and types of neurodivergence might be defined by anatomical phenotypes that we might be able to see by eye, to detect with brain imaging or to find at post-mortem, or by functional* phenotypes that we might be able to detect by observing an individual's behaviour, by testing their cognitive abilities or by using functional brain imaging (see Chapter 2). Accordingly, I have split this chapter into two main parts. In the first, I describe many of the abnormal anatomical or structural phenotypes that can emerge during brain development. We can confidently describe such changes as pathological since they are demonstrably different, or not present at all, in the brains of most people and many will have detectable negative consequences for how an individual's brain functions in some way. Occasionally, the identification and description of abnormal anatomical or structural phenotypes might be possible with the naked eye. An extreme example would be babies born with substantial parts of the brain and skull missing. More often, we need to bring additional methods to bear, such as brain imaging (Figure 2.11, Chapter 2) in the case of living individuals or detailed measurements at a macroscopic or microscopic level in the case of post-mortem studies. Examples would be detection on brain scans of dense localized deposits of calcium in the young brain or the microscopic finding at post-mortem of abnormal collections of misplaced cells.

The second part of the chapter describes many of the most common disordered or neurodivergent ways in which brains can function. Since brain function is so broad and multifaceted, it is no surprise that the ways in which we describe the brain's functional phenotypes, be they advantageous or debilitating, range from the highly specific to the rather vague. For example, seizures are specifically defined debilitating phenotypes, or symptoms if you prefer. In seizures, the brain undergoes periods of deranged activity that can cause involuntary, incapacitating movements of the body that are clearly abnormal. The term seizure is used universally and carries an agreed meaning. On the other hand, phenotypes associated with neurodivergent behaviours can be much harder to define, describe and apply consistently. This is completely understandable given their complexity and the fact that they are often describing aspects of a person's behaviour in the context of their social setting. An extreme example might be one of the indicators of neurodivergence in children given by the United Kingdom's National Health Service: 'not talking as much as other children'. Phenotypes described like this are going to be very hard to define with precision and consistency, certainly much harder than measuring the size of someone's brain or observing seizures.

Anatomical Abnormalities

Many congenital neurodevelopmental abnormalities of brain anatomy are detectable with the naked eye or with brain imaging methods at, or shortly after, birth. In fact, we have the ability to detect some of them before birth, for example, severe defects of the formation of

* The way I use the term functional corresponds with the way most others use it today. Sometimes, however, particularly in older literature, you might see the term used in a more specific way to describe psychiatric symptoms for which, it is claimed, no biological (or 'organic') cause exists, which is an unsustainable view. Difficult as it might be to identify the underlying causes of complex perceptual and behavioural phenotypes, they must reside somewhere in the brain's structure, biochemistry and/or activity even if it is often impossible to identify where using current technology.

the embryonic **neural tube** (we call these neural tube defects). The names of some have *cephaly*, meaning head, as their root. Others use the more specific root *encephaly*, meaning brain, rather than the broader root *cephaly*. I will point out where this subtle difference in nomenclature might be confusing. Also, as you will see, neither cephalic nor encephalic abnormalities necessarily affect the whole head or brain. As I go through all these abnormalities, do bear in mind that it is very unusual for any of them to occur in isolation. Usually, they co-occur as syndromes with other structural and functional abnormalities.

Anencephaly (Figure 3.1) is a severe and relatively common type of neural tube defect (spina bifida, the other main type of neural tube defect, is perhaps better known; it affects the spinal cord). Anencephaly probably affects about one in every 250 pregnancies, most of which end in miscarriage, and about one in every 5,000 births. Neural tube defects such as anencephaly have their origins during neurulation which, as I described in Chapter 1, follows hot on the heels of **neural induction** and is the process by which the relatively flat **neural plate** folds up to become the neural tube. If this process goes awry in a way that prevents the neural tube from closing completely, some of the tube remains open. In a normal embryo, the neural tube does not close at the same time along its entire length: this is emphasized by the common description of closure as zippering, which stresses the progressive nature of the sealing of the neural tube along its length. Anencephaly occurs when zippering fails to close the anterior part of the neural tube, where the brain should form. The exposed anterior **neuroectoderm** soon degenerates. Babies with anencephaly are born missing at least some portions of their brain and skull. They can be very large portions, but there is considerable variation in the exact details. In the most common form, holoanencephaly (*holo* is Greek for whole), the entire brain is absent and the newborn usually only survives for a day or so; in meroanencephaly (*mero* is Greek for partial), there is some limited development of the brain and skull and survival for longer is possible. **Exencephaly** is a disorder in which the brain develops before birth without a surrounding skull (*ex* is Greek for out, so exencephaly literally means out of the head). It often precedes the formation of anencephaly, explaining why you might see some authors talking about the exencephaly–anencephaly sequence. Individuals with exencephaly that is not a precursor of anencephaly are usually stillborn. An encephalocele (meaning a brain hernia) is another type of neural tube defect that is rarer than anencephaly. In this disorder, the skull forms abnormally and brain tissue protrudes through as a sac covered by skin or other membrane.

Another two conditions, aprosencephaly and atelencephaly, are extremely rare. In Chapter 1, I said that we divide the **forebrain**, or prosencephalon, into two parts: the **telencephalon**, which generates structures including the cerebral cortex and **basal ganglia**, and the **diencephalon**, which generates structures including the **thalamus**. In aprosencephaly, the entire forebrain is absent or malformed; in atelencephaly, only the telencephalon is affected (the Greek prefixes *a*- or *an*- mean not). There is a spectrum of possibilities between the two, all of which will result in early demise.

Many structural abnormalities are associated with enlarged **ventricles**; we use the word **ventriculomegaly** to describe ventricular enlargement of some sort (Figure 3.1). The nature and cause of ventriculomegaly is variable. For example, in the rare condition known as colpocephaly, it is specifically the ventricles at the back of the brain that are enlarged and the brain tissue surrounding them is correspondingly underdeveloped. The name

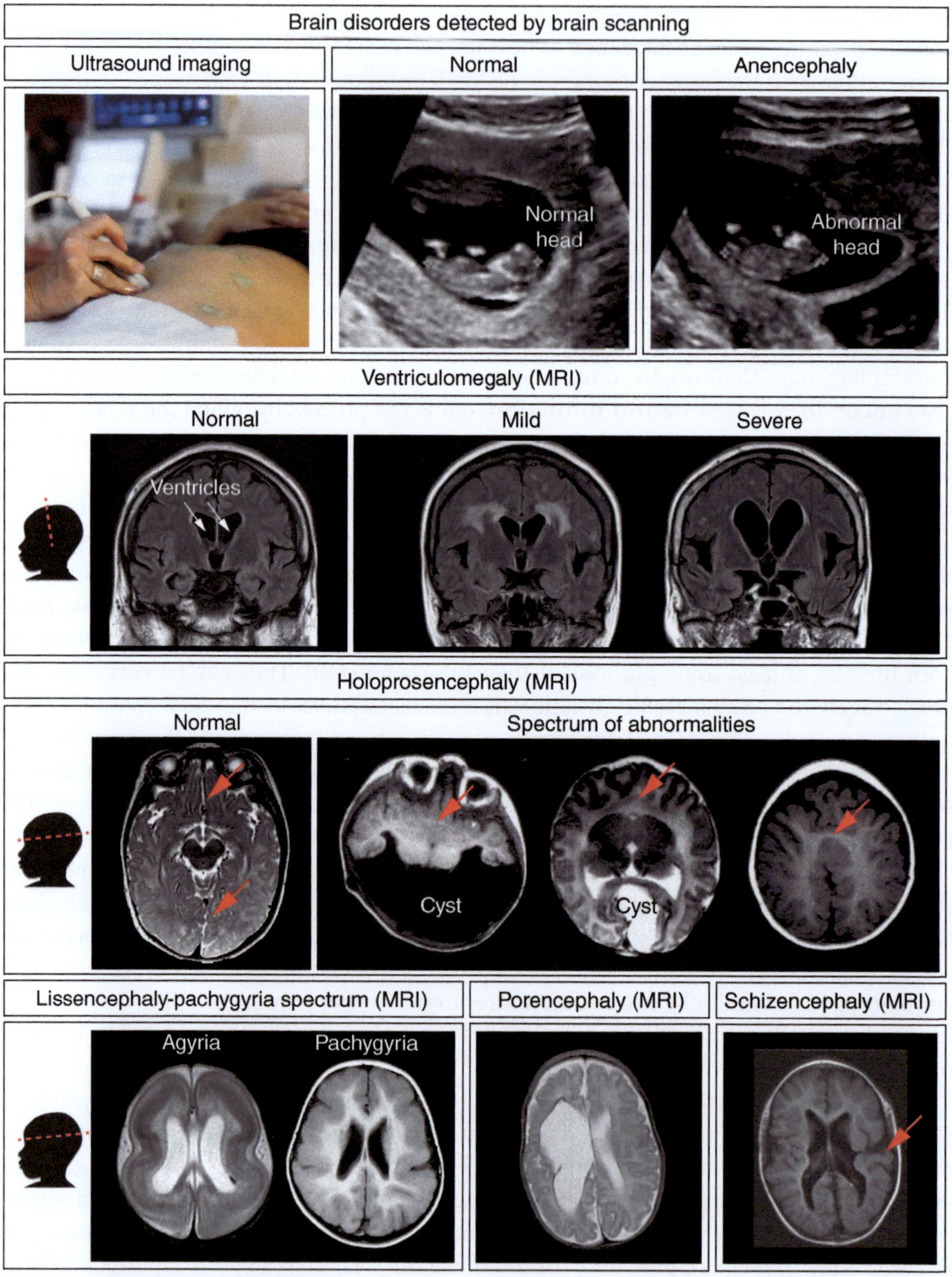

Figure 3.1 **Anatomical abnormalities of brain development.** Ultrasound scanning is commonly used to monitor the development of the fetus. In the example shown here, a mother is carrying twins: the one on the left looks normal and is lying with its head to the right; the other is lying the same way but has anencephaly and lacks a normal head. The rest of the pictures are structural magnetic resonance imaging (MRI) scans of the brains of children (Figure 2.11). Examples of ventriculomegaly illustrate a mild case, with enlargement of the ventricle on the right side relative to that on the left, and a severe case. Examples of holoprosencephaly illustrate the spectrum of

Figure 3.1 (Continued) abnormalities observed: red arrows indicate clear separation of the two hemispheres in the normal brain and regions where separation has failed in the abnormal ones; two of the brains have large posterior cysts. Lissencephaly-pachygyria spectrum examples show diffuse agyria (that is, the brain has a smooth appearance) and pachygyria (note the unusual thickness of the cerebral cortical grey matter). In the example of porencephaly, the abnormality is more severe on the left. A red arrow marks an abnormal cleft in the right cerebral hemisphere in a case of schizencephaly. *Images are reproduced with permission from Lee Langman et al.*[58] *(top row), Hahn and Plawner*[59] *(3rd row), Koenig et al.*[60] *(bottom left), Hino-Fukuyo et al.*[61] *(bottom right); and from Iida et al.*[62] *(2nd row), under a Creative Commons Attribution 4.0 International License.*

colpocephaly stands for hollow (*colpo-* in Greek) head (-cephaly), which is a rather extreme description given what actually happens. This is an example of a name ending with -cephaly, in contrast to the names of the disorders in the previous and following paragraphs, which are -*en*cephalies. In the case of colpocephaly, the use of the shorter, less specific root is historical; the name colpencephaly would have done just as well. Colpocephaly can be relatively mild compared to many other cephalic disorders; some of those affected have survived to adulthood, with increased likelihood of other abnormalities of brain structure and function.

Ventriculomegaly might be associated with **hydrocephalus**, which is an accumulation of cerebrospinal fluid in the ventricles of the brain (although the name literally means it is in the head, it is actually in the brain; note that the name ends in -us rather than -y for historical reasons and nothing else). A key feature of hydrocephalus is that the accumulation of cerebrospinal fluid causes increased pressure in the ventricles. This increased pressure will enlarge the ventricles in many cases and it risks damaging surrounding brain tissue. Be careful not to confuse hydrocephalus and ventriculomegaly. Ventriculomegaly and hydrocephalus have similar incidences in the population, affecting about one in every 1,000 newborns, but some cases of ventriculomegaly result from disorders other than hydrocephalus, and hydrocephalus does not always cause ventriculomegaly. For example, ventriculomegaly can result from a loss – or a failure to develop – of brain tissue, leaving a larger space in the brain for the ventricles and their cerebrospinal fluid. If ventriculomegaly is caused this way, the cerebrospinal fluid might be at normal pressure.

Holoprosencephaly is a failure of prosencephalic development in which the early single-chambered **prosencephalon** does not divide properly into two telencephalic vesicles (Figure 3.1). It is associated with abnormalities of face development. It probably affects about one in every 250 pregnancies, most of which end in miscarriage, and about one in every 10,000 births. Holoprosencephaly describes a spectrum of defects that range from extremely severe, in which the prosencephalon does not divide at all and there are major facial malformations such as a single eye where the root of the nose would normally be, to mild, in which apparently normal brain development is associated with cleft lip. Mild cases can survive to adulthood, although they are likely to have other abnormalities of brain structure and function.

Lissencephaly is a rare abnormality, affecting about one in 100,000 births, in which the cerebral cortex develops without normal folds (that is, without gyri and sulci) (Figure 3.1). Lissencephaly means smooth brain, but there is considerable variation in severity from no gyri (agyria) to abnormal gyri that might be broader than usual (which we call **pachygyria**;

pachy- is Greek for thick). Some refer to this variation as the lissencephaly–pachygyria spectrum. Prognosis depends on severity: most children fail to develop much after birth and many die within a few years. In a condition that is the opposite of lissencephaly, the cerebral cortex develops too many folds. We call this polymicrogyria, and it affects about one in every 10,000 births (Figure 3.2). It might be present on only one or on both sides and varies widely in severity.

Porencephaly, meaning holes in the brain, is a very rare abnormality in which cavities (or cysts) are present in the cerebral hemispheres (Figure 3.1). Hydranencephaly is an extreme form of porencephaly. **Schizencephaly**, meaning divided brain, is another very rare abnormality in which abnormal clefts are present in one or both cerebral hemispheres (Figure 3.1). Some authors include it under the heading of porencephaly. The prognosis for these conditions depends on their severity, which is very variable.

Megalencephaly (synonymous with macrencephaly) describes abnormal enlargement of the brain, which is often malfunctioning. Hemimegalencephaly is a rare form in which only one side of the brain is abnormally large (an example is shown in the middle of Figure 3.2). Here, we need to pay careful attention to nomenclature. Megalencephaly, which ends with the suffix -*en*cephaly and, therefore, refers to the brain, is not the same as macrocephaly, which ends with the suffix -cephaly and is a large head. Both megalencephaly and macrocephaly can be defined statistically based on measurements of size, and so their incidences depend on the agreed threshold above which the terms are applied. Megalencephaly is associated with a much higher likelihood of additional abnormalities of brain structure and function than macrocephaly. Whereas macrocephaly is a feature of numerous syndromes, many people with macrocephaly have neither an enlarged nor a disordered brain.

Micrencephaly and microcephaly are essentially the opposite of megalencephaly and macrocephaly: micrencephaly signifies an unusually small brain; microcephaly signifies an unusually small head. Like macrocephaly, microcephaly is definable statistically and is strongly associated with micrencephaly and many other abnormalities of brain structure and function.

There are a number of other cephalic (head) abnormalities in which it is primarily the skull that is deformed. In some cases, this is because of premature sealing of one or more of the joints between a child's skull bones (which we call sutures). These sutures are normally left unsealed long enough to allow the brain to grow. We call this collection of abnormalities craniosynostoses. There are various types of craniosynostosis arising from the premature sealing of different sutures, with names such as oxycephaly, plagiocephaly, scaphocephaly and trigonocephaly. If untreated, they are likely to have an effect on the developing brain. Many cases of skull deformities arise due to reasons other than premature sealing of sutures. They are often benign and can, for example, be caused simply by the way a baby lies when asleep.

Other structural abnormalities often have names indicating what brain structure is affected followed by the type of abnormality that has occurred. Frequent types of abnormality include **hypoplasia** (an abnormally small structure), **agenesis** (a failure of the structure to form at all), dysplasia (disorganized formation) or **atrophy** (wasting away of

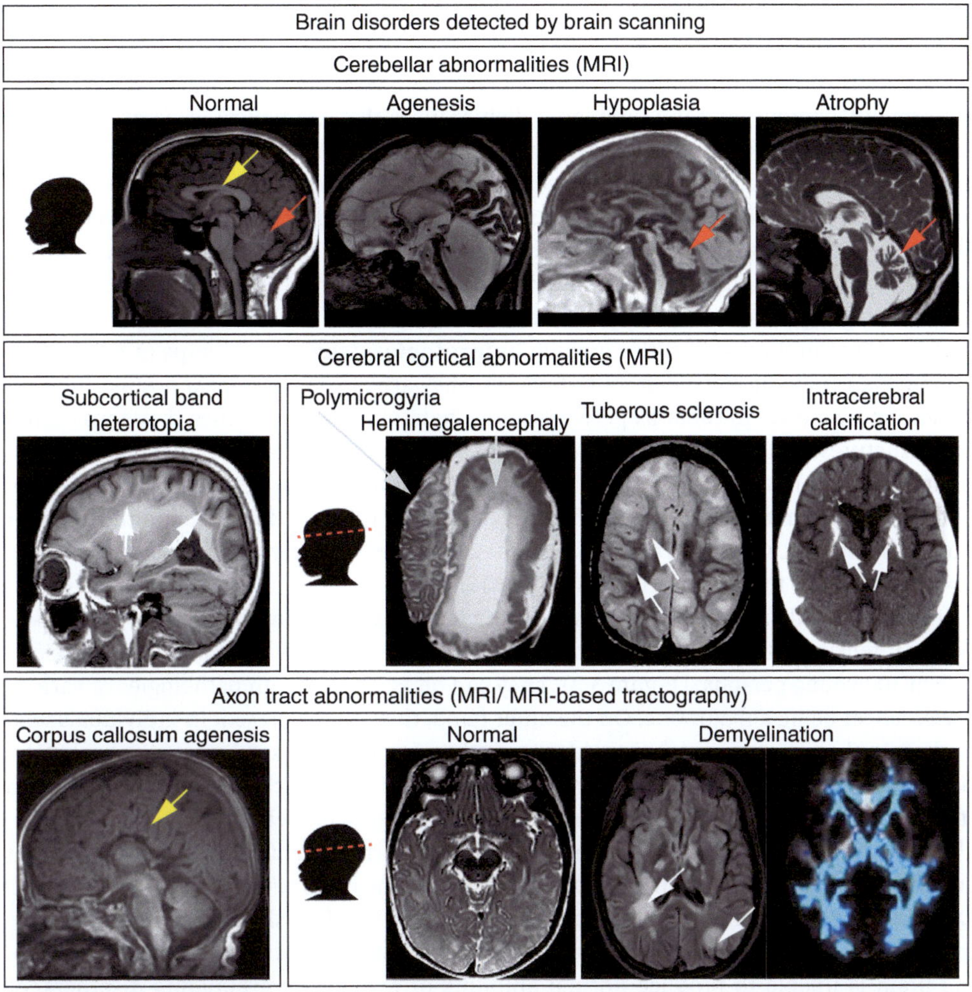

Figure 3.2 **Anatomical abnormalities of brain development continued.** Magnetic resonance imaging (MRI), including tractography (Figure 2.11), used to reveal structural disorders in the brains of children: cerebellar agenesis, hypoplasia and atrophy (red arrows point to cerebellum); subcortical band heterotopia (white arrows point to a band of tissue looking like cortical grey matter interspersed between the normal cortex and the lateral ventricle); polymicrogyria and hemimegalencephaly in the same brain; tuberous sclerosis (white arrows indicate appearance typical of cortical tubers in the subcortical white matter); intracerebral calcification (white arrows); corpus callosum agenesis (yellow arrows in pictures at top and bottom left indicate corpus callosum or where it should be); demyelination (examples shown by white arrows). In the tractography image, blue indicates regions, which are mainly in white matter, of greater diffusivity than normal, indicating the breakdown of diffusion barriers due to a loss of white matter integrity in a case of familial demyelination. *Images are reproduced Goncalves et al.*[63] */ with permission of Springer Nature. (middle row right); and Poretti and Boltshauser*[64] */ Springer Nature / CC BY 4.0. (top row), Leventer et al.*[65] */ with permission of Taylor & Francis. (middle row left), Kosmeri et al.*[66] */ with permission of Hippokratio General Hospital of Thessaloniki (bottom row left), Galardi et al.*[67] */ MDPI / CC BY 4.0. (bottom row middle), Gharaylou et al.*[68] */ Frontiers Media S.A / CC BY 4.0. (bottom row right).*

the structure after its formation). For example, cerebellar hypoplasia is a condition in which the cerebellum has not developed to a normal size (Figure 3.2).

Cerebellar abnormalities affect around one in every 5,000 children (Figure 3.2). Structural abnormalities underlying the functional disturbances that are typical of cerebellar problems, such as **ataxia** (see below), are often detectable using brain imaging methods. As well as cerebellar hypoplasia, they include cerebellar agenesis (very rare), cerebellar dysplasia and cerebellar atrophy. **Corpus callosum** disorders also affect about one in every 5,000 children (Figure 3.2). The diagnosis often rests on brain images from children who might have various sorts of functional symptoms. Like cerebellar disorders, the corpus callosum can suffer hypoplasia, agenesis, dysplasia and atrophy.

Other structural abnormalities include **heterotopias**, which are clumps of cells in the wrong place (heterotopic means out of place). They are rare. Most appear around the ventricles, and we refer to them as periventricular heterotopias. Sometimes they form band-like structures underneath the grey matter of the cerebral cortex, in which case we call them subcortical band heterotopias (Figure 3.2). Heterotopias appear as nodules in magnetic resonance images, where they can appear similar to the nodules seen in the disorder tuberous sclerosis, which are actually benign tumours rather than heterotopias (Figure 3.2; I will discuss this further in Chapters 6 and 8).

Intracranial calcifications of childhood, which show up on neuroimaging as small spots whose intensity differs from the surrounding tissue, are relatively common (approximately one in 100 young persons) (Figure 3.2). They can be associated with pathologies of various sorts, cancer being a major one. On the other hand, they can be present in individuals who show no other abnormalities (in which case they are labelled as physiological calcifications). Both pathological and physiological intracranial calcifications of childhood can involve many parts for the brain, from the forebrain to the cerebellum.

Dysmyelinating disorders are rare abnormalities of formation of the **myelin** sheaths that surround and allow the effective function of axons in the brain (Figure 1.9 and Chapter 2). Diagnoses often rely on magnetic resonance imaging in children showing functional symptoms. An example of a dysmyelinating disorder is hypomyelination, in which the developing brain is unable to produce normal amounts of myelin. *De*myelinating disorders, of which multiple sclerosis is the most common, involve the degeneration of previously normal myelin; they are more common in adults (Figure 3.2). That said, many dysmyelinating disorders involve demyelination. Disorders of myelin fall under the broader heading of **leukodystrophies**, meaning abnormal white (*leuko-* in Greek) matter growth or degeneration (-dystrophy). The underlying pathologies of the leukodystrophies (of which there are over 50) can involve not only the cells in the white matter that make myelin (the **oligodendrocytes**) but others too, including **astrocytes** and **microglia**.

Functional Disorders

In this section, I will focus on situations where the brain's functional characteristics have developed in ways that most people would agree is abnormal. These include, for example, disturbance of a person's bodily movements, such as ataxia or seizures, and severe forms of mental disturbance affecting a person's perception or mood. Functional phenotypes often fall into one or other of the two great historical disciplines of neurology or psychiatry. Broadly speaking, neurology is concerned with disorders affecting neural systems that allow us to detect events in our environment and to activate and carry out movements. Psychiatry is concerned with disorders or neurodivergent conditions affecting our interpretation of information we get about the world around us and our internal thought processes and decisions, often referred to as our mental, cognitive or intellectual functions. In general, phenotypes that usually fall under the neurological heading are easier to measure, making abnormalities relatively easier to define. The symptoms of psychiatric disorders are often complex and lie on spectra of continuously variable type and severity, and their description can be subjective and liable to vary in different cultures; at the mild end, they would often become descriptions of neurodivergence, which is the subject of the next section. The distinction between neurology and psychiatry is likely to blur increasingly as we learn more about how the brain achieves its most complex functions.

The terms used to describe functional symptoms are numerous, and their definitions can vary considerably. In general, the more easily a phenotype can be measured objectively, the more universally accepted its nomenclature and definition tend to be. Two major works that attempt to standardize terms and their definitions are the *International Classification of Diseases, published by the World Health Organization* and now in its 11th edition (it is known as ICD-11)[69] and the *Diagnostic and Statistical Manual of Mental Disorders*, published by the American Psychiatric Association and now in its 5th edition (it is known as DSM-5).[70] These works have undergone major changes over the years, particularly in the ways that neurodevelopmental disorders are described and categorized. They are likely to continue to change in the future as we learn more about the conditions that we are trying to describe.[71]

As was the case for anatomical or structural abnormalities, functional symptoms rarely occur in isolation. They are often components of syndromes, accompanying other functional abnormalities and, sometimes, detectable anatomical or structural defects that might, or might not, cause them.

Abnormalities of Posture and Movement

In discussing this topic, my focus is on phenotypes, or symptoms, arising within the brain. I am concerned with the high-level systems for controlling our muscles, the systems that *initiate* and *regulate* the processes that translate the decisions we make into actions. I will not discuss symptoms originating outside the brain, for example in the spinal cord, peripheral nerves and the muscles themselves, nor in bones, joints and ligaments, all of which need to function properly for normal posture and movement.

The abnormal phenotypes discussed here are relatively common, affecting about one in every 500 births. They arise in one or both of the two major output systems that control our movements: the pyramidal system and the extrapyramidal system. The pyramidal system comprises the output from the motor cortex via the corticospinal tract (Figure 2.1, Chapter 2). This tract gets the name pyramidal because, where it passes along the front of the brain stem enroute to the spinal cord, it bulges out in a way that looks, fancifully, like a small pyramid. Abnormalities of this system lead to a loss of voluntary movement. The extrapyramidal system is multifaceted, comprising a complex assortment of neural structures, including the basal ganglia and cerebellum, and axonal tracts, many of which are outside the brain. It is concerned with the involuntary control of movements initiated through the pyramidal system. Abnormalities of extrapyramidal system components compromise a person's ability to coordinate and modulate their movements and to maintain their posture.

The symptoms that people with disorders of posture and movement might experience include the following. Rigidity, spasticity or hypertonia describe a situation where muscles are too tight or, in other words, have too much tone. The opposite, where the muscles are floppy, is **hypotonia**. Some disorders cause muscle weakness and exaggerated reflexes, which we call hyperreflexia. Akinesia, hypokinesia and bradykinesia describe absence, decreased or slow movement respectively. Dyskinesia describes uncontrolled movements, which can range in severity from mild tremors to more debilitating movements of the body, which might be slow and writhing (athetosis) or quick and jerky (**chorea**). Ataxia refers to problems with balance or the coordination of movement on attempting to carry out a task, including walking, writing or speaking. You will often see ataxia split into different types: for example, gait ataxia affects how a person walks, truncal ataxia affects a person's trunk and the term **cerebellar ataxia** might apply if the disorder arises in the cerebellum. Disorders of movement also includes problems with the brain's output to muscles of speech, swallowing or breathing. We call difficulty or absence of speaking dysarthria or anarthria respectively and difficulty swallowing dysphagia.

I also include here seizures, which are involuntary disorders of movement that originate from uncontrolled, uncoordinated bursts of electrical activity within the brain (Figure 3.3). Repeated seizures might lead to a diagnosis of epilepsy. Seizures can be of many different types and might include rhythmic jerking movements (clonic seizures) or muscular rigidness or limpness (tonic or atonic seizures respectively). Note that some seizures involve little if any abnormal movement. These are absence seizures characterized mainly by a loss of awareness and responsiveness for a short time; there might be fluttering of the eyelids or other small jerking movements of the body.

Neurodevelopmental Delay: An Overview

While no two infants develop at exactly the same rate, most develop at similar ones, and we can evaluate statistically their rates of attainment of key neurodevelopmental milestones,

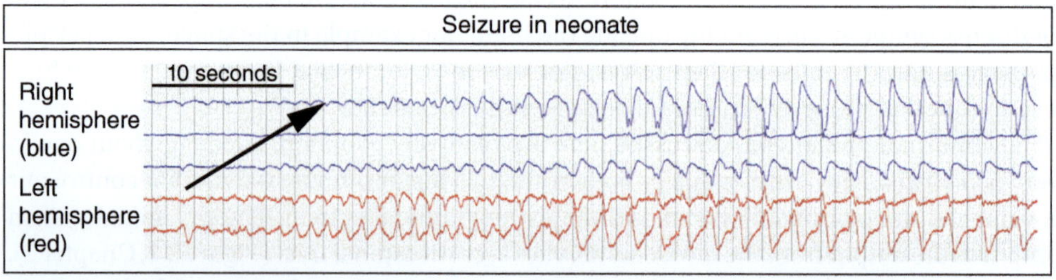

Figure 3.3 Electroencephalographic recording of a seizure. A recording from an infant in a neonatal hospital unit. A seizure begins on the left side (red trace) with rapid frequency and low amplitude. It propagates to the right side (arrow; blue trace). As it does so, amplitude increases while frequency slows. *Traces are reproduced from Ryan et al.*[72] */ Springer Nature / CC BY 4.0.*

such as crawling, walking or speaking (Table 2.1, Chapter 2). Unlike the ease with which one can monitor some aspects of an infant's growth, for example their weight or length gain, assessing the development of their brain is more complex and involves testing key skills over the first few years of life, usually under three major headings: motor, language and cognitive. Motor, or movement, skills are those that involve the body's musculature, and they separate into gross motor skills, such as crawling or walking, and fine motor skills, such as picking up an object or using a utensil to eat. Language skills split into those concerned with understanding others (receptive language) and those concerned with expressing oneself (expressive language). Cognitive (or thinking) skills are those that require the child to solve problems and explore their world using learning, memory and reasoning. These categories overlap: for example, fine motor control of the laryngeal muscles and/or fingers is required for communicating verbally or for signing and, therefore, the expression of language; exploration often involves movement and so on. Moreover, development in any one category affects development in others: for example, motor skill development enhances an infant's potential to explore and language development promotes their cognitive development. The main advantage of these categories is that they provide an organized framework guiding the design of specific tests to evaluate major aspects of infant brain function, which I discussed in Chapter 2.

Up to the age of about 5 years, most infants all around the world acquire these skills in a similar progressive pattern. Many factors both internal and external to the individual influence the timing of their key attainments. Some of these factors, such as cultural or economic ones, affect entire populations of infants, causing the average ages of milestone achievement to vary depending on prevailing circumstances (Figure 3.4 shows examples of variation in ages of milestone acquisition across several studies). It is, therefore, very important to compare the development of individual infants with the average of the population from which they come. If, following appropriate comparisons, an infant is developing skills at a rate that is significantly below the average, they are likely to attract a diagnosis of neurodevelopmental delay. Note that definitions of what is significantly below the average and what is neurodevelopmental delay tend to vary in the literature, adding uncertainty to reports of its incidence, which might be anywhere from about one to about ten per 100 infants. As you would expect, some cases are much more severe and therefore clear-cut than others, and some associate with structural abnormalities of the brain that are likely to be causative and with other functional disorders. It is unusual to apply the term neurodevelopmental delay to children over 5 years old because factors including cultural differences increasingly influence the particular skills that older children acquire and when they acquire them. Older children are more likely to undergo assessments designed to make diagnoses that are more specific.

Note that neurodevelopmental delay is different from neurodevelopmental regression, which is less common and applies if an infant achieves developmental milestones in the first few years of life only to lose them again later. Developmental regression might develop, in older children, into a very variable and uncommon condition known as childhood dementia. Whereas dementia is usually associated with the elderly, it can appear at any stage of life.

Most children with neurodevelopmental delay show a global delay in the acquisition of milestones. Less commonly, delay might be limited mainly to specific aspects of

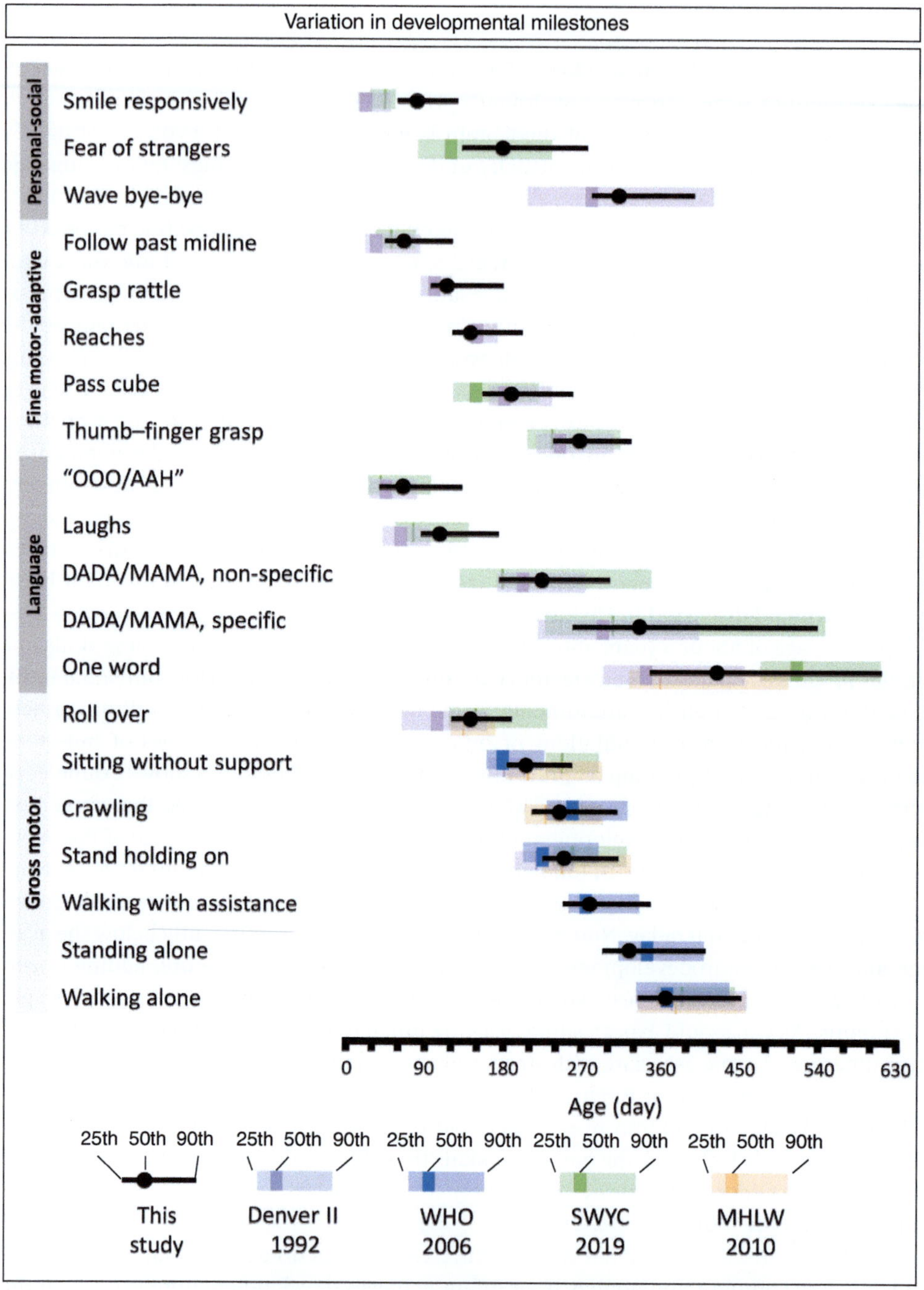

Figure 3.4 Milestone variation. The distributions of age at achievement of milestones in a study of 16,627 Japanese infants (referred to in the figure as 'this study') and studies carried out in: Denver, USA (Denver II 1992); Ghana, India, Norway, Oman and the USA by the World Health Organisation (WHO 2006); north-eastern USA (Survey of Well-Being of Young Children, SWYC 2019); Japan (National Growth Survey on preschool children 2010 conducted by Ministry of Health, Labour and Welfare, MHLW 2010). For the 20 milestones in 4 domains (personal–social, fine motor–adaptive, language and gross motor), distributions of age at milestone achievement are displayed as 25th, 50th and 90th percentile values (meaning the ages by which 25%, 50% and 90% of infants in the study attained each milestone). *Graph is reproduced from Matsubara et al.*[73] */ Elsevier / CC BY 4.0.*

development, leading to dissociation between the rates at which different attributes are acquired. Least common are extreme cases in which the order in which key milestones are achieved is completely reversed. I will give some example case studies in the next section.

Types of Neurodevelopmental Delay: Three Case Studies

To illustrate the differences between several types of neurodevelopmental delay seen by clinicians, I will compare the stories of three unrelated children, all aged about 2 years and growing up in a high-income country. They are based on real cases. In all three examples, the children are brought to the doctors because they are not yet talking when all their friends are. Of all the possible ways in which a child might differ from their peers, this one is likely to be particularly obvious and alarming to parents and often prompts them to seek professional help. The first case illustrates the most commonly encountered type of delay, the third the least.

Child 1: The doctor establishes that, although this child is not yet talking, he has been babbling since he was about one year old and, very recently, he has started saying Mama and Dada appropriately. Despite the fact that he is starting to use verbal communication, he has reached his milestones late. Babbling is achieved by roughly 50% of infants at six months and using Mama and Dada is achieved by roughly 50% of infants at a year. On further questioning, it turns out this child first sat unaided at one year and has only just started to walk. About 50% of infants would sit unaided at around 6 months and 90% would be walking by about 15 months, so gross motor development also appears significantly delayed. Regarding other behaviours, his visuomotor skills are delayed: the child first reached for objects at one year, which 90% would be doing by about six months, and has only recently started to release objects from his grasp, which most would do by one year. This child is described as having global neurodevelopmental delay.

Child 2: As with Child 1, this child started to babble late, at one year, and has just started using Mama and Dada. However, unlike the previous case, this child's gross motor development is proceeding at a normal rate: he walked at one year, ran at 18 months and can now jump. Regarding visuomotor skills, the child intentionally replaced objects before a year and started scribbling and imitating pencil strokes soon after, which are again on the average. Essentially, this child has a selective disorder of language. This is selective (or dissociated) neurodevelopmental delay.

Child 3: Unlike the previous two children, the order in which Child 3 achieved some milestones is atypical. As with the previous examples, he started to babble late and does not use Mama and Dada. He attained other skills at typical ages, however: he walked at one year, ran at 18 months and can now jump; he intentionally replaced objects before a year and started scribbling and imitating pencil strokes soon after. So far, he is like Child 2. What makes this child unusual is that he can recognize all the letters in the alphabet, whereas the average age at which his peers would do so is five years. Although infants can be trained to recognize letters, in this case the parents say they did not do this and the child must have taught himself. Despite having this advanced ability, the child does not use gesture language, such as pointing or waving, and has only just started following gesture commands by the parents, which would typically occur at one year. So, his development of speech and non-verbal communication is delayed, but he has reached some neurodevelopmental milestones unusually early. This is a significant deviation from the normal order of milestone attainment. Some higher-level milestones have been accomplished before lower-level ones. This can be described as neurodevelopmental delay with deviation.

You should remember that these terms are descriptive, that milestone acquisition is highly variable and it is unlikely that any of the three children would be diagnosed with a specific disorder at this stage on the basis of these findings alone. They might be monitored more closely than usual.

Intellectual Disability

This term is even harder to define unambiguously than neurodevelopmental delay, mainly because it is used at ages where behaviours are much more complex than during the first few years of life. It describes a broad range of phenotypes of variable severity, from profound in all circumstances to very mild and context-dependent; the latter is commonly associated with neurodivergent conditions rather than with disorder. There is not even universal agreement on the best name. Pejorative ones such as mental retardation or mental deficiency have fallen from favour, although you will still see them used. *The International Classification of Diseases* (ICD-11) proposes the term 'disorders of intellectual development'.[69] I prefer the commonly used term 'intellectual disability'. If I were to say *disorders* of intellectual development, this might be fine in this section, which has the heading 'Functional *disorders*', but would make less sense when it comes to discussing intellectual disability in the context of neurodivergence, which, as I said in the Introduction, I prefer to avoid describing as disorder. It seems fair to say that some people categorized as neurodivergent might have some sort of intellectual disability, even though it might be relatively mild and only create difficulties in some circumstances, for example in particular social settings or when attempting to carry out specific types of tasks.

Intellectual disability is sometimes defined based on measures of intellectual function and adaptive behaviour. A person's overall intellectual function – or, to put it more simply, intelligence – is commonly measured by tests taken by individuals that give a score called an Intelligence Quotient or IQ (Figure 2.10 in Chapter 2).[74,75] Although measuring IQ is an effective way to generate quantitative data that will identify individuals as being statistically different to the normal population, many with an intellectual disability have an IQ in the normal range. Their disabilities might be in their adaptive behaviours – which are learned conceptual, social and practical skills that individuals use to meet the demands of everyday life in their society at their age. Adaptive behaviours are commonly measured by interviews with trained professionals, with all the potential for subjectivity that this implies (Figure 3.5). Researchers have shown that measures of intelligence and adaptive behaviour are not particularly strongly correlated, suggesting that they might be assessing separate, independent attributes of the brain.[76] This illustrates one of the many complexities of defining functional brain phenotypes: where does one stop breaking them down into an increasing number of increasingly specific entities? Unlike phenotypes such as a person's height or weight (easy to define and measure) or the volume of their brain (easy to define, a bit tricker but nonetheless possible to measure), there is a vast array of ways in which we might define functional brain phenotypes. The degree of resolution with which I am defining them – as neurodevelopmental delay, intellectual disability and so on – is suitable for the purposes of this book but, in many areas of research, more fine-grained phenotyping (often referred to as deep phenotyping) is required. This might exploit modern machine-learning approaches to analyse data on many aspects of functional development from large numbers of individuals.

The global prevalence of individuals labelled as intellectually disabled is somewhere in the region of one to three per 100; the vast majority fall into a mild category. Intellectual disability often follows from neurodevelopmental delay. The range of methods for assessing an older child's intellectual ability is broader than is possible in the first few years after birth, simply because of the increase in skills and knowledge acquired with age. The authors of ICD-11[69] and DSM-5,[70] in common with many others, emphasize the desirability of providing some degree of objectivity to the diagnosis of intellectual disability by using quantifiable tests wherever possible. This is not always possible, however, as the authors of ICD-11 point out: 'Where appropriately normed and standardized tests are not available, diagnosis ... requires greater reliance on clinical judgment based on appropriate assessment of comparable behavioural indicators'.

The difficulties inherent in defining the term intellectual disability are significant. Multiple tests are required, some of which are highly subjective, and definitions tend to include numerous caveats and exceptions, often couched in ambiguous terms and leaving much to the judgement of individual medical professionals. Here, I will limit myself to quoting one example, in this case from DSM-5.[70]

> *Intellectual disability involves impairments of general mental abilities that impact adaptive functioning in three domains, or areas. These domains determine how well an individual copes with everyday tasks:*
>
> - *The conceptual domain includes skills in language, reading, writing, math, reasoning, knowledge and memory.*
> - *The social domain refers to empathy, social judgment, interpersonal communication skills, the ability to make and retain friendships and similar capacities.*
> - *The practical domain centres on self-management in areas such as personal care, job responsibilities, money management, recreation and organizing school and work tasks.*

As you can see, many of these are characteristics that vary greatly between all of us, even within specific populations, societies, cultures, families and so on. Setting the boundary where variation becomes disability is very hard.

Disorders of Communication

The ability to communicate effectively through language, usually written and spoken but also nonverbal, is central to most collective human activities. There are many ways in which our ability to communicate can be affected. Above, I ran through symptoms that might arise if the brain has difficulty with the systems that output signals to the muscles that allow us to talk or write or communicate in non-verbal ways. In addition,

Adaptive behaviour assessment

Item no	Item	Suggested modification	Need for modification
I. Communication domain			
40	Says month and day of birthday when asked	Approximate answer, or DOB in relation to festival	Recording birth, for example according to the constellations or in relation to festivals
29	Says first and last name when asked	Can say initials or father's name	Concept of last name may not be practiced in all communities and Father's name is often used as last/ surname
II. Daily living skills domain			
06	Feeds self with spoon; may spill	Uses fingers to eat Chapati/ Dosa appropriately	Family may not regularly use spoon at meal times.
33	Finds and uses appropriate public restroom for his or her gender.	Finds and uses appropriate place	Restrooms may not be available
3	Clears unbreakable items from own place at table	Shows appropriate care handling crockery	All these items may depend not only on availability of appliances but also the differences in culture and gender role. Most of these domestic tasks are not generally expected of male members of family, and going out, or being employed is not expected of female family members, especially
8	Helps prepare foods that require mixing and cooking (for example cake or cookie mixes, macaroni and cheese, etc.)	Can help prepare simple meals like instant noodles, dosas from available batter or poha (beaten rice)	
38	Travels at least 5–10 miles to unfamiliar destination (i.e., bikes. uses public transportation or drives self).	Can find a way of reaching the place with or without escort.	
39	Earns money at a part time job (i.e., at least 10 hours a week) for one year.	Is able to do a job assigned to him/her	
III. Socialization domain			
37	Goes on group dates	Goes out with friends	Dating is not considered appropriate behavior in most cultures and may not be a task to learn
38	Goes on single dates		
31	Goes places with friends in evening without adult supervision (for example, to a concert, lecture, sporting event, movie, etc).		

Figure 3.5 **Adaptive behaviour**. Adaptive behaviour refers to the range of social, practical and conceptual skills that individuals learn and perform in their everyday lives (examples are given here). Social skills include how we interact with each other, including how we take responsibility, our self-esteem, our personal gullibility, our ability to solve social problems and our ability to

Figure 3.5 (Continued) follow rules, laws and so on. Practical skills include how we take care of ourselves personally, our occupational skills, our organisation of schedules, routines, travel, safety, use of money and many others. Conceptual skills include our literacy and use of language and numbers and our ability to self-direct. There are many tests used to measure adaptive behaviour. They rate a set of specific behaviours on standardized scales, relying heavily on answers elicited during interviews with carers. Examples include assessments like the Adaptive Behaviour Scale (ABS), Diagnostic Adaptive Behaviour Scale (DABS), Vineland Adaptive Behaviour Scales, Scales of Independent Behaviour and the Behaviour Assessment System for Children, all of which evaluate different aspects of an individual's social, conceptual and practical life skills depending on their age and developmental stage. There are differences in the primary objectives of some. For example, the key difference between ABS and DABS is that ABS is a broader assessment tool that measures adaptive skills across a wider age range and in various life domains while DABS is specifically designed to pinpoint whether an individual has significant limitations in adaptive behaviour. ABS is not necessarily focused on diagnosing intellectual disability, whereas DABS is used primarily for diagnosing intellectual disability. In terms of content, ABS, for example, assesses adaptive behaviour on a 241-item Adaptive Behaviour Scale with General Adaptive Composite scores providing an overall measure of adaptive development in ten skill areas: Communication (speech, language and non-verbal skills); Community Use (interest in activities outside the home and ability to recognize community locations); Health and Safety (caution and an ability to avoid physical danger); Leisure (play and the ability to follow rules); Self-care (eating, toileting and bathing behaviours); Self-direction (self-control, following directions and making choices); Functional Pre-academics (letter recognition, counting and drawing simple shapes); Home Living (helping with household tasks and caring for personal possessions); Social (getting along with other people, using manners, assisting others and recognizing emotions); Motor (locomotion skills and manipulation of the environment). A critically important consideration is that not all items in these tests are appropriate in all countries and cultures. For example, the table reproduced here indicates items in the Vineland system that are difficult to use in some settings in India. *Table is reproduced with permission from Manohari et al.*[77] */ Sage Publications.*

communication can be affected by disruption of the parts of the brain responsible for understanding and formulating language prior to outputting it, including so-called speech centres such as Broca's area in the cerebral cortex (Figure 2.2, Chapter 2). Symptoms of this include aphasia (no speech), dysphasia (disrupted speech) or stammering (also known as stuttering or, less snappily but more descriptively, childhood-onset fluency disorder). Communication can, of course, be influenced more broadly by how our brain functions at a high perceptual and cognitive level, which I will move on to discuss below. The communication of people with disorders of perception or mood and neurodivergent individuals can show symptoms such as echolalia, which is the inappropriate or meaningless repetition of words just after someone else has said them.

Disorders of Perception: An Overview

In the present context, perception describes the brain's ability both to sense the world around it – using sight, hearing, touch and so on (as in "I perceive there is a person in front of me") – and to place interpretations on what it perceives (as in "by the look on their face, I perceive this person is friendly"). Disordered perception is very common in the human population, affecting as many as one in 30 of us. This is only a rough approximation because definitions of these disorders vary around the world, as do methods for recording their occurrence. These symptoms often appear in adolescence or early adulthood. Because of this, some experts do not consider them as neurodevelopmental disorders. DSM-5, for example, considers that the term neurodevelopmental disorders should only apply if symptoms appear during childhood.[70] While this might be pragmatic in the context of the clinical management of these symptoms, most researchers working on their causes agree that they almost certainly originate at least in part from events that occur prenatally, in childhood or in adolescent development, even if symptoms are not manifest until adulthood.

In a disordered state, the brain might have perceptual deficits or construct distorted perceptions of things around it, for example objects, noises and smells; or it might think that it is perceiving things that do not exist, which we call sensory deceptions; or it might interpret incorrectly the meaning of the thing it detects, which we refer to as delusional perception.

Perceptual deficits would include overall loss or impairment of one or more type(s) of sensory ability, for example complete or partial blindness or deafness, or specific loss of a particular aspect of sensory ability, such as a specific inability to recognize faces (we call this prosopagnosia). Distorted perceptions include changes in intensity or type. In hyperaesthesia, things are sensed with greater intensity than usual; hypoaesthesia is the opposite. In synaesthesia, which is usually considered neurodivergence rather than disorder, things that most people would perceive to be of a particular type, such as a sound, would elicit a sensation of a different type, for example a colour. Sensory deceptions include hallucinations, which is the sensation of non-existent things such as sounds or voices, objects or people, touches or movements and smells or tastes. A common definition of delusion, as used in a clinical context, is a belief that is clearly wrong; a delusional perception, mentioned above, is a particular type of delusion triggered by the detection of something (for example, a belief that aliens have landed triggered by seeing a green bus). People with paranoid delusions have a deep mistrust of others who they believe are persecuting or trying to harm them. Most definitions have their caveats and definitions of delusion probably have more than most; they immediately raise the question of how to define 'clearly wrong'. After all, many commonly held beliefs guide the ways in which we live our lives and many people consider some are clearly wrong. Fascinating though it is to consider these matters, I will not delve deeper into them here.

To illustrate some of the ways in which perceptions can become disordered or unusual, I will describe a few case histories from the last two centuries in the next section.

Disorders of Perception: Case Histories

Reproduced in Figure 3.6 are some pages from John Haslam's 1810 book *Illustrations of Madness*.[78] In them you can read one of the earliest descriptions of hallucinatory and delusional perceptions. The patient was James Tilly Matthews and, as Haslam points out, his belief that the 'Air Loom' existed was a delusion.

Hallucinations and delusions can manifest in many ways that have been described in numerous accounts over the last two centuries. For example, in 1923, French psychiatrists Joseph Capgras and Jean Reboul-Lachaux provided an early description of a type of delusion in which people think that individuals or groups have been replaced by doubles or imposters.[79] They described a 53-year-old woman who believed she was famous, wealthy and of royal descent. She 'transformed everyone in her entourage, even those closest to her,

Historic description of schizophrenia

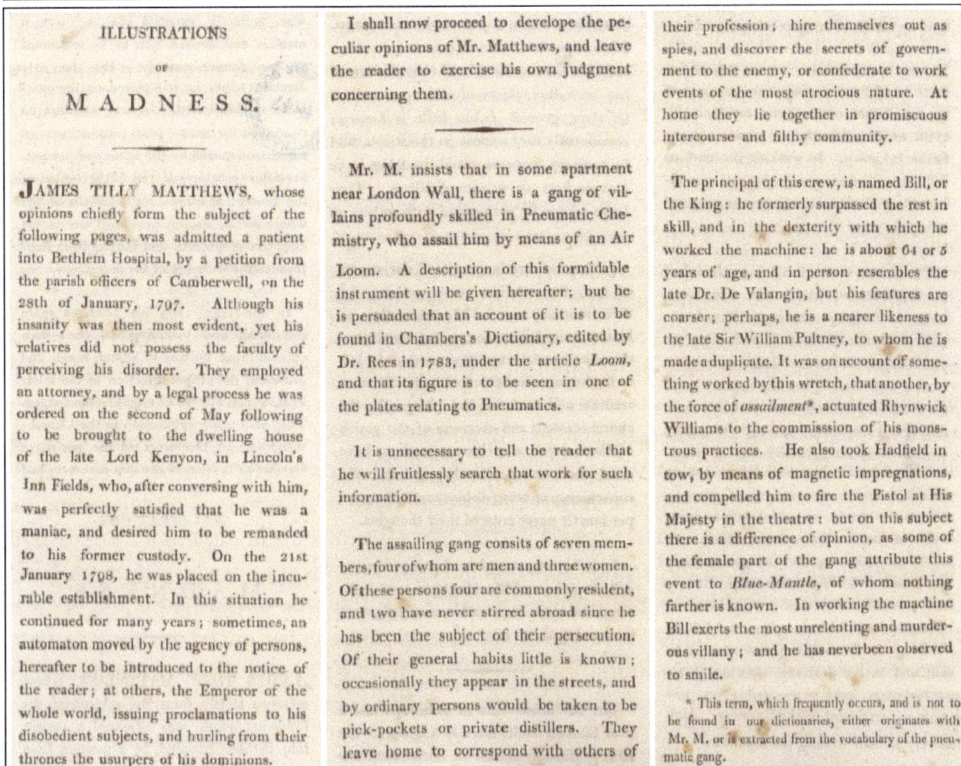

ILLUSTRATIONS

OF

MADNESS.

JAMES TILLY MATTHEWS, whose opinions chiefly form the subject of the following pages, was admitted a patient into Bethlem Hospital, by a petition from the parish officers of Camberwell, on the 28th of January, 1797. Although his insanity was then most evident, yet his relatives did not possess the faculty of perceiving his disorder. They employed an attorney, and by a legal process he was ordered on the second of May following to be brought to the dwelling house of the late Lord Kenyon, in Lincoln's Inn Fields, who, after conversing with him, was perfectly satisfied that he was a maniac, and desired him to be remanded to his former custody. On the 21st January 1798, he was placed on the incurable establishment. In this situation he continued for many years; sometimes, an automaton moved by the agency of persons, hereafter to be introduced to the notice of the reader; at others, the Emperor of the whole world, issuing proclamations to his disobedient subjects, and hurling from their thrones the usurpers of his dominions.

I shall now proceed to develope the peculiar opinions of Mr. Matthews, and leave the reader to exercise his own judgment concerning them.

Mr. M. insists that in some apartment near London Wall, there is a gang of villains profoundly skilled in Pneumatic Chemistry, who assail him by means of an Air Loom. A description of this formidable instrument will be given hereafter; but he is persuaded that an account of it is to be found in Chambers's Dictionary, edited by Dr. Rees in 1783, under the article *Loom*, and that its figure is to be seen in one of the plates relating to Pneumatics.

It is unnecessary to tell the reader that he will fruitlessly search that work for such information.

The assailing gang consits of seven members, four of whom are men and three women. Of these persons four are commonly resident, and two have never stirred abroad since he has been the subject of their persecution. Of their general habits little is known; occasionally they appear in the streets, and by ordinary persons would be taken to be pick-pockets or private distillers. They leave home to correspond with others of their profession; hire themselves out as spies, and discover the secrets of government to the enemy, or confederate to work events of the most atrocious nature. At home they lie together in promiscuous intercourse and filthy community.

The principal of this crew, is named Bill, or the King: he formerly surpassed the rest in skill, and in the dexterity with which he worked the machine: he is about 64 or 5 years of age, and in person resembles the late Dr. De Valangin, but his features are coarser; perhaps, he is a nearer likeness to the late Sir William Pultney, to whom he is made a duplicate. It was on account of something worked by this wretch, that another, by the force of *assailment**, actuated Rhynwick Williams to the commission of his monstrous practices. He also took Hadfield in tow, by means of magnetic impregnations, and compelled him to fire the Pistol at His Majesty in the theatre: but on this subject there is a difference of opinion, as some of the female part of the gang attribute this event to *Blue-Mantle*, of whom nothing farther is known. In working the machine Bill exerts the most unrelenting and murderous villany; and he has never been observed to smile.

* This term, which frequently occurs, and is not to be found in our dictionaries, either originates with Mr. M. or is extracted from the vocabulary of the pneumatic gang.

Figure 3.6 An early case of perceptual disorder. Pages from John Haslam's 1810 book *Illustrations of Madness* provide one of the earliest descriptions of hallucinatory and delusional perceptions. The drawing of the Air Loom mentioned in these pages was a figment of the patient's imagination. A reconstruction (bottom right) by modern-day artist Rod Dickinson (The Air Loom, A Human Influencing Machine, 2002, Rod Dickinson) was on display at Bethlem Hospital which, along with other sites mentioned, is in London, UK. *John Haslam / Wikimedia commons / public domain /* https://commons.wikimedia.org/wiki/File:Airloom.gif */ Last accessed on October 21, 2025. Rod Dickinson / https://theairloom.org/info.php / last accessed on Nov 04, 2025.*

such as her husband and daughter, into various and numerous doubles'. Three of her four children had died, yet she believed that they had been abducted and that her only surviving child was replaced by a lookalike.

These accounts illustrate features of common types of delusions, namely paranoid or persecutory and grandiose delusions. Other common types include delusional jealousy (a belief that a sexual partner is being unfaithful, even when it is impossible) and somatic delusions (a belief that something is wrong with a part of the body or that a part is missing).

More recently, in a 2022 report, Roberto Bottini and colleagues working in Italy, described a case of synaesthesia in a blind individual, illustrating the point that this condition extends to involve all sensory modalities and not just vision.[80] The case is of CB: 'a 40-year-old male, born totally blind by sighted parents due to congenital retinopathy caused by maternal rubella during pregnancy. CB reports a series of synesthetic experiences, involving numbers, letters, days, months, texture and space, that he has been experiencing since he can remember. CB's parents, as well as his twin sister, also born congenitally blind, have never experienced any form of synaesthesia'. In CB, number-related synaesthesia is strongest, as follows: 'Whenever CB hears, writes, reads or thinks about a whole number (i.e., 1, 2, 3, etc.) he experiences it as having a shape (i.e., a small cube of different sizes depending on the number) and very specific texture properties. For example, numbers 1 and 2 are experienced as being as scratchy as cardboard, while number 3 is as soft as velvet. Numbers 4 and 5 are as smooth as plastic, so too numbers 6 and 7, but the type of plastic he perceives is different from numbers 4 and 5. Numbers 10, 11 and 12 are experienced as being smooth and cold like metal. Numbers containing the same digit (e.g., 17 and 27) are experienced as being similar in texture. For instance, numbers 9 and 19 are both soft as velvet but differ in the quality of this softness, with 9 being softer than 19. This also translates into a sort of regularity among numbers, by which, for example, numbers between 21 and 29, or 31 and 39, present similar characteristics to numbers 1 and 9. This sort of regularity is typically encountered in other cases of visual synaesthesia, such as colour–number synaesthesia'.

Disorders of Mood or Emotion

In a clinical context, mood usually describes the overall level of excitement or happiness that a person feels. Our mood pervades all aspects of our thinking and affects our general behaviour including our emotions and reactions to events. As we all know, our mood fluctuates all the time, often in response to external events. It is relatively common for individuals to experience changes of mood severe enough to become debilitating.

Mood disorders can be complicated to describe, particularly if they involve the types of experiences that most of us have in our daily lives. These types of experiences might include a feeling of irritability or tension, difficulty concentrating, a sense of nausea or panic, palpitations of the heart, trembling, sweating and difficulty sleeping (insomnia). Who has not felt all of these things at some time or other? Clinicians might diagnose a disorder based on these symptoms if they are of such magnitude and persistence that they interfere in a significant way with an individual's ability to lead their life. More extreme symptoms that might make diagnosis easier include self-harming behaviours such as cutting, burning, poisoning or starving oneself or suicide, mutism (the inability to talk in some situations), phobias and panic attacks. Possible diagnoses include depression or anxiety. You might sometimes see depression and anxiety described as symptoms. This is an area in which the distinction between symptoms and diagnoses is blurred.

The opposite of depression is mania. Its symptoms include a feeling of extreme happiness or elation, rapid talking and exaggerated self-importance, easy distractibility, a lack of requirement for sleep and engaging in impractically ambitious projects. Like depression and anxiety, mania varies greatly in severity, both between different people and at different times in the same person. Some individuals experience both periods of depression and periods of mania.

Disorders of perception, mood or emotion are features of what we commonly refer to as mental illness, a term that is very broad and hard to define. As I pointed out above, the symptoms are extremely debilitating for some, interfering with their ability to live their lives. They lead to diagnoses of specific psychiatric disorders that I will describe in Part II. Other people might have milder symptoms, possibly in only some particular situations and not others. These might be disabling or might bring unexpected benefits and are usually described as components of neurodivergence rather than disorder.

Neurodivergence

There is a plethora of perceptual, cognitive and behavioural phenotypes that differ sufficiently in some people that we label them as divergent, but not enough that we think of them as disordered. Some of these are essentially mild versions of phenotypes I have already described. I have left others to this section because they are particularly strongly associated with neurodivergence rather than disorder. As we move to higher and higher levels of cognition, many terms apply to increasingly broad ranges of behavioural features, be they normal or typical, disordered or divergent.

One of those applied to at least 2–3 per 100 of the population is hyperactivity. We might – but we do not – use the term hyperactivity to describe some symptoms of some of the movement disorders I described earlier in this chapter: for example, seizures, chorea or tremor. When people in a healthcare setting talk about someone being hyperactive, they are referring to the fact that they are restlessly busy, often in an impulsive way that lacks focus or concentration, making them easily distractible and sometimes aggressive. They might fidget or move around more than is usual, wander or talk more than is appropriate for a situation, or have difficulty joining in calm activities such as reading. Attention deficit is another term often associated with hyperactivity to describe the neurodivergent behaviours of some individuals. Vaguer and even harder to define are behaviours that might create difficulties for neurodivergent individuals in certain social contexts, where society deems them to be unusually disruptive or to conduct themselves poorly. For example, DSM-5 defines a category of Disruptive, Impulse Control and Conduct Disorders.[70]

Neurodivergent people often have a learning disability of some type. Clearly, some serious brain disorders, for example some of those discussed earlier that affect the overall anatomy or structure of the brain, will inevitably impair overall learning, potentially drastically. This is not what people usually mean when they talk about learning disabilities. The disabilities associated with neurodivergence can be quite specific and might not be immediately obvious. For example, dyslexia is relatively common, affecting about 1 in 10 of us.

People with dyslexia find it hard, to an extent that varies from person to person, to read, write and spell words. Other examples are dysgraphia, which is difficulty specifically with writing, or dyscalculia, which is difficulty with arithmetic.

Other phenotypes that we often consider under a heading of neurodivergence are obsessive and compulsive behaviours. Obsessions and compulsions mean much as they do in everyday use. They typically involve having repeated intrusive thoughts that create a sense of anxiety. As they increase in severity, they can drive a person to carry out the same acts repeatedly, which temporarily relieves the anxiety. Stimming is a term used to describe a person's production of repetitive actions, including body movements or noises (stimming is short for self-stimulation and the action is called a stim). Stimming is thought to have a self-soothing function. Examples include the act of rocking backwards and forwards repeatedly while sitting or standing, finger or toe tapping and so on. All of us do these things some of the time but if someone does them a lot it might contribute to them being considered neurodivergent. The word tic is used to describe uncontrolled fast, repetitive muscle movement, such as blinking or finger-clicking, or the generation of repetitive sounds or words (occasionally offensive ones); the main difference between tics and stims is that tics do not appear to have a self-soothing function. Tics are quite common in children. They often appear between ages 5–10 years and usually go away spontaneously with time.

The behavioural phenotypes that can lead to a diagnosis of neurodivergence often involve types of complex interactions with other people in social settings. For example, some people might find social situations in general very stressful, they might place unusual reliance on strict routines for doing things, they might often find it difficult to understand the thoughts or feelings of others, they might come across as acting in a detached or even rude manner, they might have difficulty realising when someone is saying something that should not be taken literally or they might be unusually observant of small details. There are many other examples that I could give. None are easily defined and measured. As I discussed for the phenotypes above, a critical issue for clinicians making a diagnosis is the extent to which they create difficulties for a person. For researchers looking for mechanistic explanations for defined conditions, the imprecision inherent in defining these phenotypes and, therefore, the diagnoses that result from them add to the difficulty of discovering underlying causes. Not being clear on what exactly is being caused makes it hard to figure out what is causing it! Hopefully, approaches such as deep phenotyping, which I mentioned earlier in this chapter in the context of intellectual disability, will help resolve some of the difficulties.

Summary

- Phenotypes or traits are observable characteristics; they can be anatomical or functional, with the latter including perceptual, cognitive and behavioural traits.
- Many distinct anatomical abnormalities of prenatal brain development have been described, some of which can be severe and incompatible with life; some are detected before birth by imaging the fetus, others are identified soon after birth.

- Phenotypes associated with functional disorders are generally detected after birth and might include difficulty with movement, seizures, delayed development, intellectual disability, communication difficulties, disordered perception, mood or emotion.
- Phenotypes associated with neurodivergence can overlap those associated with disorder and, in addition, often include hyperactivity, learning disability and obsessive behaviour.
- Some of the phenotypes that are associated with disorder, and most of those associated with neurodivergence can be very hard to define precisely, presenting a challenge for researchers attempting to place individuals into discrete diagnostic categories.

Part II

Ultimate Causes of Neurodiversity and Neurodevelopmental Disorder

4

Genetic Variation

Genetic variation refers to the differences that exist among us all in the deoxyribonucleic acid (DNA) that our cells contain. Most genetic variations cause no detectable differences among us; I will explain why in this chapter. Some variations do have an impact, however, by causing – or more often predisposing – individuals to develop disorder or neurodivergence. In this chapter, I will explain how genetic variation arises and the circumstances in which it is likely to affect a person.

The Nature of DNA

An essential property of DNA is its ability to self-replicate. DNA also makes its own breeding grounds by directing the construction of cells and organisms within which to undergo self-replication. The structure of DNA not only allows it to carry out these two major functions but also gives it the leeway to change and evolve.

DNA molecules are very large polymers made up of repeating subunits, or monomers, which we call nucleotides. There are four of them. Each type contains one of the following nucleobases (more commonly referred to simply as bases): cytosine (C), guanine (G), adenine (A) or thymine (T). Nucleotides join in chains many billions long; we usually call these chains strands. Each base along a DNA strand forms chemical bonds with a specific complementary base in another strand that lies beside it (the strands sometimes come apart, as I will explain in a moment). We call the bases that are linked in this way 'base pairs': C pairs with G and A pairs with T. I have illustrated these principles in Figure 4.1 and you can see more details of the bases in Chapter 8 (Figure 8.6). Most of the time, the two linked complementary strands are twisted together (hence, the common description of DNA as a double helix). The complementary DNA strands sometimes separate, for example during DNA replication or ribonucleic acid (RNA) synthesis. RNA synthesis, known as **transcription**, is the first step in specifying the molecules needed for constructing cells and organisms.

The order of the four bases along the strands of DNA constitutes a code for the cell to use. These uses fall into two major categories. (i) The identification by other molecules in the

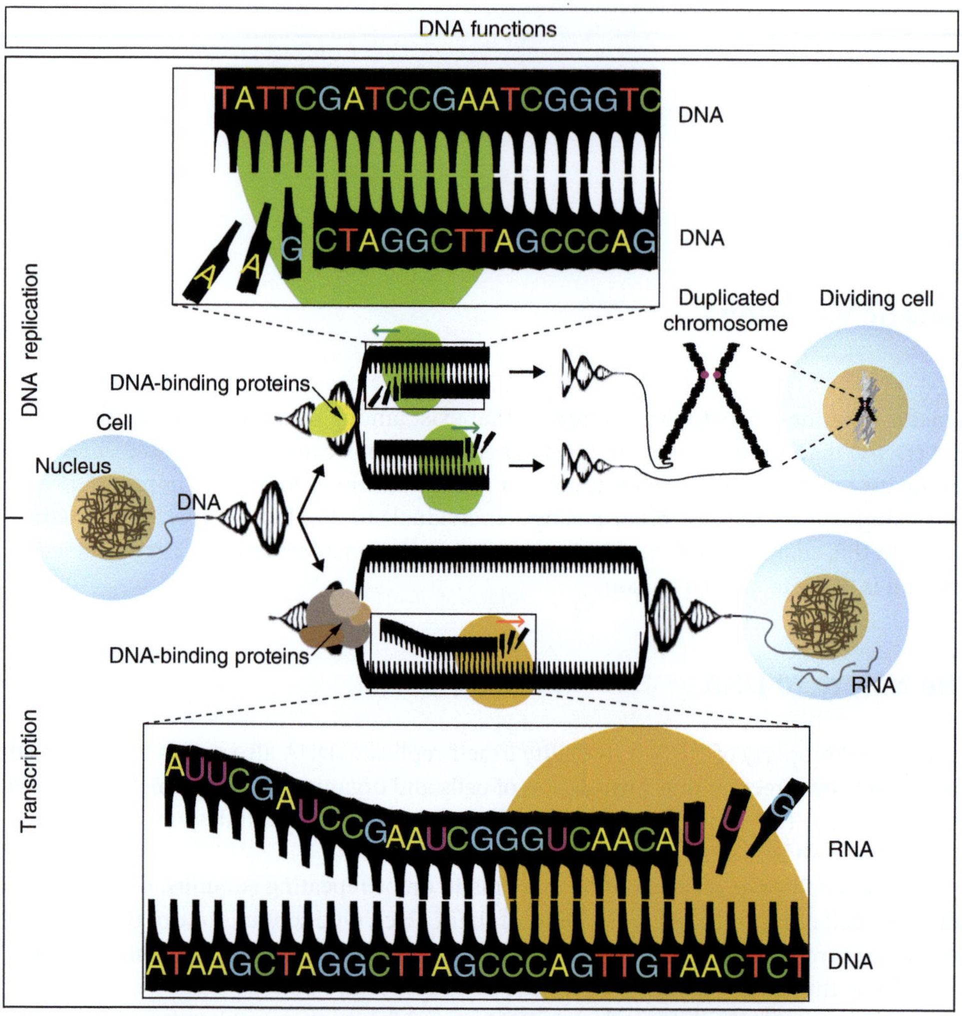

Figure 4.1 The essential activities of DNA. Here, I have depicted DNA and RNA at different magnifications, ranging from a line to a sequence of named bases. The nuclei of most of our cells contain a tangle of 46 very long DNA molecules, or chromosomes. Their total length is about 2 m per cell, an arresting value given that most cells are only about 0.00002 m in diameter. Most of the time, DNA comprises two complementary strands of four bases (A, T, C and G) wound around each other in a double helix. The two strands come apart under the influence of specialized DNA-binding proteins during DNA replication and RNA synthesis (transcription). In DNA replication, the two separated strands act as templates for the synthesis of new strands by DNA polymerase proteins, which move along the template strands adding new complementary bases (green arrows). The two new double strands wind back up to make a duplicated chromosome comprising two sister chromatids linked at regions we call **centromeres** (purple dots). The duplicated chromosomes line up in the centre of a dividing cell and each sister chromatid of each chromosome is drawn into a different daughter cell. In transcription, DNA-binding proteins initiate the movement of RNA polymerase proteins along the template strand (orange arrow) producing RNA (using uracil, U, instead of thymine, T). After transcription, the original DNA strands come back together. We call the region where the DNA strands temporarily separate a transcription bubble.

cell of specific places in the DNA where they can interact with it to control its function (these molecules include the DNA-binding proteins shown in Figure 4.1). (ii) The provision of instructions to guide the construction of molecules that the cell needs to stay alive, to reproduce itself and its DNA and to make it into a cell of a specific type equipped for specific tasks. Crucially for DNA's ability to evolve, its ability to do these things is not constrained by its exact length and sequence of bases: DNA can live with variation in both of these parameters.

Our cells contain DNA in two locations, nuclear and mitochondrial. Nuclear DNA is what people are usually referring to when they talk about our genome. Normally, our **somatic cells** (meaning the cells of our body apart from sperm and eggs, which we refer to collectively as **germ cells**) each contain about 6.4 billion base pairs of DNA in 46 very long DNA molecules: our chromosomes. The complement of 46 chromosomes in our somatic cells comprises 44 **autosomes**, which come in pairs that share similarities, and two **sex chromosomes**, either two X chromosomes in females or an X and a Y chromosome in males. Paired chromosomes other than the X and Y share almost the same length and sequence of base pairs, whereas lengths and sequences vary greatly from pair to pair and between X and Y. Each of our parents gives us one of each pair of autosomes and one of our sex chromosomes. We describe the normal complement of two of each autosome and sex chromosome as diploid (*diploos* is Greek for double); our germ cells are haploid, containing only one of each autosome and one sex chromosome, either X or Y (*haploos* is Greek for single).

We inherit mitochondrial DNA from our mother (rarely if ever our father).[81,82] Mitochondria are tiny structures providing energy to the cells that contain them. Each of our cells contain thousands of them. It is likely that they evolved as independent organisms and eukaryotic cells (meaning cells like ours, which have a nucleus enclosed in a nuclear membrane) incorporated them several billion years ago. This would help explain why they carry their own genome. Mitochondrial genomes are contained in circular molecules of DNA called mitochondrial chromosomes. Each molecule contains about 16.6 thousand base pairs and each of our cells can contain many thousands of copies of them, with several copies in each mitochondrion.

The DNA that humans have inherited has been replicating for billions of years, from the time when simple lifeforms emerged to the relatively recent few million years of human evolution. DNA evolves because it has the potential not only to replicate itself but also to change. Whereas most changes have little or no effect on the organism that contains them, some might affect – in a good or a bad way – that individual's chances of surviving or reproducing. Those that are advantageous are likely to spread through subsequent generations and those that are disadvantageous are not. This is the basis of evolution by natural selection. Human geneticists are often particularly interested in studying whether and how populations of people rid themselves of potentially harmful changes, a process they refer to as purifying selection.

While DNA's ability to change is good news for the evolution of species by natural selection, it is not necessarily good news for individuals who might develop with a disadvantageous change. Understanding how such changes happen and what their consequences might be for individuals is a crucial part – although only a part – of understanding the origins of human diversity and disorder.

How Variation in DNA Comes About

DNA Molecules Are Prone to Develop Errors

Our nuclear genome replicates itself every time one of our cells divides. Adult humans are made up of somewhere around 30 trillion cells, several hundred billion of which are in your brain. Making a person involves a truly enormous amount of DNA replication. To try to gain a sense of its magnitude, imagine making 30 trillion copies of a bank note. If you stacked them, one on top of the other, the stack would reach 8 times further into space than the distance to the moon. It is unsurprising that significant numbers of errors creep into this number of copies. Mistakes occur in two main ways.

Some errors occur spontaneously. One reason is that the four bases can, and occasionally do, pair incorrectly when DNA is replicating. The four bases – A, T, C and G – occur in slightly different chemical forms, some of which are common while others are rare. Incorrect pairing often involves the rare forms but even the common forms are capable of forming mismatches. DNA is flexible enough to tolerate the occasional mismatch (often referred to as wobble base pairs) and such errors are a natural risk during DNA replication.

Other errors happen when harmful factors damage DNA. DNA is susceptible to damage because some of the chemical bonds that hold it together are not very strong. Harmful factors can arise either within the body (endogenous factors) or externally (exogenous factors).

Examples of endogenous factors include the highly reactive forms of oxygen known as **reactive oxygen species** (often abbreviated to ROS), which are the products of normal metabolism and can damage DNA's bases. Fortunately, cells have evolved a way of countering the potential harm that reactive oxygen species can cause. They produce an enzyme that helps to break down reactive oxygen species, called superoxide dismutase, and distribute it to their cytoplasm and mitochondria.

Examples of exogenous factors include viruses and ionizing radiation (which is radiation, such as radioactivity, with sufficient energy to knock electrons from molecules such as DNA). We are exposed to viruses and to at least some background level of ionizing radiation all the time. Viruses replicate by harnessing aspects of their host cell's biochemistry and in doing so they can inflict damage on the cell's DNA. Ionizing radiation can damage DNA molecules either directly by smashing into them or indirectly by raising the levels of reactive oxygen species.

The rates at which errors occur in nuclear DNA range from tens of thousands to a million per cell per day.[83,84] These rates are scarily high, but evolution has equipped cells with DNA repair mechanisms that correct most of the mistakes.[85,86] I have summarized these mechanisms in Figures 4.2 and 4.3. The repair mechanisms illustrated here are important for minimizing errors that might cause developmental disorders. They involve a large array of specialized proteins for recognizing, removing and replacing faults. Many of them operate in cells that are replicating their DNA in preparation for cell division, which is when errors tend to occur. Note that there are other mechanisms not illustrated here, which operate on types of DNA damage implicated particularly in the development of cancers.

As shown in Figures 4.2 and 4.3, wherever possible cells use an undamaged strand of DNA (either from the same or the **sister chromatid**) as a template to accurately

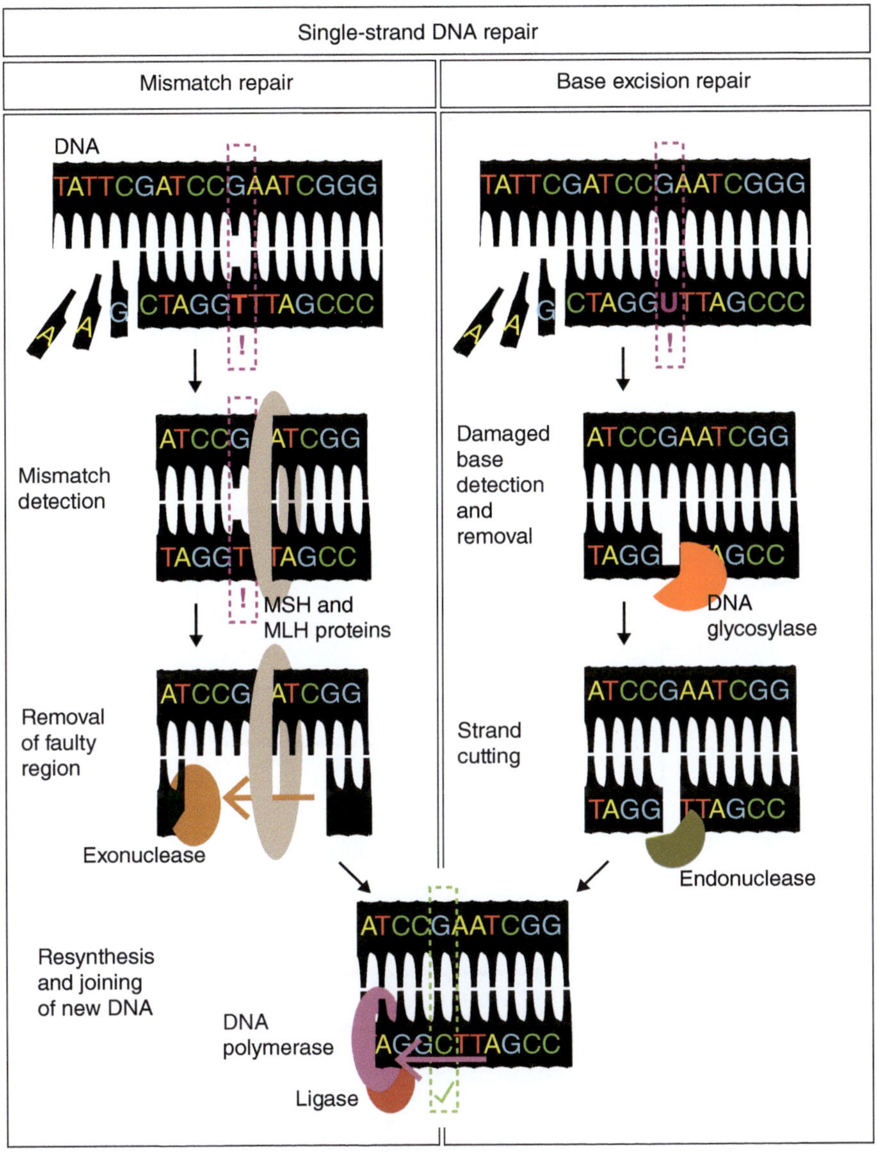

Figure 4.2 **How DNA repairs itself.** Mismatch repair: As DNA replicates, it can incorporate mismatched bases, here a T instead of a C. A group of proteins (called MSH or MLH followed by a number) detect such mismatches and correct them by working together with exonucleases, which cut out DNA bases, DNA polymerases, which make new DNA, and ligases, which join the new DNA into the existing strand. Base excision repair: Sometimes harmful events such as exposure to reactive oxygen species damages the bases. In this case, a chemical change to a C has turned it into a uracil (U), which DNA does not contain (RNA does). Many other changes are possible. DNA glycosylases and endonucleases cut out the damaged base. Polymerases and ligases replace the removed base with a correct one.

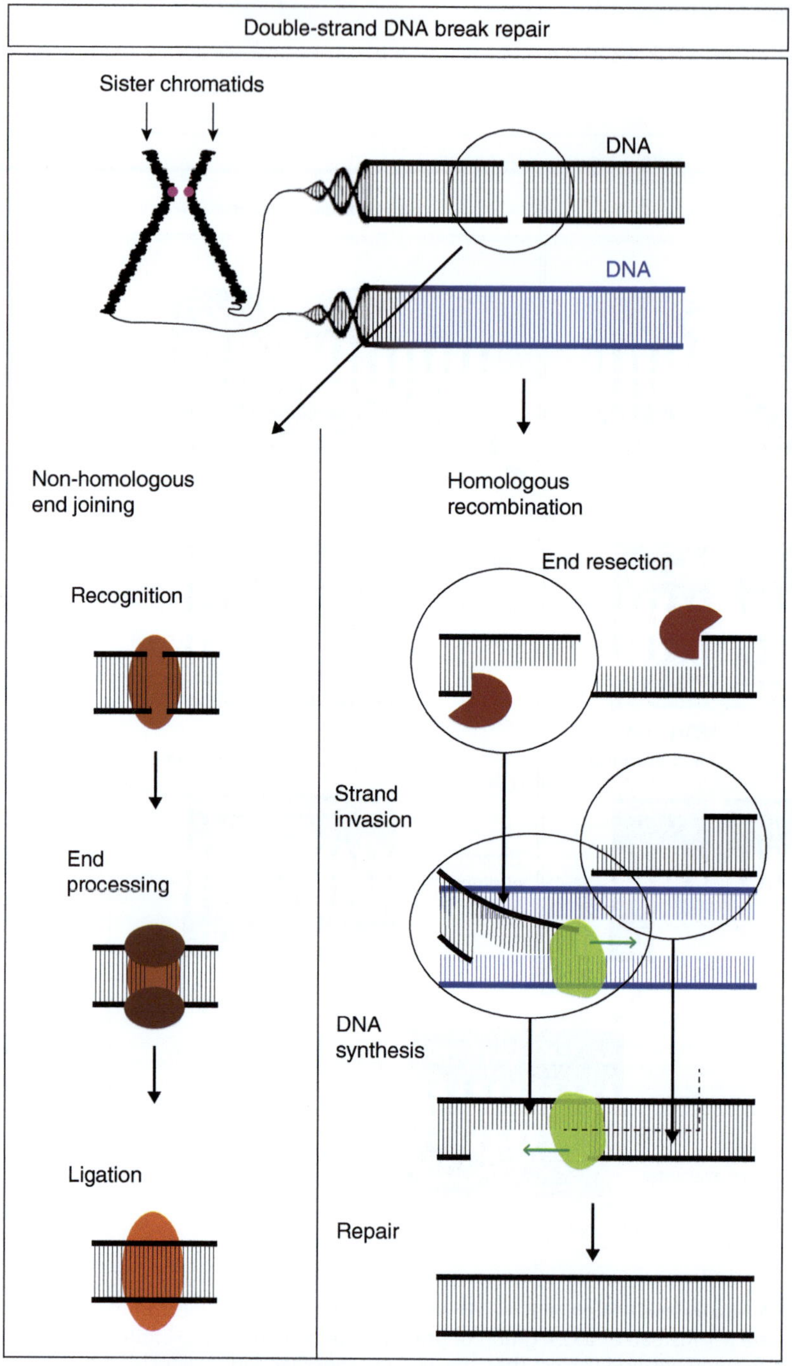

Figure 4.3 **How DNA repairs itself, continued.** Double-strand DNA break repair: Breaks across both strands of a DNA molecule can produce rearrangements of the genome. While such rearrangements are integral to the normal production of germ cells, as I will explain later in this chapter, they are hazardous in other circumstances. Cells have two main ways of repairing these breaks. In non-homologous end joining, a group of proteins (called XRCC followed by a number) work together with exonucleases and endonucleases, polymerases and ligases to recognize the break, process the separated ends and join the DNA. In homologous recombination, specialized proteins resect bases on either side of the break to generate lengths of single-stranded DNA. These then invade the sister chromatid (look again at Figure 4.1 if you have forgotten what the sister chromatid is: I have made its DNA blue here), separating its strands, finding and binding complementary DNA. This binding initiates the synthesis of new DNA that returns to bind to complementary DNA in the original chromatid, thereby bridging the break with single-stranded DNA, which acts as a template for the generation of double-stranded DNA.

resynthesize DNA. If they cannot do this, cells can continue DNA replication through a damaged region, sometimes attempting to correct it in the process (we call this translesion synthesis); it is more likely than other mechanisms to perpetuate errors.

Some Errors in DNA Remain Uncorrected, Creating Mutations

Despite the repair mechanisms I have just outlined, some errors do get through and are then transmitted through all subsequent cell replications in that individual and, potentially, to their offspring. If a change in the sequence of DNA bases caused by a spontaneous error of DNA replication or DNA damage remains uncorrected by the cell's DNA repair mechanisms, we call it a mutation. Any descendant cells produced by the mutant cell will inherit that mutation. We usually reserve the term mutation to mean a change of this type, rather than a temporary flaw that the cell's DNA repair mechanisms succeed in correcting.

When a new mutation emerges for the first time, we call it a **de novo mutation** (*de novo* is Latin for afresh). If a de novo mutation occurs after the **zygote** has divided then only some of that person's cells will carry that mutation; the earlier any specific mutation arises, the greater will be the number of cells carrying it. Researchers have estimated that up to the eight-cell stage, the very early embryo accumulates de novo mutations at a rate of 2–3 per cell division.[87] This number drops as the cells' DNA repair mechanisms mature. Most of these mutations have no detectable effect, for reasons I will discuss below. Sometimes, however, they can affect the individual in some discernible way, causing disorders, including neurodevelopmental ones, or diseases such as cancer (which is not addressed in this book).

We use the term genetic mosaic to describe individual organisms (human or non-human) made of cells with different genomes. As you can see from what I have just said, we are all genetic mosaics because we all have some relatively small amount of variation among the genomes contained in our cells.[87,88] Much of the time, we focus on the far more numerous similarities between the DNA in our cells; for example, I remember my schoolteachers saying that all our cells contain the 'same DNA'. It is true that we can often ignore our genetic mosaicism, but it is best to remember its existence since it can matter. Even **monozygotic twins** that have arisen from the splitting of one zygote will show genetic mosaicism within and between them.[89] Although we often call them identical twins, strictly speaking they are not. Nevertheless, their DNA will share many more similarities than the DNA of any other two individuals, including non-monozygotic siblings, and their study can therefore shed light on the relative contributions of genetic and non-genetic factors to how we develop (more on this in Chapter 6).

There are some important differences in the way mitochondrial DNA replicates and its susceptibility to mutations. Mitochondrial DNA replicates continuously even when cells are not dividing: we call this relaxed replication. Mitochondrial DNA is relatively short-lived, having a half-life of about 7–10 days, so the amount of replication that it undergoes in the lifetime of just one single cell, let alone of all the cells in the body, is potentially enormous. Mitochondria have less efficient DNA-repair mechanisms and a local environment that is more likely to induce faulty replication. Together, these differences result in mitochondrial DNA accumulating more errors that persist than nuclear DNA does.[90–92]

Mutations Can Pass to the Next Generation

Some de novo mutations occur in the cells that make a person's germ cells. We call these cells the germ cell **lineage** (the term lineage describes the succession of cell divisions producing any given cell or type of cell). If a mutation occurs in a person's germ cell lineage, then it might pass to their children. The chances of this happening depend on how many germ cells it affects, which depends in part on how early in the lineage the mutation occurred. The incidence of de novo mutations in germlines increases with parental age. In fathers, this is probably in large part because their sperm are produced by an ever-increasing number of cell divisions across the reproductive age range, so copy errors are more likely in older men.[93,94] In older mothers, the process that creates their eggs are more prone to errors involving entire chromosomes. On average, humans pass about 60 de novo mutations to their children. We know this from studies that have compared the genomes of thousands of so-called trios, which comprise two parents and at least one child. De novo mutations that pass from an individual to the next generation are inherited. If they persist through subsequent generations (which they might or might not, depending on what effects, if any, they have), then they increase the amount of DNA sequence variation in the human population as a whole.

Mutations Can Persist with Different Frequencies in the Population

Based on what I have said so far, you can see how every individual's genome comes to be unique. If we were to compare the sequences of the base pairs in the nuclear DNA of any two randomly selected individuals, we would find on average about 6 million differences between them. The vast majority of these differences would fall into a category of genetic differences known as **polymorphisms**. We say that a region of the genome (which can be of any size from one base to many hundreds of thousands) is polymorphic if we find two or more different sequences (or sequence variants) in that region in the population. Geneticists generally limit the use of the term polymorphism to sequence variants that are present in significant numbers of people, defined as more than 1% of whatever population is under consideration.

It is important to distinguish between the terms: genetic polymorphism and mutation. They do not describe exactly the same thing. Essentially, mutations can become polymorphisms, although only some do. A genetic polymorphism is a naturally occurring relatively frequent (>1%) variant form of the DNA sequence in a given region of the genome that has arisen in an unknown way at some time in the past. The pool of genetic polymorphisms present in the present-day human population was – and still is – fed by de novo mutations that manage to persist through multiple generations and to spread through a population so that the proportion of people that now inherit them rises above 1%. We refer to mutations that are no longer new, but are not frequent enough to call polymorphisms, simply as mutations or inherited mutations. De novo or inherited mutations that have overtly harmful effects are likely to remain too rare to become polymorphisms in most populations.

For reasons that I will explain later, most polymorphisms have no detectable consequences. However, some do exert effects, which can be either positive or negative for the individual. The effects might be relatively subtle: for example, they might affect a person's likelihood of developing a disorder caused primarily by other factors. They might predispose to, rather than cause, disorder or they might protect against it. Some polymorphisms with potentially harmful effects persist in a population because they also confer advantages under certain circumstances. The balance of advantages versus disadvantages conferred by a polymorphism might vary between populations in different parts of the world. For example, a generally disadvantageous polymorphism might have the potential to be advantageous by protecting against harmful factors prevalent in only some groups of people or in only some geographical regions. Such polymorphisms are likely to persist at a higher frequency among people exposed to the harmful factors, whatever they happen to be. The take-home message is that whether a mutation becomes a polymorphism and, if so, how frequent it is are complicated matters.

Categorizing Types of Genetic Variation

Nuclear DNA Variants Can Be of Many Different Sizes

One way of categorizing genetic variants, be they mutations or polymorphisms, is according to the length of DNA they involve. The categories I describe here are commonly used but you will not find strict definitions universally applied in the literature. Variants that are anywhere from a single base up to about 1,000 contiguous base pairs are often called small-scale sequence variants. The commonest mutations involve single isolated base pairs, which might be altered, duplicated or missing; we call them point mutations.

Variants that are larger than small-scale sequence variants, from about 1,000 base pairs up to about 5 million base pairs, are often called genomic structural variants. Some variants are big enough to be visible when looking at the chromosomes down a microscope, a process called **karyotyping** (Figure 4.4). You will find them referred to as chromosomal variants (somewhat confusingly because, at the end of the day, all types of variation occur within the chromosomes). This categorization according to size is a good place to start. We can then overlay further information based on what has happened in the mutant or polymorphic region.

Nuclear DNA Variants Can Arise in Many Different Ways and Be of Many Different Types

We think that most point mutations arise during DNA replication, for reasons outlined above; factors in our environment, such as chemicals that damage DNA (mutagens) or ionizing radiation, can increase their frequency. Variants involving only one base pair are of the following two main types.

In the first, the base pair at a specific location in some people is different to that in others: we call these single nucleotide variants (or SNVs for short). If a single nucleotide variant is frequent enough to be a polymorphism, we call it a **single nucleotide polymorphism**, or SNP (often pronounced 'snip').

In the second, a particular base pair is present at a specific location in some people and absent in others. Since this might have originated from the insertion or the deletion of the

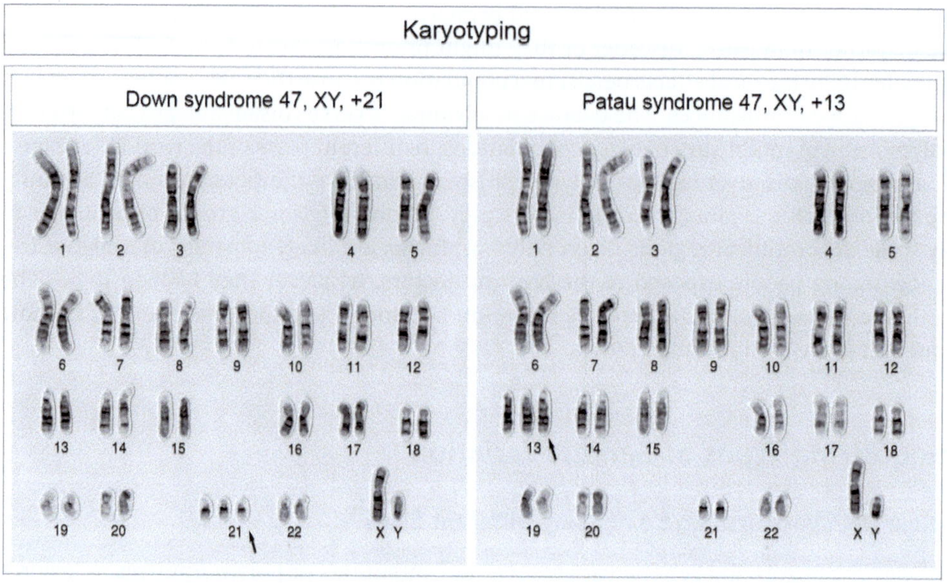

Figure 4.4 Karyotyping. Karyotyping involves the creation of a chart of chromosomes isolated from cells that are in the process of dividing. It can be done on a variety of cell types, blood cells being a common one. Live cells in the sample are grown in the laboratory and are given a chemical to stop them during division, when their chromosomes are condensed. They must be condensed for this to work; between divisions, the chromosomes are long tangled up DNA molecules (as illustrated in Figure 4.1). The chromosomes can be stained in a variety of ways before being photographed, revealing stereotypical and diagnostically useful banding patterns. The images are laid out in order, as in these pictures. The person on the left has an extra chromosome 21 (+21) and therefore 47 chromosomes in total (including the X and Y chromosomes); this is Down syndrome. The person on the right has an extra chromosome 13; this is Patau syndrome. *Reproduced from Wessex Reg .Genetics Centre / Wellcome Collection / CC BY 4.0.*

base pair at that location at some time in the DNA molecule's ancestry, and we might not know which occurred, we often call these variants indels (*in*sertion or *del*etion variants). They are more common in regions of the genome that do not code for proteins.

Insertion or deletion variants that are more than one base pair are also often referred to as indels, as long as they are less than about 1,000 base pairs long (there is no absolute agreement on the exact upper limit of what we call an indel). There is partial overlap between the use of the word indel and that of another commonly used descriptive term: **copy number variation** (**CNV**). In this term, the word copy refers to a stretch of DNA. We should have two copies of each stretch of DNA for autosomes – which are paired, remember – and the X chromosomes in XX females. But this is not always the case, as the term 'copy number variation' implies; the number of copies of stretches of DNA can vary from person to person. Stretches of DNA can be deleted, which would reduce the number of copies of that region in a person's genome. The reduction would be either to one, if it affects one chromosome, or to none, if it affects both chromosomes. Alternatively, stretches of the genome can be replicated, sometimes many times.

So, we apply the term CNV over a much broader range of variant sizes than the term indel. At one extreme are CNVs made up of multiple copies of just two or three base pairs;

we often call these repetitive, or repeat, sequences. At the other extreme are CNVs that involve the replication of sequences millions of base pairs long. CNVs of all lengths are remarkably frequent, and they make a major contribution to human genetic variation. Current evidence indicates that up to about 10% of the human genome contributes to CNVs.[95–97] The regions where CNVs occur are not scattered evenly throughout the genome, but rather they occur in hotspots enriched for repeated sequences, which favour their creation.

Indels and CNVs can form in several ways. How any particular one has come about is hard to define with certainty and the likely mechanisms are sometimes complex, with both the insertion and deletion of bases possibly contributing. Two of these mechanisms of formation, polymerase slippage and single strand annealing, are illustrated in Figure 4.5. Other possible causes are unequal crossover, where crossover refers to the process by which parts of chromosomes switch, mainly during meiosis in the germ cell lineages (as I will describe later in this chapter). Finally, some indels and CNVs result from events that include the movement from one place in the genome to another of mobile, highly repetitive genetic sequences called transposable elements.[98,99] They are very common, making up almost 50% of the human genome, and their existence constitutes yet another reason for thinking of the sequences of DNA molecules as being somewhat fluid and prone to change.

Large chromosomal variants are relatively common, affecting close to 1% of all live births (this is an overall approximation; the incidence varies with factors such as maternal age). They create large-scale changes in a person's genome. Many cause disorders; the fact that some do not is surprising, perhaps, given their size. In some chromosomal variants, there is an abnormal number of chromosomes (there are examples of this in Figure 4.4). Normally, our somatic cells carry a diploid complement of 46 paired chromosomes. Variation from this is called aneuploidy. A person might have an extra chromosome (called **trisomy**) or, rarely, more than one extra chromosome (for example, the presence of two extra chromosomes is tetrasomy). Alternatively, a person might be missing a chromosome (which is called monosomy). Almost all aneuploidies that cause disorders arise de novo.

Other chromosomal variants involve only a part of a chromosome. Some of these abnormalities involve either the deletion or the duplication of a portion of a chromosome. In some cases, a part of a chromosome moves to a new location in the same chromosome or joins with another; this is called a **translocation** (Figure 4.6). Sometimes parts from two chromosomes exchange, and we call it a reciprocal translocation. If the exchange involves no net loss or gain of DNA (or, perhaps we should hedge our bets and say no *major* net loss or gain, since we cannot be sure there is none at all), we say it is a balanced translocation. Sometimes a whole chromosome joins with another, which is called a Robertsonian translocation. In other cases, a part of a chromosome breaks off and reattaches itself the opposite way around; this is called an inversion. Another rare abnormality is the formation of a ring chromosome when the ends of a breakaway portion of a chromosome join to make a circle.

Mitochondrial DNA Is Particularly Prone to Variation

Most mitochondrial variants are the result of either point mutations or larger-scale deletions. They probably form through the types of mechanism already described for nuclear DNA. As explained earlier, the state of our mitochondrial DNA is even more fluid than that

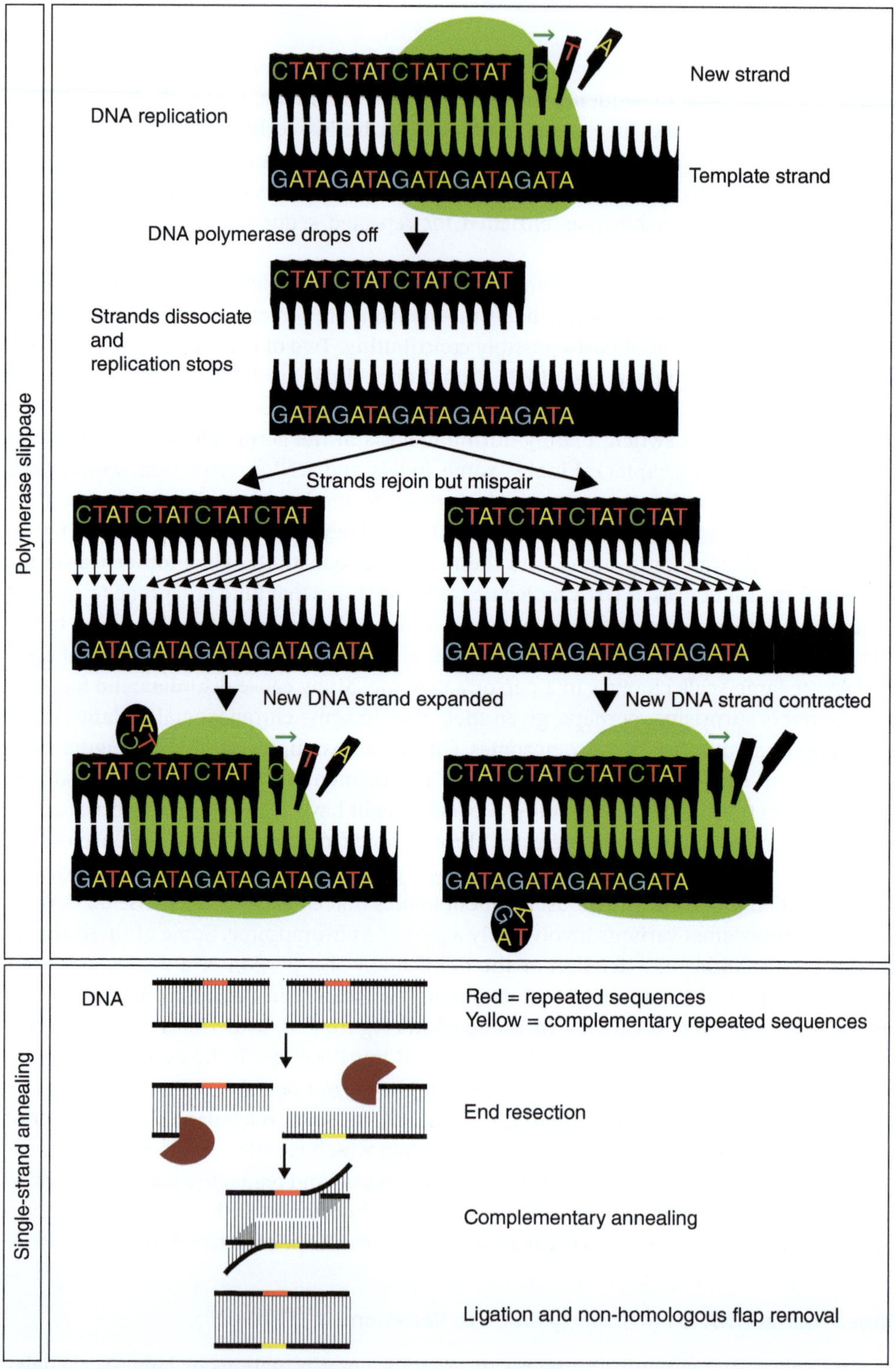

Figure 4.5 Indel and copy number variation (CNV) formation. This is most likely to occur in regions where DNA sequences are highly repetitive, with the same short sequence repeated over and over again. We think that most indels and CNVs result from polymerase slippage (also known as replication slippage) during DNA replication. In this process, the DNA polymerase complex falls

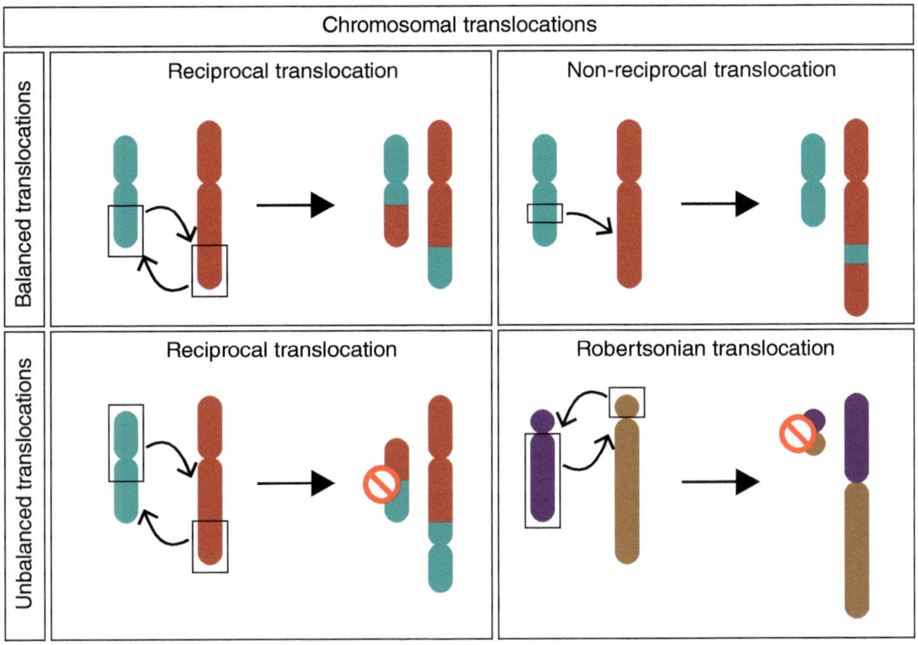

Figure 4.6 **Nuclear chromosomal translocations.** Here, I have depicted nuclear chromosomes in a stylized manner, using a different length and colour to distinguish them. The constriction in each is the centromere, which divides chromosomes into two arms. In balanced translocations, there is no (or at least very little) loss of genetic material overall. In reciprocal translocations, parts of two chromosomes swap places; in non-reciprocal translocations, a part of one chromosome moves into another. In unbalanced translocations, genetic material is lost. This can occur in reciprocal translocations that result from a region with a centromere swapping places with a region that has none, resulting in one chromosome with two centromeres and one that is likely to be lost since it has no centromere. In Robertsonian translocations, chromosomes whose arms are very different in length (called acrocentric) exchange material. The long arm of one swaps with the short arm of another and the relatively tiny product from the two short arms is lost.

Figure 4.5 (Continued) off, the two strands dissociate and replication stops. When the two DNA strands pair up again, they sometimes do it incorrectly because repeated sequences (here GATA on the template strand) anneal to complementary bases in the wrong place. There are two possibilities. On the left, bases on the partly synthesized (new) strand have annealed to bases too far back along the template strand and the bases on the new strand that should have annealed there are looped out. Once the new strand is complete, it will be longer than the template. On the right, the opposite has happened: bases on the partly complete strand have annealed to bases too far along the template strand and the bases on the template strand that should have been annealed are looped out. Once the new strand is complete, it will be shorter than the template. Some indels and CNVs result from single-strand annealing, which can occur during a cell's attempt to repair a break of both strands in one of its DNA molecules (have a look back to Figure 4.3). Here, the processing, by end resection, of a double-strand break that has occurred between two repeated sequences exposes two complementary sequences (one coloured red and the other yellow) which can anneal. Repair of the break is completed by subsequent removal of the non-homologous flaps. The upshot is that only one repeat remains from a stretch of DNA that originally contained two repeat sequences with non-repeat sequence between them.

of our nuclear DNA. The number of mitochondrial DNA molecules and any variants tend to vary among our cells more than our nuclear DNA does.[90] Where there is more than one variant of the mitochondrial DNA molecules in a cell, we call this is a heteroplasmic state. If all the mitochondrial DNA molecules in a cell are the same, even if they all carry a mutation, then we call this a homoplasmic state. You may be wondering, given that there are so many mitochondrial DNA molecules per cell, each with the potential to accumulate variants, whether it is likely that many cells are truly 100% homoplasmic, with *all* mitochondrial molecules identical. You could be right, but it is hard to be sure since detecting variants present in low proportions is very difficult. It seems reasonable to accept that when we use the term homoplasmic state, or homoplasmy, we actually mean a homoplasmic state as far as we know; when we use the term heteroplasmic state, or heteroplasmy, we mean that we can detect differences.

Why Only Some Genetic Variants Affect People

A Variant's Location in the Genome Matters

The overall picture painted by research over the past decades, as summarized above, is that DNA molecules are malleable. They can change in a single individual and in a population. Some of these changes might be detrimental to the individual, but more often they are not. Where in the genome changes occur is very important in determining whether they affect the individual. To appreciate why this is the case, I need to explain what the various parts of our genomic DNA do … and whether they do anything at all.

First, some definitions of terms. In genetics, the term locus (plural loci) means the position of a stretch of DNA in a chromosome. Loci are defined using letters and numbers (Figure 4.7). Some loci contain genes, a term that can be surprisingly hard to define precisely. The difficulty arises in part because the term was coined before DNA and how it works was discovered. Different people, including some scientists, use it in slightly different ways. Sometimes the word is used to refer to the parts of the genome that code for proteins (protein-coding sequences). I prefer a broader definition because we now know that many active parts of the genome do not work by encoding protein. A succinct definition might be a stretch of DNA that affects an individual's characteristics. Those characteristics might be of any type and magnitude. They might be manifest in a person's appearance or behaviour. Biologists often refer to an individual's characteristics as traits or phenotypes, as I discussed in Chapter 2. Many stretches of DNA affect an individual's traits because they code for ribonucleic acid (RNA) molecules. Some – but only some – of the RNA molecules will go on to guide the construction of the proteins that make up the body, in a process that we call **translation**. These RNA molecules are a type called messenger RNA, or mRNA (Figure 4.8).

Some genes code for RNA molecules that are not translated to make proteins (Figure 4.9). These are called **non-coding RNAs**, or ncRNAs (non-coding here means that they do not code for protein). Some of these ncRNAs, namely transfer RNA (tRNA) and ribosomal RNA (rRNA), carry out functions essential for translation. Others have regulatory functions: some are very small, such as microRNAs (miRNAs); others are longer, such as long

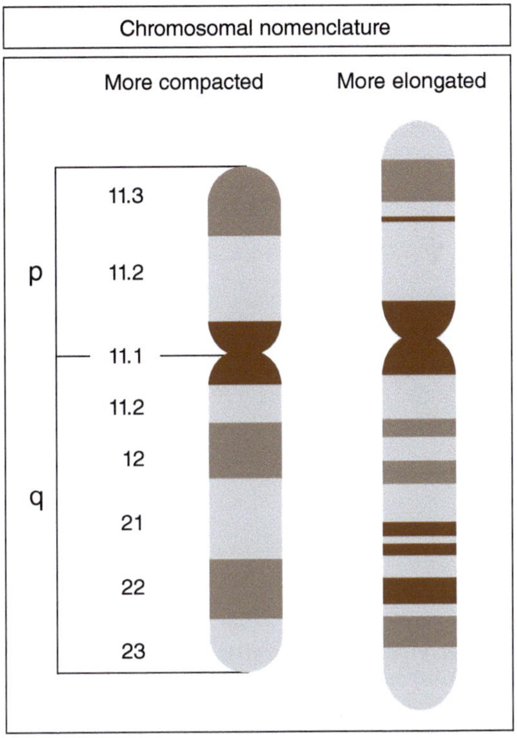

Figure 4.7 **Describing chromosomal locations.** To pinpoint loci on chromosomes, geneticists give them an address. This often uses a standardized method related to the bands seen on stained chromosomes (see Figure 4.4 on karyotyping). First, they state the number of the chromosome (the one shown here is 18) or whether it is X or Y. Then they give a letter to denote the arm of the chromosome (p for short arm, q for long arm). Then they give more numbers to define in which band or bands the locus is located: for example, 18p11.2 or 18p11.2–p11.3. The type of staining and the particular stage of cell division at which the chromosomes are seen defines the pattern and number of bands that can be identified. In general, early in division the chromosomes are more elongated and more bands are seen; as cell division progresses, the chromosomes become more compacted and fewer bands can be resolved. Bands are usually designated by two digits sometimes followed by a decimal point and one or more additional digits, which represent sub-bands within a light or dark area. The numbers increase with distance from the centromere. These are the basic conventions. Additional terms are added to zoom in on specific individual base pairs at a locus if, for example, a geneticist wants to define the exact position and nature of a particular base pair change (I will not go into details here). Also, now that human genomes have been sequenced, researchers can provide a more specific address in terms of nucleotides if required. Different researchers often present slightly different values for the molecular location of the same sequence because interpreting the enormously long sequence of the human genome is not entirely straightforward and can be done with a range of methods that produce small differences.

non-coding RNAs (lncRNAs). These and other ncRNAs can affect an individual's traits in a variety of ways, for example by interacting with the mRNAs that code for proteins. In general, our understanding of ncRNAs lags behind that of mRNAs, but research into this complex world is extensive and knowledge is growing rapidly.

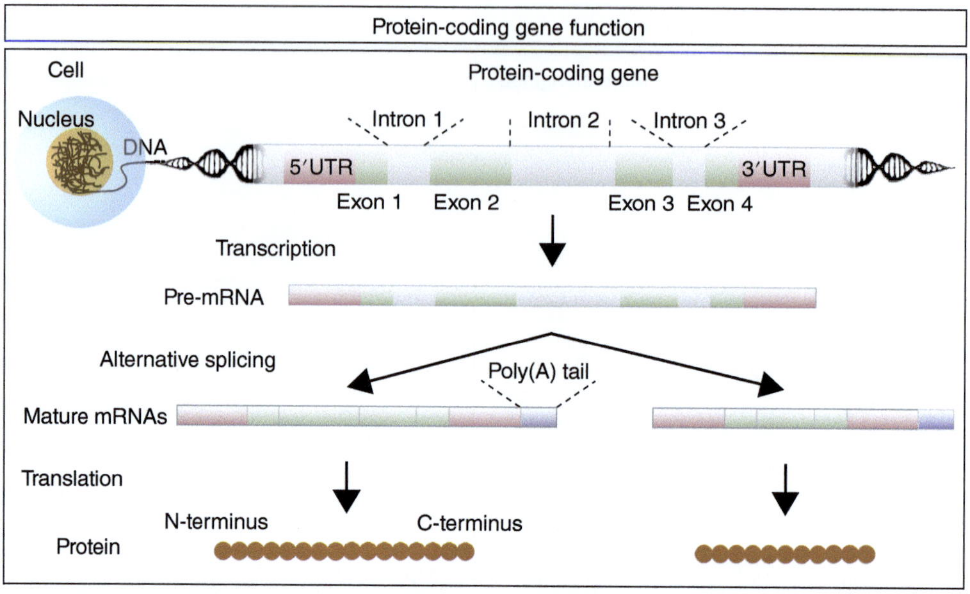

Figure 4.8 The structure of protein-coding genes and mRNAs. Protein-coding genes contain sequences that code for protein (exons; green) and intervening sequences that do not (introns; grey). Regions in red are transcribed but untranslated (they are called the 5′UTR and 3′UTR, where UTR stands for untranslated region and 5′ is the end from which transcription and translation start). Transcription produces pre-mRNAs. Note that nucleic acids can only be synthesized in the 5′–3′ direction and we sometimes refer to the 5′ end as being upstream and the 3′ end as being downstream. The cell uses a molecular complex called the spliceosome to remove the introns and very often – in about 90% of our protein-coding genes – splices the exons together in different combinations to produce alternative mature mRNAs and hence alternative proteins from the same gene. During mRNA processing, a string of A bases is added to the end of the molecule: this is called polyadenylation and it adds a polyA tail. Longer tails are associated with longer survival of the mRNA, whereas short tails are found in short-lived mRNAs. Mature mRNAs are then exported from the nucleus, which involves RNA-binding proteins, and translated on ribosomes in the cytoplasm. The N-terminus marks the beginning of the protein and the C-terminus marks the end. These terms come from the molecular structure at the two ends: there is a free nitrogen-based group at the N-terminus and a free carbon-based group at the C-terminus.

Earlier in this chapter I said that the function of the DNA code includes two major components. One is the provision of instructions to guide the construction of molecules that the cell needs; this function has been my focus in the previous paragraphs. Another is the provision of specific sequences that can be identified by other molecules that need to interact with specific sites in the DNA to control its function. These sites include **promoters**, **enhancers**, silencers and insulators, which are known collectively as **regulatory elements** (Figure 4.10). Promoters lie close to the sequences they regulate, whereas enhancers, silencers and insulators can be very distant, sometimes hundreds of thousands of bases away.

Now we come to a critical question about the many elements in our genome that carry out all these important functions: how much of our genomes do they occupy? When the human genome was sequenced in the 1990s, it was discovered, to the surprise of many, that if you add up all the stretches of DNA that code for proteins, they occupy less than about 2% of the genome. Most of the rest, excluding the regulatory elements and so on, was described by some as junk, which has turned out to be somewhat pejorative given that important functions have been ascribed to at least some of it, as I have discussed above. Our genomes contain somewhere in the region of 20,000 protein-coding genes and, in addition, there might be more than 100,000 other genes that do not code for protein – it is very hard to estimate this.[100] Nevertheless, it is the case that, as yet, we have not identified functions for most of our genome. Most variants, which arise through chance, are likely to emerge in parts of our genomes whose alteration has no detectable effects on the individual, simply because those parts are so much bigger.

Do bear in mind, however, that the jury is still out on the critical functions, if any, of the vast swathes of apparently non-functional DNA in our genomes. For example, a major study called the Encyclopaedia of DNA Elements (ENCODE) project suggests that most of it is likely to have some functional importance. For now, we should keep an open mind. At the very least, changes in these regions might contribute to gradual evolutionary change over long periods, which is a fascinating topic but for a different book.[101]

When Variants Arise and How They Affect Cells Matter

As I said above, de novo mutations occur in the DNA of our cells throughout life. The earlier that mutations arise, the greater the proportion of a person's cells they have the potential to affect. Whether and to what extent they spread depends on how they influence the cells that carry them. As we have seen in the previous section, many mutations have no effects and can spread unhindered. At the other extreme, if mutations prevent cells from replicating or surviving, they will be lost, taking their mutation with them, and there might be no consequences. If mutations have an opposite effect, inducing the overproduction of cells, then cancer is a possible long-term outcome (I will not be discussing cancer in this book).

Sometimes, however, the effects of mutations are less extreme. For example, they might not be overtly detrimental to cells but might predispose cells to develop in a divergent or disordered way in response to other factors, such as those in their environment or the presence of other predisposing variants in that same genome. Alternatively, mutations might affect genes that are not essential for the mutated cell's immediate survival and replication but are important later, in which case there might be consequences either for that person at some stage of their life or for their offspring.

How Genetic Variants Can Affect People

Variants in Protein-coding Regions Can Affect the Proteins Produced in Many Different Ways

If a variant affects the protein-coding region(s) of one or more gene(s), then the possible outcomes are numerous. I will start by considering the possible effects of point mutations.

The insertion or deletion (that is, an indel) of a single base in a protein-coding DNA sequence can have a big effect if it messes up the cell's translation of the corresponding

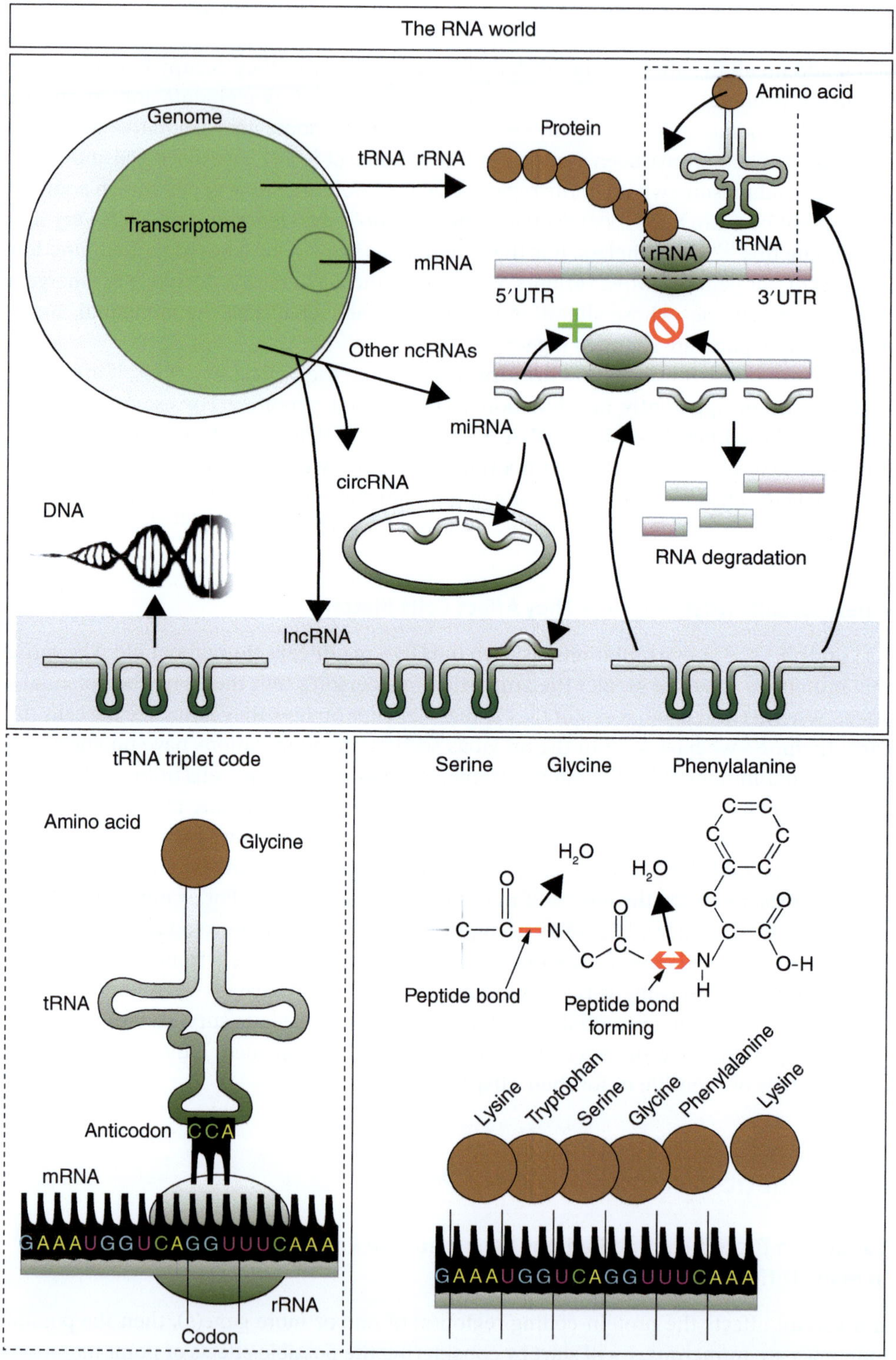

Figure 4.9 The world of RNA molecules. Most of the human genome (about 75%) codes for RNAs that are not translated into proteins, termed non-coding RNAs (ncRNAs). Only about 2% code for mRNAs that are translated into proteins. Many RNAs have essential roles in regulating the activities of cells and their genomes in a variety of ways, some of which are illustrated here.

mRNA. To understand how this can happen, you first need to know how translation works to produce proteins.

Proteins are long chains of 20 different amino acids (which are small organic compounds comprising mainly carbon, hydrogen, oxygen and nitrogen atoms). The number and order of amino acids along the protein chain is specified by the sequence of bases in the mRNA (which has been made by transcription from the gene that codes for that protein). When a cell translates mRNA to protein, it reads the sequence of bases in the mRNA molecules three at a time, starting at a specified point in the sequence, usually marked by the sequence AUG. The reading is done by a set of translator molecules, including those in the ribosomes and a collection of transfer RNAs (tRNAs). Each transfer RNA molecule contains a sequence able to recognize – because it is complementary to – a specific triplet of bases in the mRNA sequence. This set of translator molecules moves along the mRNA from one end to the other, dealing with each new triplet of bases as it goes (these triplets are called **codons**). As each new codon is encountered, a transfer RNA molecule containing the specific triplet of bases complementary to the new codon (called an anticodon) is called into action. Transfer RNA molecules, which link specific anticodons and amino acids together, carry their specific amino acid load to the ribosome, where it is added to the growing protein (see Figure 4.9). Some codons do not encode amino acids; instead, they bring synthesis to a halt and are called stop codons. Stop codons are needed because only some stretches of mRNA molecules specify protein, and the system needs to know where those stretches end.

Figure 4.9 (Continued) Transfer RNAs (tRNAs) use their anticodons to recognize specific base triplets in mRNA (called codons) and bring the amino acids defined by the triplets to the ribosomes, where they are added to the growing protein during translation. Ribosomes contain ribosomal RNA (rRNA) and proteins. The enlargement of the area outlined with a dashed line shows the amino acid glycine being brought to the codon GGU (you can easily find online tables listing codons that specify each of 20 amino acids and those that signal translation to stop). Bottom right: as amino acids are brought by transfer RNAs to the growing protein, they join by forming peptide bonds between the carbon (C)-based group at the end of the preceding amino acid and the nitrogen (N)-based group in the new amino acid (water is formed when these bonds are made). Here, glycine has been added, phenylalanine (codon UUC) is being added and lysine (codon AAA) will be added next to the -COOH (carbon-based) group of phenylalanine. Other ncRNAs include microRNAs (miRNAs), so-called because they are typically only 22 nucleotides long, circular RNAs (circRNAs) and long non-coding RNAs (lncRNAs). Much research is currently devoted to elucidating their diverse functions and mechanisms of action, defining their clinical relevance and discovering their potential uses as biomarkers or therapeutic targets. Some of the functions we know about are illustrated here. The binding of miRNAs (there are probably many thousands of different types of them in humans) to particular complementary sequences in the coding sequences and/or the 5′UTR and/or 3′UTR of mRNAs can either enhance translation (this often happens if binding is to the 5′UTR) or inhibit it, sometimes by causing mRNA degradation (inhibition is typical if binding is to the coding sequence or 3′UTR). Some degree of mismatch of the complementary sequences is tolerated in the binding, which is why the miRNAs are not perfectly lined up against the mRNAs in this diagram. Both circRNAs and lncRNAs can bind miRNAs, inhibiting their actions by sequestering them away from mRNAs (we often say they act as molecular sponges, although by sequestering specific complementary miRNAs they act more selectively than your average sponge). The actions of lncRNAs are extensive: they can act on DNA, affecting its compaction for example (as occurs in X chromosome inactivation, which I will come to later in this chapter); they can interact with mRNAs, for example increasing their stability or blocking miRNA binding; they can interact with proteins, for example affecting where they end up in the cell. We are probably only scratching the surface of this world of intricate control mechanisms.

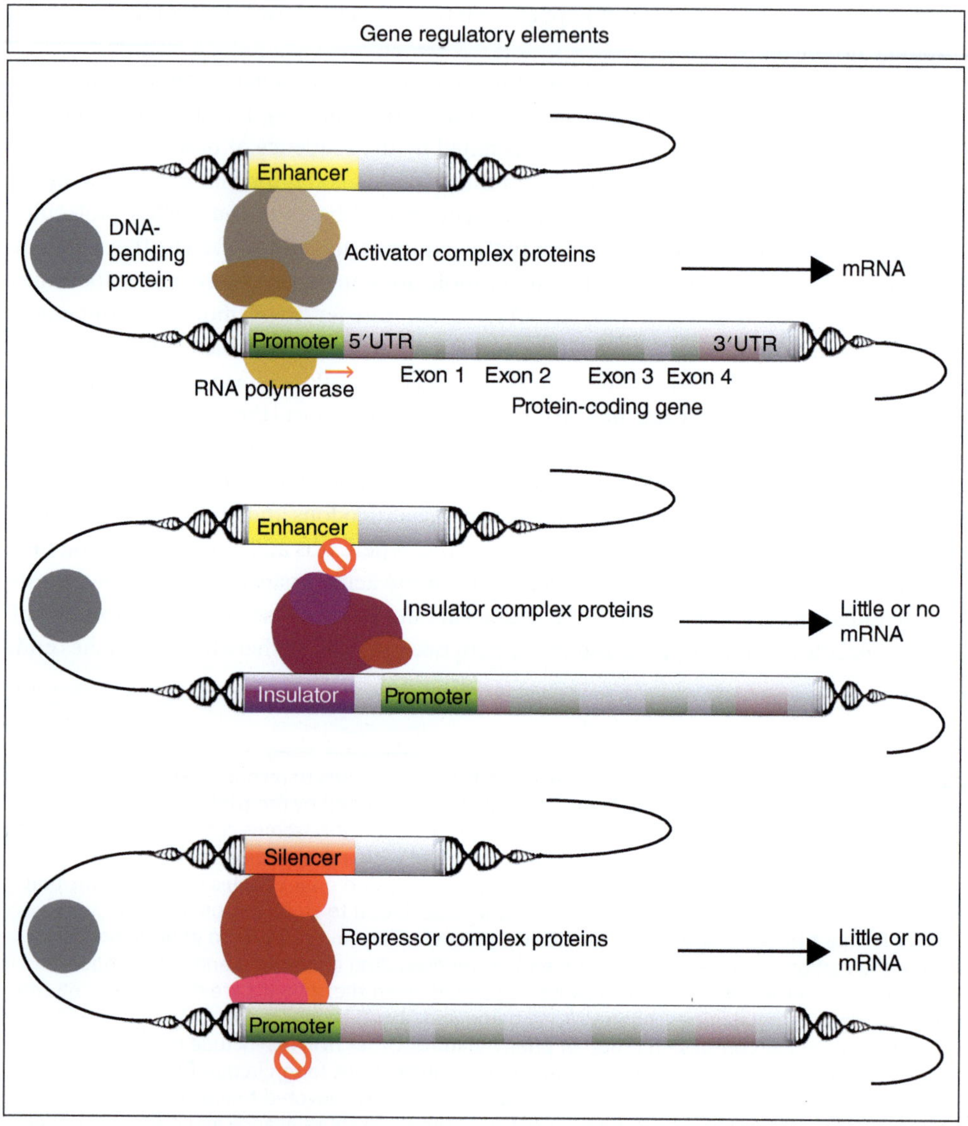

Figure 4.10 **Gene regulatory elements.** Gene regulatory elements are DNA sequences that are recognized and bound by proteins to control transcription from specific regions of the genome, such as those coding for mRNAs (as is illustrated here). Some of these proteins bind some elements and not others because they recognize specific DNA sequences; which of these proteins a cell contains will affect which elements are active. Other proteins are involved at many or all elements: for example, RNA polymerase is needed ubiquitously to make mRNA. Gene regulatory elements include promoters, enhancers, insulators and silencers. Promoters are found near the point at which the transcription of a gene would start and they initiate the process by binding RNA polymerases, among other proteins. Enhancers are more distant elements that are bound more selectively by specific proteins, such as lineage-specific transcription factors, that recognize particular sequences. Many protein co-factors are involved, creating regulatory complexes. Enhancers and promoters interact via their activator protein complexes; DNA bending proteins allow this to happen. Insulators block this interaction between enhancer and promoter and prevent the spread of transcriptional activity. Silencers bind proteins that repress promoters and thereby transcription.

Back, then, to the single base indel. Hopefully you can see that a triplet-based process of translation will be disorganized by a single base indel because everything that happens after the indel will be knocked out of register. If the cell is reading three bases at a time, and one base somewhere along the sequence is removed or added, the translator molecules will read the triplet code incorrectly after the indel: they might add the wrong amino acids to the protein; or they might stop its synthesis too early if the shift has created a premature stop codon; or they might allow its synthesis to continue beyond where it should if the shift has removed the normal stop codon. In the jargon of the field, we say that such mutations alter the reading frame of the gene, and we refer to them as frameshift mutations.

If a point mutation in a protein-coding DNA sequence substitutes one base with another (a single nucleotide variant, or SNV, see above), then this will not cause the reading frame to shift but it might, or it might not, have other consequences. Certainly, it will change the triplet of bases that the translator reads at the SNV, but it might have no consequences because there is redundancy in the code. For most amino acids, several different codon sequences (which normally are the same at the first two bases, differing at the third one) can bring that same amino acid into the growing protein. This means an SNV might be inconsequential if all it does is to change which codon triggers the addition of the same amino acid to the protein. We call these changes, which we can think of as being silent, synonymous substitutions. Alternatively, an SNV might involve a nonsynonymous substitution. A nonsynonymous substitution might change which amino acid is added at the corresponding position in the protein, which we call a missense mutation. Such changes might affect the function of the protein, depending where they are and what the function of that part of the protein is, or they might have little or no effect. The latter is more likely if the substitution is by an amino acid with similar biochemical properties, which we call a conservative substitution. Alternatively, a nonsynonymous substitution might create a premature stop codon; we call this a nonsense mutation. In many cases, the cell destroys mRNAs carrying nonsense mutations (a process known as nonsense-mediated decay[102]). Sometimes this decay does not happen, particularly if the premature stop occurs in the final exon of a gene. If a protein were to be made from this truncated mRNA, it would be too short, which would likely significantly affect its function. It might lose functions or it might gain new ones, including the possibility that it might interfere with the functions of normal protein made from an unaffected copy of the same gene on the paired chromosome (if there is one, which would not be the case for a gene on the X or Y of an XY male, for example). If a nonsynonymous substitution were to change a stop codon to an amino acid encoding codon, translation would continue into the normally untranslated region; we call these read-through mutations.

What about variants in protein-coding regions that involve more than one base? Small-scale indels and CNVs that involve numbers of base pairs that are not multiples of three will cause frameshift mutations (for example, in Figure 4.5, I illustrated the expansion or contraction through polymerase slippage of a repeating four base GATA sequence). Deletions that are multiples of three base pairs might cause a missense mutation since sequences following the deletion will remain in frame. Insertions that are multiples of three bases might result in the introduction of additional amino acids into the protein; sequences following the insertion will remain in frame. CNVs comprising multiple repeats of even a few bases in the protein-coding sequence of a gene are likely to disrupt its function, for example by creating proteins that are unstable and likely to degrade rapidly or are unable to function properly for other reasons, such as a loss or a distortion of their functionally important regions. They are likely to have consequences for thc individual: for example, the insertion into protein-coding sequences of a repeated three-base-pair sequence, known as a **trinucleotide repeat expansion**, causes some human neurodevelopmental disorders (more on these in Part III).

Variants might interfere with the production of proteins not by directly altering the protein-coding sequences in the genome themselves, as I outlined in the previous section, but by affecting mechanisms that control which bits of protein-coding sequence are transcribed into mRNA. Remember from Figure 4.8 that the protein-coding regions of most protein-coding genes (which we call exons) are split across multiple stretches of DNA, and they are separated by regions that do not code for protein (which we call introns). When transcription occurs, the RNA first produced – which we call precursor mRNA – contains both exons and introns. The latter are later cut out and the exons joined together in a process called **splicing**. This is accomplished by molecular complexes celled spliceosomes, and splicing can often produce many different mRNAs containing different combinations of exons from the same gene. We call this process alternative splicing, and we call the products splice variants. The functions of different splice variants of the same gene can differ in important ways. The spliceosome carries out splicing by recognizing specific sequences at the junction of exons and introns. If mutations affect these sites, then the mature mRNA might retain introns or have exons spliced out of it, either of which might affect the function of the protein produced by that mRNA.

Variants in Regions That Do Not Code for Protein Can Affect the Individual for Many Reasons

Variants of all sizes, from point mutations upwards, can occur in genes or in parts of genes that do not code for protein but are important for other reasons. For example, they might affect non-coding RNA (Figure 4.9). Or they might affect the functions of the regulatory elements that control how other genes work, in other words, promoters, enhancers, silencers or insulators (Figure 4.10). Why would such effects be important? Remember that not all genes are active in all cells at all times. Multicellular organisms comprise many different, often highly specialized, cell types that develop because different sets of genes are active in different cells (in the jargon of the field, we say that different sets of genes are 'expressed' in different cells). Gene expression must be orchestrated exquisitely to ensure that cells develop, survive and function optimally. If these control mechanisms go wrong due to mutations of gene regulatory mechanisms illustrated in Figures 4.9 and 4.10, divergent and disordered development and function are likely outcomes.

Quite how critical gene regulatory regions in the genome are in determining an organism's traits can be appreciated by considering what it is that makes humans different from other species. Roughly half of our genome is similar to that of other animals and even plants; unsurprisingly, the greatest similarity (about 95%) is with the genomes of our closest living relatives, chimpanzees and bonobos. The proteins that our genes encode are 99% similar between humans and chimpanzees. The exact percentages depend how the calculations are done, but these numbers make the point that the major differences in appearance and behaviour between us and other species are unlikely to be caused solely, or even mainly, by differences in the proteins our DNA encodes. Instead, they are more likely to be due to differences in how the protein-coding genes are used, in other words to differences in their expression. This view has been strengthened by analysis of the genomic regions that are unique to humans. These regions are most likely to make us different from all other species, even chimpanzees and bonobos, and are referred to as human accelerated regions (often abbreviated to HARs).[103–105] The vast majority of human accelerated regions – about 92% – map to genomic regions that do not code for proteins, and there is evidence that at least some of these regions contain gene regulatory elements.

Large Variants Involving Protein-coding or Non-coding Regions Are Likely to Affect the Individual

Large CNVs can have effects in many ways, depending on their nature as well as their location. In general, a large CNV is more likely to cause a clinical disorder than a small one. On the other hand, small CNVs are more common than large CNVs, so, all together, they end up producing roughly similar numbers of clinical disorders.

Large CNVs might result in there being either more or less copies than normal of one or more entire gene(s); we refer to these as increases or decreases in gene dosage. The effects of altered gene dosage are not easy to predict and depend on many factors. For example, in the case of changes in the dosage of protein-coding genes, whether and, if so, to what extent the encoded protein levels are altered in the cell might depend on whether compensatory mechanisms kick in. Some genes are known to autoregulate their own levels of protein production, putting production up if levels in the cell fall or turning production down if levels are too high. For genes where this mechanism applies, the cell could mitigate the effects of the loss of one functional copy of a gene by making the other work harder to produce the required protein, and it could mitigate the effect of having too many copies by making each copy work less. Large CNVs can also bring normally distant genes closer together, which might affect how they work, and can even result in two normally separate genes combining to make a fusion gene with a new function.

As with large CNVs, large-scale chromosomal deletions or duplications will reduce or increase the numbers of copies of the genes in the affected chromosomal region or on the affected chromosome, in other words they affect gene dosage. The effects of other chromosomal abnormalities such as inversions and translocations depend on exactly what has happened. If the abnormalities have left the person with a normal amount of genetic material, albeit rearranged, then this might not affect that individual (unless the rearrangements have broken critical genes). For example, balanced translocations (Figure 4.6) generally do not affect the individual carrying them. A bigger risk is for the children of people with such abnormalities, for reasons explained later in this chapter.

How Changes in Proteins Might Affect People

The Effects of Changes in Proteins Depend on Many Things

Most of the genetic variants that we know cause clinically defined disorders affect the exons of protein-coding genes (you will often see the collection of all exons in the genome described as the **exome**). I have described above the different ways in which variants of protein-coding genes can alter the structure and therefore potentially the function of the proteins those genes produce. The effects of altering a protein depend on many variables, including the nature of the change in the protein (for example, how big the change is, where the change is and what has changed), whether cells can compensate for having the altered form, when the protein is needed and what biochemical processes it is involved in.

The amino acid sequences, the three-dimensional structures and the functions of proteins are interlinked. For example, the amino acid sequence of a protein affects how it folds into a three-dimensional structure, which in turn affects how it interacts with other molecules and therefore how it functions (more on folding in Chapter 7, Figure 7.12). It is not easy to predict how a protein folds from its sequence of amino acids, although approaches based on the use of artificial intelligence have made great progress in recent years.[106]

Most proteins have regions within them whose functions we either know from experimental work on those regions themselves or can infer because they are similar to regions with known function in other proteins. We often call these regions **domains,** and their identification can give important clues to the function of the protein as a whole. Many proteins have multiple functionally important domains that do different things, which complicates matters: variants affecting different domains of the same protein can have different effects. There is rarely a simple way of predicting exactly how a protein works, making it difficult to predict what the effects of any given variant would be. As I alluded to earlier, for example, while many variants reduce the ability of the protein to function (we call this loss-of-function), they might less commonly have the opposite effect, increasing or adding to the protein's activity (we call this gain-of-function). It is also possible for a variant form of a protein produced from one of the two copies of a gene to interfere with the normal protein produced from the other copy, usually by blocking its actions. Usually, we have to resort to using empirical, experimental approaches to test the effects of variants. For example, researchers might examine the effects of introducing mutations found in patients into human brain cells grown in culture, using systems such as those outlined in Figure 1.1 in Chapter 1.

If we include splice variants, there are many tens of thousands of different proteins, and they probably all do something at least slightly different. Many of these proteins have multiple functions in possibly diverse cellular processes. In many cases, we know at least some of the things they do, but there is much still to learn about some. In the following sections, I will briefly consider the types of processes that variants can affect. I will expand on them when I come to describe the proximate causes of specific disorders in Part III.

Changes in Some Proteins Affect DNA Replication, Repair or 3-dimensional State

Variants that interfere with the functions of proteins that work on our DNA often harm people in some way. This is because they have the potential to affect many, or even most, cellular processes. Some of these proteins are involved in DNA replication and repair (Figures 4.1–4.3). Variants that affect the ability of these types of protein to work properly have the potential to undermine the most fundamental function of cells, namely to reproduce their DNA with great fidelity. Other proteins, which we call **transcription factors,** control which genes a given cell expresses and at what levels by binding to regulatory elements (Figure 4.10 and see next section).

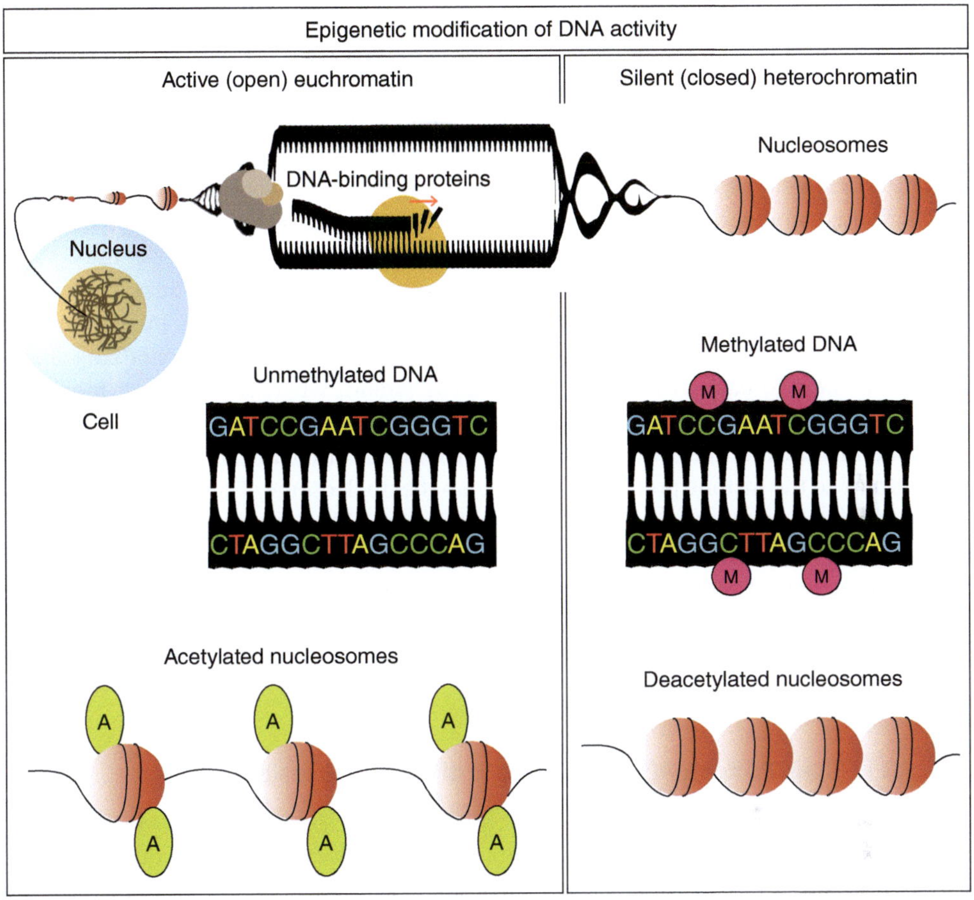

Figure 4.11 Epigenetic modification of nuclear DNA function. Broadly speaking, the nuclear DNA of cells that are not dividing contains regions that are in one of two main states, either active or silent. The active regions of a cell's DNA are open for transcription whereas the silent regions are not. DNA winds itself around proteins called histones to form structures called nucleosomes. If the nucleosome proteins are acetylated (achieved by histone acetyltransferase enzymes adding acetyl groups to them), the DNA becomes relatively free of nucleosomes and can be transcribed. Deacetylation of the nucleosomes (achieved by histone deacetylase enzymes) and methylation of DNA itself (achieved by DNA methyltransferase enzymes adding methyl groups to mainly C bases) results in inactive DNA tightly wound into nucleosomes.

For a gene to be transcribed, the DNA containing it must be in in a suitable state. In all eukaryotes, nuclear DNA is often bundled up into a tightly wound supercoiled structure called **chromatin**. Chromatin is organized by protein complexes called **nucleosomes**, which are made up of proteins called **histones**, around which the DNA of each chromosome winds itself.[107] Further winding and the actions of additional stabilizing and supporting proteins allows the chromatin to be compressed into the chromosomes. Transcription cannot occur in the densely compacted regions of a chromatin supercoil; these regions are called heterochromatin. Before these regions can express their genes, the chromatin must be loosened or relaxed, in which case it is known as euchromatin (Figure 4.11).

Proteins that either regulate the presence or absence of methyl groups (made of one carbon and three hydrogen atoms) on DNA itself or modify the histone proteins it is wound around can loosen the DNA to allow transcription. The alteration of gene expression by chromatin modification (which we refer to as chromatin remodelling) with no change in DNA sequence is often called **epigenetics** (*epi* is Greek for over or near). Beware, however, some use the term more broadly to include the study of any factors that affect how DNA works.

DNA methylation occurs on most bases of the C variety (converting them from cytosine to methylcytosine) provided they are followed in the DNA sequence by bases of the G variety. These are called CG dinucleotides, or more commonly CpG dinucleotides. If a region contains large numbers of CpG dinucleotides, it is known as a CpG island. By the way, the reason that a little 'p' is placed in the nomenclature is to indicate that we are talking about C's and G's next to each other in the same DNA strand, where a phosphate molecule (denoted p) joins them, and not C's paired with G's on opposite strands.

Histone modification involves the addition or removal of chemical groups, such as an acetyl group, to the histone proteins that make up the nucleosome. These modifications to the histone proteins change their interactions with the DNA. For example, the addition of acetyl groups loosens the attraction between the histones and the DNA. All these processes require the actions of proteins, such as the histones themselves, the enzymes that modify them and the enzymes that are involved in DNA methylation. Variants affecting the functions of these proteins can cause severe neurodevelopmental disorders by having widespread effects on transcription.

Changes in Some Proteins Affect the Control of Gene Expression

The term gene expression, which I introduced in an earlier section, is widely used in biology. It is a convenient term to cover a complex process, but when we say that genes are expressed or we discuss their levels of expression, there can be a risk of ambiguity. We might be referring to amounts of particular RNAs or of particular proteins (if the genes in question are protein coding), or maybe both. The amount of each mRNA or protein in a cell is the result of a dynamic equilibrium between the rate at which cells make mRNAs or proteins and the rate at which they degrade them. Different sets of proteins regulate these processes (making it unwise to assume, as is sometimes done, that mRNA measurements will necessarily act as a reliable proxy for protein measurements or vice versa).

The precise choreography of gene expression from euchromatic regions of the chromosomes is regulated at the DNA level by transcription factors that act on the regulatory elements associated with each gene (Figure 4.10). These transcription factors can either activate or repress gene expression. Activation occurs when specific activator proteins bind to specific promoters and enhancers, which they recognize by their particular sequence of bases. They initiate transcription by attracting co-factors and RNA polymerase, which is required for all mRNA synthesis, to create large molecular activation complexes. Repression occurs when specific repressor proteins bind to specific silencers or insulators – again, recognizing them by their particular sequences of bases – and interfere with the activation complexes. The same sequences are often found in the regulatory elements of many genes, allowing the same activator and repressor proteins to affect the expression of multiple genes. Variants that disrupt these transcription and factors and associated co-factors are likely to have countless effects by influencing the expression of hundreds or even thousands of other genes (which we sometimes refer to as downstream genes). They are, therefore, likely to have numerous detectable effects on the individual.

Regarding protein synthesis, processing and degradation, many protein-coding genes produce proteins essential for ribosomes, translation initiation complexes, protein processing (also called posttranslational modification) and protein degradation by two major routes: the ubiquitin–proteasome pathway and lysosomal proteolysis. I will say more about these last two in Chapter 7. Posttranslational modifications are essential for the structures and functions of proteins and can include protein cleavage or the addition of functionally important molecular units such as acetyl, phosphoryl, glycosyl and methyl groups (the previous section contains an example of how this can affect histone proteins; Figure 4.11). Genetic variants that interfere with processes such as these have the potential to impact the activities of many proteins and are liable, therefore, to affect the individual in some detectable way.

Changes in Some Proteins Affect the Structure of Cells and Tissues

The main or exclusive functions of many proteins is to maintain the structure and integrity of cells and tissues. The importance of such relatively inert proteins, often called **structural proteins**, is most obvious outside the brain: for example, the collagens are structural proteins that make a large contribution to your skin, the structural protein keratin forms your hair and nails and so on. Structural proteins are also essential in the brain. Here are some examples. Tau and tubulin proteins provide organization and shape to cells by contributing to their internal **cytoskeleton** (as the name implies, these are the microscopic scaffold structures inside cells). They are also components of structures called **mitotic spindles** that split the genetic material of dividing cells between their two daughter cells. **Myelin** proteins contribute to the myelin sheath that surrounds the axons of nerve cells to ensure that they transmit electrical impulses efficiently. Structural proteins such as collagen contribute to the complex meshwork of molecules that surround brain cells, called the **extracellular matrix**. Please note it is the proteins themselves that are structural; the structures they contribute to, such as the cytoskeleton, myelin sheath and extracellular matrix, carry out many functions in addition to support, as we shall see in Part III.

Changes in Some Proteins Affect Other Cellular Activities

One of the essential attributes of cells is their ability to move ions or larger molecules from one place to another, either within themselves or between cells. For example, the electrical activity of nerve cells is made possible by **ion channel** proteins or **ion pump** (also called ion transporter) proteins located in their cell membranes (as discussed in Chapter 2). Other proteins, including many located in cell membranes, are components of the extensive biochemical pathways that cells use to communicate with each other in a process called intercellular signalling. Intercellular signalling pathways involve sequences, or cascades, of biochemical reactions through which cells receive information in the form of molecules released from other cells and then generate responses; these pathways are called **second messenger** systems (the first messengers being the **signalling molecules** themselves).

Many of the proteins involved in a cell's metabolism – that is, the biochemical processes directed primarily towards sustaining its life – are enzymes catalyzing reactions affecting the fats and carbohydrates that, together with nucleic acids and proteins, make up our cells and tissues. These enzymes are involved in metabolic pathways that either construct (we call this anabolism) or break down (we call this catabolism) molecules in ways that are essential for cells to survive. Some proteins are involved in metabolic pathways for other reasons: for example, they might be required to traffic molecules needed for metabolism between the parts of the cell or between cells.[108] Many proteins involved in metabolic pathways are located in cellular organelles, especially the endoplasmic reticulum, Golgi apparatus, lysosomes, peroxisomes and mitochondria. Some of the proteins in the mitochondria are encoded by genes in the mitochondrial genome, others are encoded by genes in nuclear DNA. Many mitochondrial proteins are involved in the metabolic processes that generate **adenosine triphosphate (ATP)**, which is effectively a fuel for the cell to break down to obtain the energy it needs for all its biological processes (there is more on ATP in Chapter 8: Figures 8.6 and 8.7). The mitochondria are often thought of as the cell's powerhouse.

The Inheritance of DNA Variations

The Human Population Contains Many Alleles of Our Genes

An **allele** is a term for a gene variant that is inherited and persists in the population. When discussing the inheritance of DNA variations, we prefer to use the term allele rather than variant because the latter covers all forms of genetic variation, including many that do not persist in the population (for example, lethal de novo mutations). Note that alleles are not necessarily polymorphisms, as defined earlier in this chapter, because they might not be present in >1% of the population. Some alleles, particularly harmful ones, might be very rare.

Alleles might differ in just one base pair or in up to many thousands. As is the case for other organisms, the human population as a whole contains an immense pool of different alleles of our genes. We call this the gene pool. Alleles circulate in the gene pool in ways that depend on how humans move around the planet and interbreed. Our traits (or phenotypes) and their diversity are affected by which particular alleles we inherit from our parents. One example, often cited, is that our eye or hair colour is determined by which of the alleles of the genes controlling these features we inherited. New mutations add new alleles to our gene pool all the time.

Normally, our somatic cells contain two alleles of every gene on our autosomes. Females with two X chromosomes have two alleles of every gene on the X chromosome. Males with one X and one Y chromosome have one allele for every gene on the X and one allele for every gene on the Y. If an individual has two alleles that are the same, then we say that that individual is **homozygous** for that allele. If an individual has two alleles that are different, then we say that that individual is **heterozygous** for that allele. The terms homozygous and heterozygous are not usually applied to alleles on the X and Y chromosomes in XY males (I say XY males because some males can have more than one Y chromosome). This is because XY males can only have one allele of each gene; instead, we say that alleles on the X and Y in XY males are hemizygous.

Harmful Alleles Can Persist in the Human Gene Pool

Some alleles are harmful, or have the potential to be harmful depending on circumstances, as I have already discussed at length. You might think that all such alleles would vanish from the gene pool, as a result of the process of purifying selection, which I mentioned early in this chapter. It is a virtual certainty that harmful alleles have been eliminated from the gene pool in the past and we do not know about them. But there are many harmful, or potentially harmful, alleles retained in the gene pool that do not necessarily have catastrophic effects on the individual.

Many harmful alleles persist in the gene pool because they are **recessive.**[109] A recessive allele is one whose effects – be they innocuous or harmful – are masked, or overridden, by a second allele of a different type, which we call a **dominant** allele. Examples of alleles whose effects are innocuous are those for blue eyes (recessive) and brown eyes (dominant); note that in this case the two alleles are different, but both are functioning properly. Recessive alleles that are potentially harmful (for example because their products are either poorly functioning or toxic in some way) might not affect a person's phenotype if the individual carries a corresponding allele that is dominant. In other words, the harmful nature of the recessive allele might not matter to the person who has it as long as they are heterozygous with a dominant allele that is able to compensate or rescue the effects of the potentially harmful allele. We often describe a person with a recessive harmful allele as a carrier, and they might well be unaware that they are a carrier.

Sometimes harmful alleles can be either dominant or semi-dominant (semi-dominant alleles are said to show incomplete dominance). A semi-dominant allele is one that causes a detectable effect if it is heterozygous with a dominant allele, but has a greater effect if it is homozygous. In other words, its effects are partway between those of a dominant and a recessive allele. Dominant, and to a lesser extent semi-dominant, alleles are likely to reveal themselves in a person's phenotype and harmful ones are likely to be selected against in the gene pool. Nevertheless, even dominant harmful alleles are retained in the gene pool if, for example, they have their harmful effects later in life, by which time they might already have been passed to the next generation.

Another factor influencing the persistence in the gene pool of harmful alleles, be they recessive, dominant or semi-dominant, is their **penetrance** (Figure 4.12). The issue of penetrance takes us back to the opening sentence of this section in which I said that 'some alleles ... have the potential to be harmful, depending on circumstances'. In other words, some alleles might not have harmful effects in some people. The penetrance of an allele is the frequency with which it produces a detectable effect in an individual. Another word that you will see used in this context is expressivity: this means the degree to which any given genotype is expressed as a phenotype within a single individual (Figure 4.12). Expressivity is different to penetrance: it describes the impact on a person, whereas penetrance is a measure of how often a particular genotype has an impact within a population of people. For example, if a particular allele has a detectable effect on only a small proportion of people but its impact is large in those people, we would say it has low penetrance but high expressivity. Expressivity can be variable in affected individuals. Why an allele might affect different people differently is often not clear: it might depend on what alleles of other interacting genes that person carries; or on their exposure to environmental factors that might be

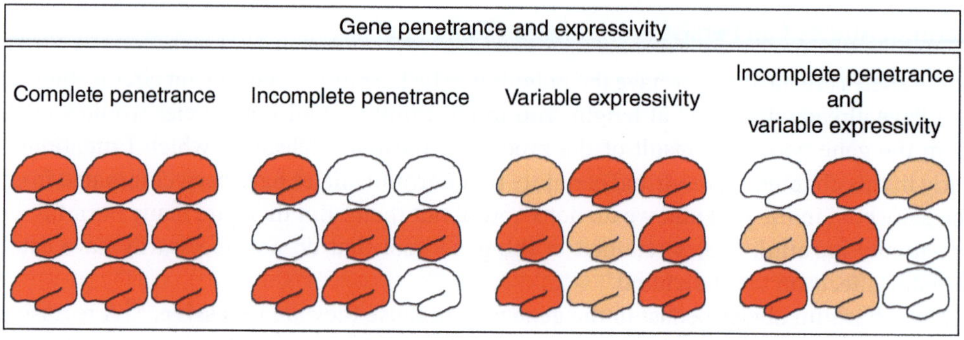

Figure 4.12 Penetrance and expressivity. Complete penetrance: all individuals are affected equally (red brains); in other words, expressivity is the same in all. Incomplete penetrance: white brains are unaffected; expressivity is the same in all affected individuals. Variable expressivity: some are affected more than others (penetrance is full since all are affected to some extent). Both partial penetrance and variable expressivity can occur together.

needed somehow to allow the effects of the allele to become manifest; or it might be because the allele affects the *chance* of a detectable effect emerging in developmental processes that are, by nature, susceptible to apparently random variation from person to person.

Genetic Factors Determine the Impact of Inheriting a Harmful Allele in Nuclear DNA

It was Gregor Mendel's work in the 19th century – long before we knew about DNA, genes and alleles – that laid the foundations of modern genetics, including our modern understanding of dominant, semi-dominant and recessive alleles described above. Mendel did not study harmful human alleles – he worked on pea plants – but the principles of Mendelian inheritance are the same no matter the species, what the alleles actually do and whether they are harmful or not. Although his work was critically important, we now know that the genetics that dictate the likelihood of inherited potentially harmful alleles having effects on the individual are often more complex than the processes he discovered.

I will start with relatively straightforward scenarios that are examples of Mendelian inheritance involving alleles of single genes on the autosomes (Figure 4.13). If one parent carries a harmful recessive allele and the other does not, there is a 50:50 (1 in 2) chance that a child will become an unaffected carrier (top left of Figure 4.13). If two people carrying harmful recessive alleles of an autosomal gene have a child, there is a 1 in 4 chance that the child will inherit two harmful recessive alleles for the gene, which will create a problem for them (top right of Figure 4.13). The chance is 1 in 4 because each parent can give either the dominant or the recessive allele to the child. To spell it out, a dominant allele from the father could combine with either a dominant or a recessive allele from the mother, and both combinations will be okay for the child or a recessive allele from the father could combine with a dominant allele from the mother, which will also be okay for the child, but if a recessive allele from the father combines with another recessive allele from the mother, this will not be okay. It is often the case that the sequences of the harmful recessive alleles from a person's mother and father are different, which is not surprising since there are

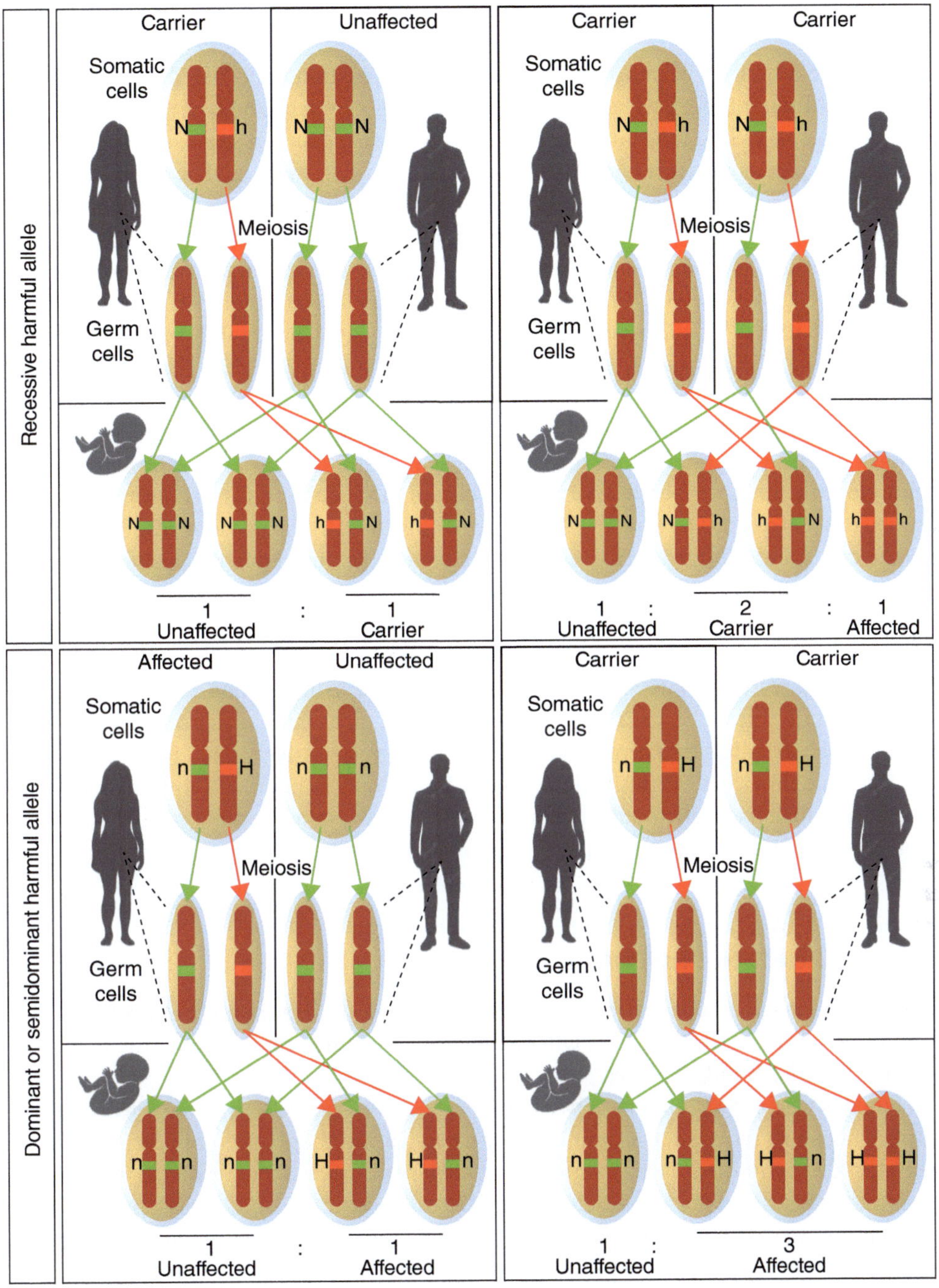

Figure 4.13 Modes of Mendelian inheritance. In each of the four panels, a pair of chromosomes (depicted as in Figure 4.6) separate into different germ cells in the mother and father (this happens during the process of meiosis). Each chromosome from one parent combines with the equivalent chromosome from the other parent to create the embryo. The top panels show two scenarios that can result in a child carrying or showing the effects of a harmful recessive allele (h; red) with frequencies that depend on whether one or both parents carry such an allele. The bottom panels show two scenarios that can result in a child showing the effects of a harmful dominant allele (H, red) with frequencies that depend on whether one or both parents carry, or show the effects of, such an allele. Green alleles are non-harmful (N if dominant; n if recessive).

many ways to produce a harmful mutation in a gene. We commonly describe this arrangement of two different recessive alleles of the same gene as compound heterozygous.

The chances of a child inheriting an allele that will affect them in some undesirable way is greater if a parent carries a harmful dominant or semi-dominant allele combined with a harmless recessive allele (Figure 4.13). If such a parent has a child with a person with two harmless alleles, then there is a 1 in 2 chance that the child will inherit the harmful allele; in other words, there is a 50:50 chance the parent with the harmful allele will give it to the child, which might cause the child problems, perhaps later in life (bottom left of Figure 4.13). If such a parent has a child with another person carrying a harmful dominant or semi-dominant allele of the same gene (a rare occurrence), then the child has a 3 in 4 chance of inheriting at least one harmful allele and has a 1 in 4 chance of inheriting two harmful alleles, which is very likely to be highly problematic or even fatal (bottom right of Figure 4.13).

Things can be a bit different for alleles on the sex chromosomes. Genes present on the X chromosome are described as **X-linked,** whereas those only on the Y chromosome are described as **Y-linked**. The sequences of most of the X and Y chromosomes and the genes they contain differ, the exceptions being short regions at their ends that contain similar genes in both, which are called pseudoautosomal regions. There are far fewer genes on the Y chromosome, which is much smaller than the X chromosome (Figure 4.4). This probably helps explain why X-linked disorders are more common than Y-linked ones. Since males who are XY carry only one copy of their X-linked and Y-linked non-pseudoautosomal genes, we cannot describe their alleles by terms such as dominant or recessive because there is no second allele for them to be dominant over or recessive to (as I pointed out earlier). If a male inherits a harmful allele of an X-linked or, less often, a Y-linked gene, then the harmful allele is likely to affect that person. I will explain what happens if a female inherits a harmful allele of an X-linked gene in the next section.

Epigenetic Factors Determine the Impact of Inheriting a Harmful Allele in Nuclear DNA

For some genes, epigenetic factors (discussed above and see Figure 4.11) affect the likelihood that inherited harmful alleles will create problems. I will first consider the potential effects of harmful X-linked alleles.

In females who are XX, there are two copies of X-linked genes, so we could describe the X-linked alleles by terms such as dominant or recessive, but there is a complication. One of a female's X chromosomes is inactivated by an epigenetic process that results in it bundling up so that it cannot work*.[110] Alleles on that chromosome will not function no matter whether they are recessive or dominant. This means, for example, that if a female is carrying a harmful recessive allele on one X chromosome, this might cause problems for her cells in which the other X chromosome is inactivated.

Inactivation of one of the X chromosomes starts to occur in cells of the early embryo. Which of the two X chromosomes is inactivated varies between cells, apparently randomly

---

* X chromosome inactivation is achieved by a very large long non-coding RNA (lncRNA) called XIST and proteins that associate with it. They coat the X chromosome to be inactivated and recruit enzymes that compact it by modifying DNA, RNA and histones.

(I say apparently because it is difficult to tell, even at the best of times, if an event is truly random). In some cells, the inactive X will have come from the father, whereas in others it will have come from the mother. This means that in XX females the proportions of cells in which a particular X-linked allele is active will vary. If the allele in question is harmful, it is likely to have a lower overall impact in females in which relatively higher proportions of cells have inactivated the X chromosome containing it; indeed, such individuals might not notice it at all, depending what the allele does.

X-linked alleles cannot pass from a father to their XY son because fathers only give Y sex chromosomes to their sons, but they can pass via their XX daughters to their XY grandsons. To spell it out, if a father passes a harmful X-linked allele to a daughter, who gets a harmless allele from their mother, the daughter might be an unaffected carrier but she can pass the harmful X-linked allele to her XY son, who might then feel its effects.

Another factor affecting the impact of inheriting a harmful allele is the fact that some genes are only active if the person inherits them from their mother and others are active only if they come from their father. We call this genomic imprinting.[111] Like X chromosome inactivation, genomic imprinting is an epigenetic process that silences genes through chromatin modifications, such as methylation. It affects a specific set of genes, amounting to about 1% of human protein-coding genes. These genes often reside close together in clusters. Why some genes are imprinted and others are not is intriguing and remains largely mysterious. It has been observed, however, that imprinted genes are expressed prominently in the brain and variants that interfere with their functions are likely to contribute to neurodevelopmental disorders in humans.

Genomic imprinting comes about as follows. As an individual develops and generates cells of ever-increasing diversity and specialization, specific chromatin modifications that contribute to these specializations tend to pass from dividing cells to their descendants. This is, effectively, an epigenetic memory allowing populations of cells to expand while retaining their identity (we call it gene bookmarking). It applies to all cell types including the germ cells. The modifications that are associated with sperm and eggs are, however, no longer appropriate in the early embryonic cells generated by conception. These embryonic cells must produce all the diverse cell types required for successful development, so they normally remove epigenetic chromatin modifications carried over from the egg and sperm. This removal of epigenetic modifications allows the embryonic cells' genomes to return to being, effectively, blank slates. We call this reprogramming. In the approximately 1% of protein-coding genes that undergo genomic imprinting, the early embryo does not remove epigenetic modifications transmitted from the sperm (in the case of paternally imprinted genes) or the egg (in the case of maternally imprinted genes). Imprinted genes are somehow protected from reprogramming.

Why does imprinting matter in the context of genetic disorders? The answer is that disorders can result if a parent gives a faulty version of a gene that they alone are capable of giving an active version of. The child will now have no properly active allele because the allele from the other parent is imprinted and therefore silent. Sometimes, a deletion of a large chromosomal region might remove both a maternally imprinted and a paternally imprinted gene (made more likely by the fact that imprinted genes are often grouped together in the genome). If the child were to inherit such a deletion, then which of the two genes would have an active allele in the child would depend on whether the deletion came from the

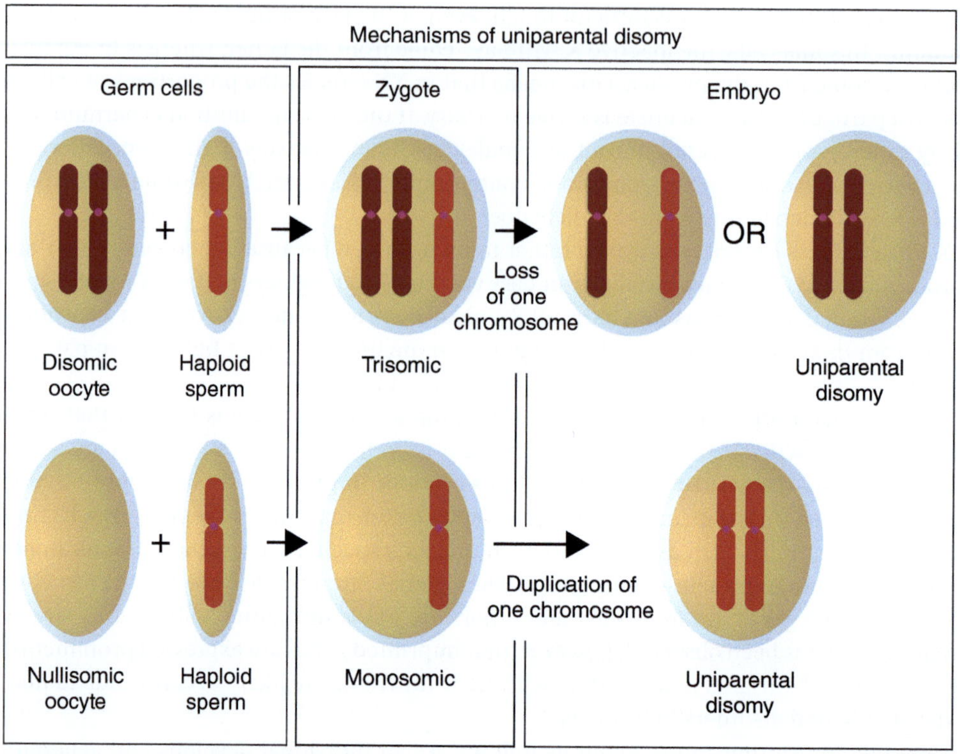

Figure 4.14 Some mechanisms leading to uniparental disomy. Due to errors in the generation of the normally haploid germ cells, they might contain an extra copy of one of the chromosomes (we call this disomic, here in an oocyte) or no copies of one of the chromosomes (nullisomic, again in an oocyte here). When fertilized by a haploid sperm, the zygote will inherit three copies of the chromosome in the top case or only one in the bottom case. Corrective chromosomal loss or duplication can occur. This either risks (top line) or guarantees (bottom line) leaving the embryo with both copies (disomy) from one parent (uniparental).

mother or the father. It is also possible for a child to receive two copies of a chromosome, or part of a chromosome, from one parent and no copies from the other parent due to chance errors either in germ cell formation or in the early embryo (we call this uniparental disomy: Figure 4.14). For most genes, this is unlikely to matter because the child has two alleles of each gene and most genes are not imprinted. As I hope you can anticipate, however, if uniparental disomy were to result in two imprinted – and therefore silent – genes coming from the same parent, then the child would lack any active allele of those two genes.

Factors Determining the Impact of Inheriting a Harmful Allele in Mitochondrial DNA

Things are a bit different when we come to the inheritance of mitochondrial disorders. First, let me point out a straightforward consequence of a point I made earlier: since we inherit our mitochondria from our mother, fathers with a mitochondrial disorder will not transmit the disorder to their children.

The next two important points to recall from what I said earlier are that cells typically contain thousands of mitochondria and that each mitochondrion carries its own mitochondrial DNA comprising mitochondrial genes with potentially different alleles. Earlier, I discussed the states of homoplasmy and heteroplasmy. Let us say that a harmful allele is present in all of a female's mitochondrial DNA molecules (a homoplasmic state), then this state will be transmitted to all of her children. You might think that this would mean that the mother and children would definitely develop a disorder, but often we find that only some are affected, an observation that is sometimes attributable to the presence or absence of interactions with other genetic or environmental factors.

Things are less predictable still if a harmful allele is present in only some of a female's mitochondrial DNA molecules (a heteroplasmic state). In this case, whether the mother or her offspring suffer a disorder is dependent on not only other genetic or environmental factors but also what proportion of mitochondrial DNA molecules carry the harmful allele in any of these individuals. Most harmful mitochondrial alleles are functionally recessive, meaning that a high proportion of mitochondrial DNA molecules must carry the harmful allele before the individual is affected. The proportion of mitochondrial DNA molecules carrying a particular allele can swing wildly from one generation to the next because of the so-called mitochondrial genetic bottleneck. This bottleneck occurs during egg development and limits the number of mitochondrial DNA molecules transferred into each developing egg. The initial selection is apparently random. The DNA molecules later replicate once they are beyond the bottleneck. The random selection at the bottleneck causes random shifts in the proportion of different alleles transmitted to the next generation. If very few (or perhaps even no) molecules carrying harmful mutations are selected, then their impact will be low or non-existent – which is good news for the offspring – and vice versa.

Causes of Diversity and Disorder Involving Multiple Variants

Many Causes of Diversity and Disorder Involve Numerous Alleles Rather Than Just One

In the previous section, I focused mainly on how alleles of single genes can affect individuals. We use the term monogenic to describe conditions that arise in this way; for all the reasons explained earlier in this chapter, any particular monogenic disorder (I will give plenty of examples of them in Chapter 6) is relatively rare. More frequently, disorders with a genetic basis involve multiple genes, and we describe them as polygenic.[112]

Sometimes polygenic disorders result from a deletion or duplication of a genomic region that includes more than one gene, or from a large chromosomal abnormality. Whereas most such variants arise de novo, some, mainly those involving relatively small deletions or duplications that nevertheless include multiple genes, can pass from one generation to the next. Their patterns of inheritance can follow Mendelian principles and their effects are usually dominant.

However, the ultimate causes of most types of polygenic disorders and neurodivergence are significantly more complex and much harder to identify. This is because they are due at least in part to the combined effects of the alleles of many genes scattered across the genome. Working out the ultimate causes of such conditions is made more difficult by the fact that they might well involve genetic and non-genetic factors. Genetic factors might increase the risk of a certain condition developing, and non-genetic factors arising either from the brain's environment or from random variation in the biochemical processes of development might be needed to trigger its emergence. Conditions caused by combinations of genetic and non-genetic causative factors are described as multifactorial. Yet another complication for research in this area is the difficulty I discussed in Chapter 3 of defining precisely the conditions whose causes we are trying to explain, particularly since they are often neurodivergent in nature and affect the highest levels of human brain function. And, as if that were not enough, there can be variation in the combinations of genetic and non-genetic factors that produce apparently similar symptoms. I will discuss ways of overcoming these challenges in the next section.

Complex polygenic and multifactorial conditions tend to run in families, in other words they are heritable, but they do not have straightforward recognizable modes of inheritance. It is worth explaining why the underlying genetics make this so. For the most part, alleles assort independently into the germ cells of the mother or father (with important exceptions discussed below) (Figure 4.15). The first reason is that, in each parent, there is apparently random variation during meiosis (the process that creates germ cell) in which of the two chromosomes of each pair is copied into any given germ cell. The second reason is that, for each pair of chromosomes, the alleles that go into each germ cell are not simply all copied from the same chromosome. Instead, the alleles on the paired chromosomes mix due to genetic recombination, which involves the breaking and recombining of pieces of DNA from the two chromosomes to produce new combinations of alleles for each gene along the chromosome. The exceptions to independent assortment occur when two genes are close enough together on the same chromosome that their likelihood of being assorted together increases; we call this genetic **linkage**. We use the term haplotype (based on the word haploid) to describe a set of alleles along a single chromosome that tend to be inherited together in this way.

The independent assortment of alleles that dominates the formation of germ cells makes it very unlikely that a person with a complex polygenic or multifactorial condition involving multiple alleles scattered across the genome will pass that same combination of alleles to their children. Whether the combination of alleles that a person inherits has an undesirable effect on them, for example by increasing their risk of developing disorder or neurodivergence, will depend on the cumulative effects of many alleles that individually make only small contributions. The inheritance of a critical number of potentially detrimental alleles is more likely if one or both parents carry a large number of them.

How We Investigate the Ultimate Causes of Diversity and Disorder Involving Multiple Variants

One factor that helps us investigate complex polygenic and multifactorial causes of diversity, be it disorder or divergence, is that many of the conditions caused in this way are relatively common compared to individual monogenic conditions. Given the variability in the exact phenotypes exhibited by individuals with a clinically defined polygenic or multifactorial condition as well as their underlying causes, studies need to involve large numbers of people so that they can measure and take account of variation.

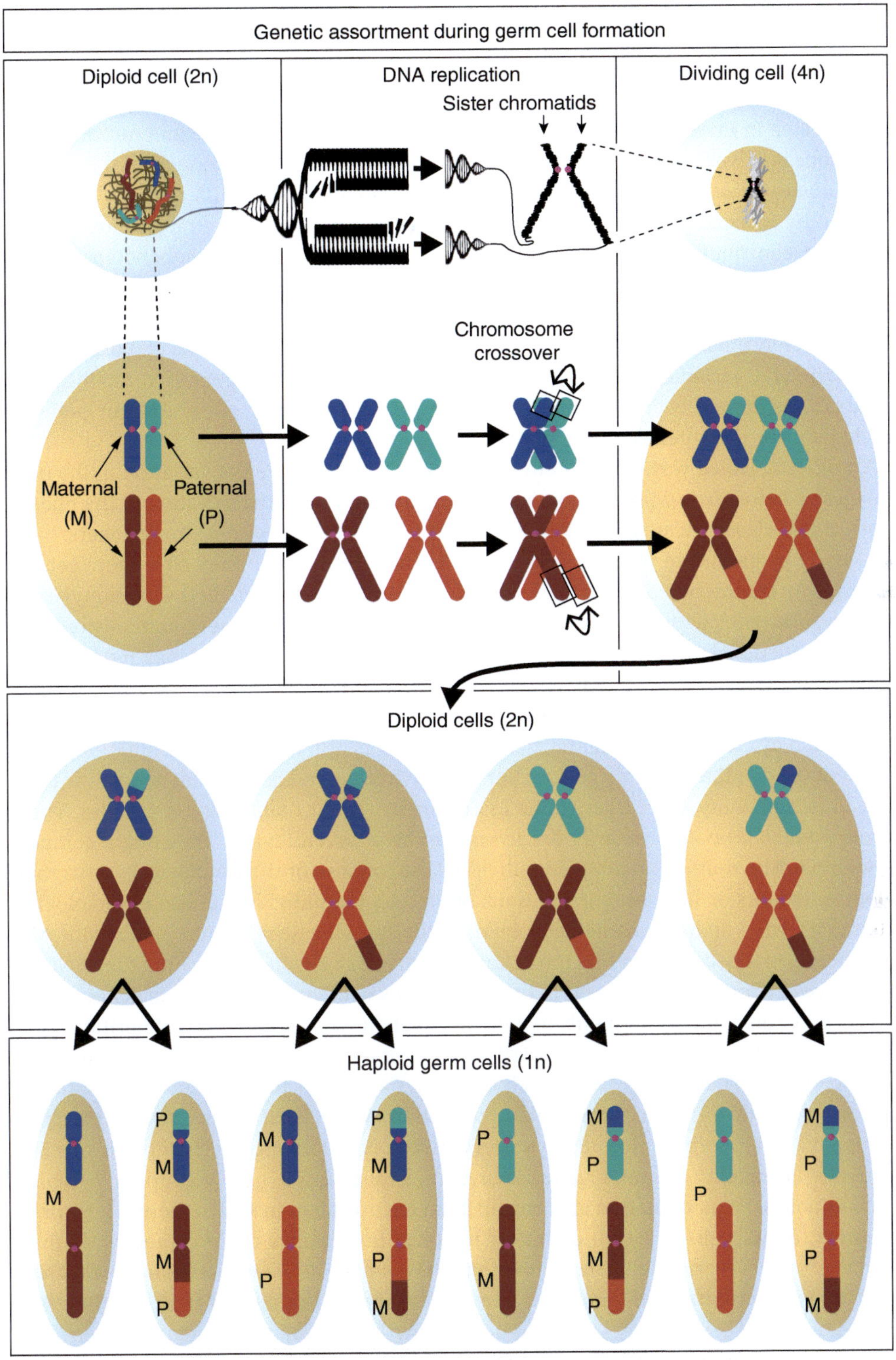

Figure 4.15 Assortment of alleles during germ cell formation. In germ cell formation, diploid cells carrying two of each type of chromosome (2n), one maternal and one paternal, duplicate their DNA, as occurs in dividing somatic cells throughout the body. Unlike in the division of somatic cells, meiosis involves chromosome crossover between parts of each pair of duplicated chromosomes such that duplicated chromosomes now have maternal and paternal parts mixed together. One of each pair of duplicated chromosomes, now of mixed parentage, moves into each daughter cell (which are diploid, or 2n) and these daughters then divide to generate haploid germ cells (1n). As you can see, these cells contain a range of mixes of maternal and paternal regions.

A first step in investigating the ultimate causes of complex polygenic and multifactorial conditions is to work out how much of the variation in any given phenotype in the population is due to variation in genetic factors, which we call the **heritability** of that phenotype. Heritability is stated on a scale from zero to one or as a percentage: numbers close to zero mean the contribution of genetic variation to the variability of that phenotype is low, whereas numbers close to one or 100% mean that most of the variation is genetic in origin. The amount of phenotypic variation that is not accounted for by genetic variation (which is the difference between the heritability value and either one or 100%) could be due to environmental variation or to the inevitable random person-to-person variability that occurs in the massively complex biochemical events that generate our brains (an engineer might describe the latter as noise).[113]

A well-established way of evaluating heritability is to compare the variation in the phenotypes of individuals who are closely related genetically with the variation in the phenotypes of less closely related individuals. The similarity of traits with a strong genetic basis is likely to be greater in more closely related individuals because they have more genetic material in common. For example, studies comparing variation in the phenotypes of monozygotic twins with variation in the phenotypes of dizygotic twins (commonly referred to as non-identical twins) or other siblings are particularly valuable since monozygotic twins have the most genetic material in common. You will often see these types of study referred to collectively as family or twin studies.[114,115] To disentangle the contributions of genetic and external environmental factors, researchers also use adoption studies in recognition of the fact that not only genetic but also environmental factors that contribute to an individual's phenotype are likely to be more similar within than between families. These studies, which do not necessarily use twins, can be of several types. They might compare how similar the traits of an adopted individual are to those of their biological versus their adoptive parents or how similar the traits of an adopted individual are to non-biological siblings growing up in the same household. In the special case of monozygotic twins who have been separated early in life, important insights can be gained by testing the effects of differences in family environment in two individuals with virtually identical genotypes.

There are some important points to understand about what estimations of heritability do and do not tell us. First, they do not tell us what proportion of a given trait is caused by genetic factors. They tell us how much of the variability in that trait in a population is caused by genetic differences across that population, which is different. Second, it is a proportion, not an absolute value, and it varies between populations. For example, if we compare the heritability of intelligence scores in a population growing up in a typical environment with a population growing up in an environment in which there is little or no variation (an unrealistic scenario, but for now just imagine we could somehow do this), then the heritability of intelligence scores would be higher in the second than the first simply because environmental variably is lower or absent in the second than the first. And finally, of course, heritability values do not say anything about which genes and environmental factors are involved in determining a trait.

A modern approach to identify genomic variants associated with particular phenotypes is using **genome wide association studies** (often abbreviated to GWAS).[112,116] It is very important to remember that revealing an association between variants and phenotypes is only a first step in the hunt for the genetic causes of phenotypes. Much additional experimental work is required to test whether variants associated with a trait do in fact contribute to causing it.[117] I will say more about this in Chapter 6.

The idea behind GWAS is that relatively common genetic variants that occur in 1% or more of the population (called polymorphisms, remember), can be present in a higher proportion of people with a particular phenotype, for example a clinical disorder, than in others. To find these associations between the frequency of particular polymorphisms and particular phenotypes, GWAS need to look at large enough populations to allow effective statistical analyses to be done. Studies might include many hundreds of thousands of people, over a million in some cases. Many GWAS look at single nucleotide polymorphisms (SNPs); some look at larger CNVs. It is very important to understand that the polymorphisms selected for study in GWAS are chosen to cover the entire genome and are themselves *unlikely to contribute* to the phenotype or trait. Instead, if a researcher identifies a polymorphism that is associated with a trait, this signals to them the presence of a nearby functionally important variant. The functionally important variant will be found close enough to the polymorphism that it tends to stay linked with it as alleles undergo assortment through the generations (we call this linkage disequilibrium). Each genome wide association study can look at thousands of polymorphisms at the same time and, since we are dealing with complex polygenic diversity, they expect to identify numerous variants associated with each condition. The effects of individual variants usually turn out to be small, with the sum of possibly thousands of them needed to contribute to causing disorder or neurodivergence.

GWAS provide information on places, or loci, in the genome that are associated with an increased chance, or risk, of a certain phenotypic trait, including disordered or neurodivergent ones. Many traits can be measured in some way, giving numbers whose values vary. For example, we can quantify the dimensions of the brain or its component parts and we can put numbers on cognitive characteristics such as intelligence quotient (IQ) measured by standard tests (see Chapter 2). We call these quantitative traits and each genomic locus that might turn out to contribute to the trait's magnitude we call a quantitative trait locus (often abbreviated to QTL). Traditionally, quantitative traits are contrasted to traits that are considered to be categorical, or discrete, which are often called qualitative traits. In reality, very few traits associated with brain development and function are likely to be completely discrete. Most will be quantitative. You might think an extreme trait such as the complete absence of a brain structure could be considered discrete – either the structure is there or it is not – but things are rarely so simple. For example, if the absent structure initially forms to some extent and degenerates later, we can ask how much of it forms and how quickly it disappears, which are in principle measurable quantities. If we look at multiple cases of such a disorder, answers to these sorts of question will almost certainly yield a range of values.

These approaches allow the generation of polygenic risk scores in individuals.[118] This is done by taking into account all a person's genetic variants that are known to be associated with a given phenotype – for example, a particular disorder – each weighted according to the nature of its effect on the phenotype. In practice, weighting involves giving each variant a number that represents the magnitude of the effect and its direction, by which I mean whether the variant makes the effect more likely (a positive number) or less likely (a negative number). For any individual, these numbers for all the relevant variants in that person's genome are added together to give a polygenic risk score for the phenotype under consideration in that person. It is important to note that polygenic risk scores are based on a person's genetics and do not take into account other factors, such as those from their external environment. This means that they do not tell us the overall likelihood that a person will develop particular traits. For example, a person might have a relatively high polygenic risk score for a particular condition, but they might be able to lower their overall likelihood of developing that condition by making changes to their way of life. Also note that a person will have different polygenic risk scores for different conditions; for each condition, the risk is calculated based on which of the alleles associated with that condition they carry. How polygenic risk scores are calculated and used is a subject of much research since it offers the opportunity to focus efforts to prevent the development of disorder on those most at risk from their genetic makeup.

Summary

- DNA molecules vary in detail from person to person; these differences make a major contribution to the phenotypic diversity of individuals.
- DNA molecules are long chains of four kinds of molecules whose order makes a code; this code guides the types of proteins and other important molecules each cell makes.
- DNA is prone to errors as it self-replicates during cell division; most DNA has no known function, and errors in these regions have no detectable effect on an individual.
- Cells can correct errors in their DNA but some – called mutations – escape and might be transmitted to subsequent generations; mutations range in size from single base changes to large chromosomal deletions or duplications.
- Our mitochondria – numerous small energy-producing organelles within each cell – each contain their own DNA, which is also prone to develop errors that can be transmitted to subsequent generations.
- Whether, and if so how, a genetic variant impacts an individual depends on factors such as its location in the genome; for example, whether it is in a region with an important activity and what that activity is.
- The nature of each variant determines the impact of its inheritance on an individual; the possible impact ranges from a major dominant effect to a subtle contribution alongside many other genetic and non-genetic factors.

5

Environmental Variation Affecting the Developing Brain

This chapter is about environmental factors, by which I mean factors that impinge on the developing brain from outside it. Outside the brain is the rest of the body and then, beyond that, the surrounding world. So environmental factors might be ones in the rest of a person's body – in other words, internal to them – or outside them altogether – their external environment. This might seem obvious, but the reason I am stopping to make you think about it is that the term is often used to mean the opposite of genetic factors, whereas some environmental factors that affect the developing brain can be traced back eventually to genetics. Such factors might result, in part or in whole, from the genetics of the affected person themselves or from the genetics of other species, most obviously viruses and bacteria, or other people.

A person's own genetics might lead to the brain being affected from outside by, for example, causing variation in biochemical events elsewhere in the body that then impact brain development. To illustrate, consider the case of bilirubin, a potentially neurotoxic substance normally found at elevated levels in the blood of newborn babies*. In some babies, it is produced at unusually high levels because of variation in the baby's genome. The high bilirubin levels are in the brain's environment, from where they can damage its development; but the cause of the high level is, at the end of the day, genetic.

The genetics of other people, ranging from the mother (from early embryogenesis onwards) to the increasing numbers of other people that an individual encounters as they grow up, might affect the brain in a wide variety of ways. For example, in early life, genetic variants affecting a pregnant mother's levels of hormones, such as thyroid hormones, which are needed by both mother and fetus, can impact fetal brain development. Later in life, the developing brain is affected by other peoples' behaviour, and genetics plays a large part in determining the way others behave. For example, the developing brain of a child will likely be adversely affected if they are brought up in an environment containing one or

* This potential neurotoxicity is why health professionals are keen to treat jaundice in newborns (jaundice is yellowing of the skin). It is usual for newborns to have too many red blood cells and they need to break some down, producing bilirubin, which is yellow. Because their livers are not yet fully developed, newborn babies are not very efficient at removing bilirubin from their blood, and occasionally it can build up to levels that risk damaging the brain (this is called kernicterus). Levels do not usually get this high and, if they do, treatment is possible.

more abusive individuals, and those individuals' tendency to abusive behaviour might have a genetic component.

In this chapter, I will first add a bit more to my discussion in Chapter 4 of how scientists disentangle environmental from genetic causes of brain phenotypes. I will then discuss some of the main discrete, or relatively discrete, identifiable environmental factors that can cause, or might contribute to, the development of brain disorder and neurodivergence, such as specific chemicals and infections. Finally, I will discuss examples of less easily defined environmental factors, such as maternal stress or socio-economic status. I will provide many examples of specific and broadly defined environmental factors in the coming sections. Figure 5.1 provides an overview of some of these. It reinforces the important point that the environment can influence brain development through protective mechanisms as well as by increasing vulnerability. There is not much we can do about our genetics, but we

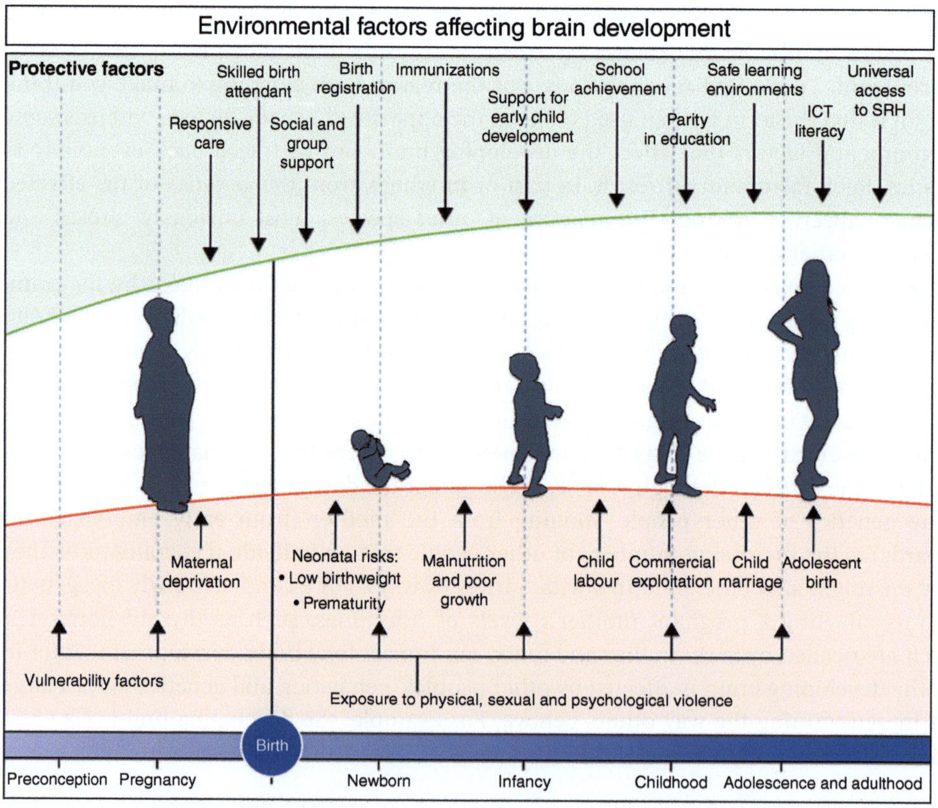

Figure 5.1 Some of the vulnerability and protective factors affecting children. The gap between the upper and lower curves represents the fact that the potential for children to avoid vulnerability factors and draw protective ones into their lives varies greatly around the world. ICT: information and communications technology; SRH: sexual and reproductive health. The figure was produced by a Commission set up by the *Lancet* journal, the World Health Organization (WHO) and the United Nations Children's Fund (UNICEF) to consider how governments, medical professionals and society as a whole can accelerate progress on child health and wellbeing strategies. *Reproduced from Clark et al.*[119] */ with permission of Elsevier.*

create much of our own environment, individually and collectively. We have the option, if we chose, to do a great deal to optimize the development of our brains not only by removing vulnerability factors but also by increasing protective ones. Sadly, due to the way we construct our geopolitical structures, this is only a realistic option for some of us.

Ways of Identifying the Effects of Environmental Variation

The Growth of Epidemiology as a Science

Many environmental factors that are liable to affect developing brains negatively, such as certain chemical toxins, smoking, alcohol, infections and so on, have been found by the methods of epidemiology. Epidemiologists study the patterns of occurrence of chosen conditions of interest in defined populations. By patterns of occurrence, I mean things like: who has the condition, who does not, what are the differences in their circumstances, lifestyles and so on. The conditions that are selected for study do not need to be harmful; they could be beneficial or protective (for example, we could study the epidemiology of happiness). Usually, however, we focus on the need to understand agents that are liable to damage us in some way individually and collectively. Epidemiologists make associations between patterns of occurrence of conditions and the distributions or spread of possible causative agents. Sometimes they can determine whether these agents are in fact causative by, for example, attempting interventions aimed at diminishing the impact of the proposed causative agent.

The critical distinguishing characteristic of epidemiology is signalled by the use of the Greek root *demos* in the word. *Demos* means the people, and epidemiology involves the study of populations rather than individuals. *Epi* means upon, so epidemiology is the study of what is upon the people. Epidemiology became a major branch of scientific research in the mid-1700s. Its rise to prominence resulted from the systematic collection by teams of physicians, often coordinated by the military or by commercial stakeholders, of information on the spread of infectious disease among large, defined populations such as soldiers in conflict zones or enslaved people. We should never forget the involvement and contribution of these compulsorily enlisted participants. The data they provided was analysed using statistical methods that were in their infancy at the time. From these ethically deeply flawed and methodologically limited beginnings, epidemiology has progressed to become the foundation of public health.

Sometimes epidemiological studies start with the hypothesis that a particular condition is due to a specific, discrete environmental agent. For example, scientists might have evidence of an association between the dumping of a toxin into the sea where people fish for food and the emergence of a disorder in the area. As we shall see below, this is what happened in Minamata City, Japan in the early 1900s. The intervention that proved the cause was to stop the dumping.

Other times, epidemiological studies might start with a less specific hypothesis. It might be more a case of trying to find out whether environmental factors have a particularly strong effect on a trait or not. As mentioned in Chapter 4, the contributions of environmental factors to phenotypes have been evaluated using family, twin and adoption studies, including studies of **monozygotic twins** separated by adoption at birth. In these separation studies, there is the potential for twins with almost identical genetic material to experience very different environments: any differences in traits are likely to be caused by environmental differences. In fact, these studies showed remarkably little difference in the neurological traits of twins brought up in different environments. Although in theory these sorts of study are an attractive way of trying to tease apart genetic and environmental influences on traits, the range of traits they can examine is limited by the number of subjects that can be recruited.

Longitudinal Epidemiological Studies

Longitudinal studies select and follow large groups of people over long periods of time, potentially many decades. They are often primarily observational, collecting repeat measurements of whatever features are of interest to the researchers at intervals from people leading their normal lives without the application of any external influence. They are useful for evaluating the relationship between risk factors and the development of disorders. They contrast with so-called cross-sectional analyses in which researchers measure many variables at a given point in time, thereby providing no information on changes with time and limiting their usefulness for examining possible causal relationships.

Many longitudinal studies follow people from birth. For example, the United Kingdom has carried out six large-scale birth cohort studies since 1946. Other examples I am familiar with come from Scotland. The ongoing Lothian Birth Cohort studies (Lothian is a region where Edinburgh is located) are based on work initiated in 1921 and 1936 and have focused much of their research on how childhood intelligence relates to intelligence and physical and mental health at older ages[†]. These studies have been able to assess the stability of intelligence throughout life and the influence of genetic and environmental factors on cognition and cognitive ageing. The *Theirworld* Edinburgh Birth Cohort study[‡] is following babies that are born preterm, which can lead to long-term disability (I will have more to say on this later in this chapter).

Recent Advances in Epidemiological Research

Genome wide association studies (GWAS) (the principles of which I described in Chapter 4) have proven to be a very powerful tool in the epidemiologist's armoury. They can provide powerful insights not only into potential genetic causes of divergence and disorder but also into the influences of environmental factors through the use of Mendelian randomization.[120] The key issue that Mendelian randomization seeks to overcome is that association does not imply causation: remember that just because two things are associated does not mean that one causes the other (for example, two events might be associated because they occur at the same time, and although one might cause the other, it might not). Mendelian randomization uses data from observational studies that are not themselves

† https://lothian-birth-cohorts.ed.ac.uk/.

‡ https://theirworld.org/projects/theirworld-edinburgh-birth-cohort/.

designed primarily to test causation. It takes these data and attempts to work out if a chosen, defined risk factor is likely to contribute to causing a particular phenotype. In the context of this chapter, I am considering risk factors that derive from a person's external or internal environment, but Mendelian randomization can be used to study potential genetic risk factors too.

Mendelian randomization exploits the wealth of statistical information that GWAS have given us on genetic variants – such as **single nucleotide polymorphisms** (SNPs) – and their associations. How does it work? This is probably easiest to explain by giving an example. Let us imagine that we want to know whether high alcohol consumption causes a particular brain phenotype. We could ask a collection of people what their alcohol consumption is, categorize them according to their answers and see if the phenotype we are interested in is more common in the high consumption group. We might find that it is, in which case we have an association. But we do not know whether the high alcohol intake causes the phenotype. The association could be explained by some other factor(s), for example underlying genetics, that we do not know about. These other factors might directly create the brain phenotype and also (in parallel) make people drink more alcohol. In other words, the unknown factor(s) could be operating on brain phenotype and alcohol intake through separate, independent mechanisms, with alcohol levels having no direct effect on the phenotype. If we can rule out such possibilities, it becomes much more likely that high alcohol intake causes the phenotype. Help is at hand if we know, from all of our genome wide association data, that certain specific variants (usually single nucleotide polymorphisms) associate statistically with people's alcohol intake and – this next point is critical – *are very unlikely to cause or affect the brain phenotype directly*. We might or we might not know why these variants associate with alcohol intake, but this does not matter. By using the fact that the behaviour (drinking alcohol heavily) is predicted by the variants, we can select groups of people with different variants and hence different levels of *predicted* alcohol intake. If this approach finds no association between predicted high alcohol intake and the brain phenotype, then high alcohol intake probably does not contribute to causing the brain phenotype. If the association remains, the likelihood that high alcohol intake causes the phenotype increases.

That is how it is done, but why does it work? How does this method help us test whether unknown factor(s) might be operating on brain phenotype and alcohol through separate, independent mechanisms? The answer is in the name: Mendelian randomization. Remember the principle, discovered by Mendel, of the independent assortment of **alleles** (Chapter 4). The point is that any variants that might, unbeknownst to us, be causing or contributing to the phenotype and also, through separate independent mechanisms, be increasing alcohol intake should assort randomly between the categories of people we defined by using our alcohol intake-predicting variants. Any such variants will be distributed randomly among the groups of people we selected based on their (predicted) alcohol intake, thereby controlling for any contributions they might make to causing the phenotype.

In practice, Mendelian randomization is more complex than the outline I have given here and is based on assumptions not all of which I have explained. It needs to be applied with care. Its judicious use is now contributing successfully to the identification of many environmental factors contributing to the causes of neurodevelopmental disorders and types of neurodivergence.

Specific Environmental Factors

The development of the brain can be adversely affected by factors coming from the external or internal environment, including harmful chemicals (an example of which – bilirubin in the internal environment – I mentioned above), infection, metabolic disorders occurring outside the brain, **autoimmunity**, malnutrition, physical trauma and preterm birth.

Exposure to Harmful Chemicals

We are surrounded by many potentially harmful chemicals. Some occur naturally, although it is often the case that human activities increase the risks they pose. For example, farming practices can increase exposure to the naturally occurring fungal molecule fumonisin, which is found particularly in corn and can cause **neural tube** defects (I will discuss how it does this in Chapter 7). Others are man-made, for example as products or by-products of the pharmaceutical and chemical industries. If our developing brains are exposed to these potential toxins in high enough concentrations, then they can cause lasting damage. The levels of exposure that must be achieved for this to happen vary from chemical to chemical and the effects they have will depend in large part on the age of the person who is exposed. In general, exposure at early stages of life is particularly risky. Early fetal exposure might cause major life-long structural defects, but even later exposure can affect the brain's functional development. Here I will give some well-documented examples of chemicals with the potential to cause problems for the developing brain.

Alcohol, produced in naturally occurring fermentation and concentrated for human consumption, is toxic to neurons and glia and people can become dependent on it. Its chronic overuse can also cause problems for the developing brain not only directly but also indirectly, for example by interfering with the gut's ability to absorb the vitamin thiamine, which is required for neurons to produce the energy they need. Alcohol is particularly harmful to the fetus and fetal alcohol spectrum disorders are relatively common. They affect around 1–5% of people in those populations that have been studied, although it is hard to measure incidence precisely due largely to variation in the criteria used to diagnose an inherently variable syndrome comprising a range, or spectrum, of phenotypes. Fetal alcohol spectrum disorders occur in people whose mothers drank alcohol at any stage during pregnancy. The quantity and frequency of consumption impact the phenotype, whose characteristics include physical, intellectual and behavioural difficulties, including abnormal facial features, problems with cognitive tasks such as learning, memory and communication and disturbances of attention and emotions.

In postnatal life, usually from adolescence onwards, researchers have observed a co-occurrence of psychiatric disorders with alcohol abuse – and, indeed, with the abuse of other substances, including legal or illegal drugs and other intoxicating chemicals. This work illustrates the difficulties that can arise when we try to ascribe cause or effect to environmental factors brought about by altered behaviour on the part of the affected individual: is the substance abuse contributing to cause the psychiatric symptoms, or vice versa, or are they independently generated manifestations of an underlying disorder? The answers are not always clear, but it is likely that all three possibilities can occur. First, substance abuse might trigger changes in brain structure and function that increase the likelihood of a person developing a psychiatric disorder, particularly in genetically predisposed individuals. Second, researchers have found that people with a psychiatric disorder tend to abuse drugs or alcohol initially as a form of self-medication but later as a dependency. Third, both substance abuse and psychiatric disorders can run in families and certain genetic variants might be risk factors for both.

The fetus can be harmed by some therapeutic drugs given to the mother to treat a wide variety of conditions, including epilepsy, cancer or inflammation and recreational drugs such as cocaine. The extent of the effects depends in part on when during pregnancy the drugs were taken and at what doses, and probably also on whether the genetics of the embryo or fetus puts it at increased risk. Some therapeutic drugs interfere with folate levels in the blood or with folate metabolism; folate levels can also be low due to inadequate levels in the diet (see below: I will explain why folate levels matter in Chapter 7).

Other examples of chemicals that can harm the developing brain include heavy metals such as mercury, lead or arsenic. Once again, the effects depend on the nature of the exposure: for example, when in life it occurred and at what level. One well-documented episode of mercury poisoning was uncovered in Japan in the 1950s. A particularly neurotoxic form of mercury, methylmercury, had been discharged over decades from a chemical manufacturing plant into the sea off Minamata City, where it contaminated seafood and then humans. This caused a condition known as Minamata disease in which affected adults experienced numerous neurological symptoms, with paralysis, delirium, coma and death in extreme cases, and the effects on the fetus resulted in children suffering brain degeneration, intellectual disability, difficulty walking and speaking and other symptoms.

Infections

Many viral, bacterial or parasitic infections have the potential to affect the developing fetus. Most reach the fetus via the placenta; some take other routes, such as via the mother's reproductive tract. The risk to the members of a population of becoming infected with a specific viral, bacterial or parasitic pathogen varies enormously with the following: geographical location and local climate, which affect the types of pathogens and vectors present (vectors transmit infective agents and include living organisms such as insects and other animals or non-living vectors such the surfaces of objects); air, water, food and sanitation quality, which affect transmission; the availability of antimicrobials to treat infections and limit their long-term consequences. For an individual, the timing of infection is an important factor affecting outcome. For example, the chikungunya virus (named after a word in the Kimakonde language of Tanzania meaning 'to become contorted'), which is spread by mosquitos, is only likely to result in a child developing intellectual disability if infection is passed from a mother with a high level of virus in her blood to her baby around the time of birth. If infection occurs at other times, it is less likely to have an impact. Other factors affecting outcomes include an individual's general health and level of immunity. For example, worldwide vaccination programmes have massively reduced the incidence of poliomyelitis, which has the potential to attack the brain and leave children with lifelong weakness and paralysis.

Infections during pregnancy or of infants and young children can have long-term effects on brain development. Many pathogens, including Zika virus (named after the Zika forest in Uganda), mumps virus, rubella virus, chickenpox virus, cytomegalovirus, syphilis bacteria and toxoplasmosis parasite, can cause congenital structural defects of the brain, such as abnormal gyrification, **anencephaly**, brain **atrophy**, cerebellar **hypoplasia**, cerebral calcifications, **hydrocephalus**, **megalencephaly** or **micrencephaly** and **ventriculomegaly** (I discussed these phenotypes in Chapter 3). They and other pathogens, such as influenza virus and human immunodeficiency virus, can cause functional symptoms such as altered muscle tone, eating disorders, hyperreflexia, intellectual disability, learning disability, neurodevelopmental delay and sleep problems (see Chapter 3). Measles in children and adolescents can sometimes be followed by the progressive deterioration of aspects of the brain function they had attained before infection.

Some infective agents, such as human immunodeficiency virus, might affect brain development by interfering with a person's immune system. In another possible example of this, Streptococcus bacteria have been implicated in triggering the long-term development of obsessive, compulsive and other neurodivergent behaviours in a condition known as paediatric autoimmune neuropsychiatric disorders associated with Streptococcus infections. Whether the underlying cause involves autoimmunity remains uncertain and controversy surrounds many aspects of the diagnosis.

Effects from Metabolic Disorders Occurring Outside the Brain

Here I will discuss how metabolic disorders arising in a person's internal environment but outside their brain can impact their brain's development. I will not discuss genetic variants whose effects include disruption of the metabolic activities of cells throughout the body including brain cells themselves. Their effects on brain cells are likely to be both direct, by actions within the brain itself, and also indirect, by affecting the internal environment surrounding the brain. In such conditions, it can be difficult to disentangle the origins of the brain phenotypes that result. I will discuss these sorts of situations in later chapters.

Conditions such as diabetes affect hundreds of millions of people worldwide, and numbers are rising. Epidemiological studies have shown that children and adolescents with these conditions have a significantly higher risk of developing mental health problems, including disordered eating patterns, self-harming behaviours, sleep disturbance, symptoms of depression and anxiety. The discovery of this association has given rise to an area of clinical research and practice known as metabolic psychiatry. One success in this area has been the use of dietary changes to alleviate symptoms, which lends weight to the hypothesis that the association is causative. In other words, it seems that changes in the body's general metabolic state can have knock-on effects on the brain. Since the brain uses about 20% of the oxygen and calories taken in by the body, metabolic disorders limiting the availability of these essential ingredients for the brain is very likely to affect its development and function. The details of how the brain is impacted at a cellular level are not well worked out. It is very likely that other risk factors, including genetic ones, act directly on the brain to determine its level of resilience to the consequences of metabolic disorders arising from outside it.

Autoimmunity

In autoimmune disorders, the immune system, which has evolved to defend the body by attacking foreign potentially harmful substances or organisms, attacks its own normal brain cells instead. The reasons it does this are usually unknown, but might in some instances follow infections, as I mentioned above. It results in inflammation of the brain, referred to as autoimmune encephalitis. The attack can target proteins in neurons, including synaptic proteins, **ion channels** and intracellular receptors. Sometimes it is not the neurons that are attacked but other cells, such as those that make the brain's blood vessels.

Malnutrition

Malnutrition in fetal and postnatal life covers a wide range of possible dietary deficiencies, from a lack of a specific substance or a defined set of substances to more generalized malnutrition. Examples of specific nutritional deficiencies include a shortage of iodine, zinc or folic acid in the diet; there are many others. Iodine is essential for making thyroid hormone, and thyroid hormone is essential for normal brain development, as it regulates particularly the timing of critical processes such as **precursor cell** migration, maturation, synapse formation and **neurotransmitter** synthesis. Why zinc is so important for brain growth and function is not yet clear. Folic acid deficiency is associated with developmental defects described in Chapter 7. Generalized malnutrition can be associated with other environmental factors, such as socio-economic status (see below) or conditions whose causes have a strong genetic component, for example the eating disorder anorexia nervosa. This is another example of the point I made earlier, that some environmental factors have origins that include genetics.

Physical Trauma

Physical trauma arising in a wide variety of ways can affect our developing brains at macroscopic or microscopic levels. For example, it might result from mechanical impact on the brain before, during or after birth. It might result from high-energy ionizing radiation or X-rays (Chapter 4). The trauma might go unnoticed at the time or it might induce pain. Our body's pain receptors, called nociceptors, which are nerve endings in skin, muscle, joints, bone and organs, detect harmful (or noxious) events and our brains interpret their signals as the unpleasant emotional and sensory experience we describe as pain. Newborn babies respond to noxious stimuli in a way that make us think that their brains create a sense of pain, but whether their personal experience is the same as that of an older child or adult is impossible to tell since they cannot describe their feelings using the richness offered by verbal language.[121,122]

Preterm Birth

There is evidence that being born too early increases a person's risk of developing conditions such as cerebral palsy, neurodevelopmental delay, intellectual disability and neurodivergence and of suffering mental health problems during adolescence and adulthood.[123–125] Brain scans have been reported as showing differences in the brains of preterm babies. Not all studies have found these associations with preterm birth, reflecting the acknowledged

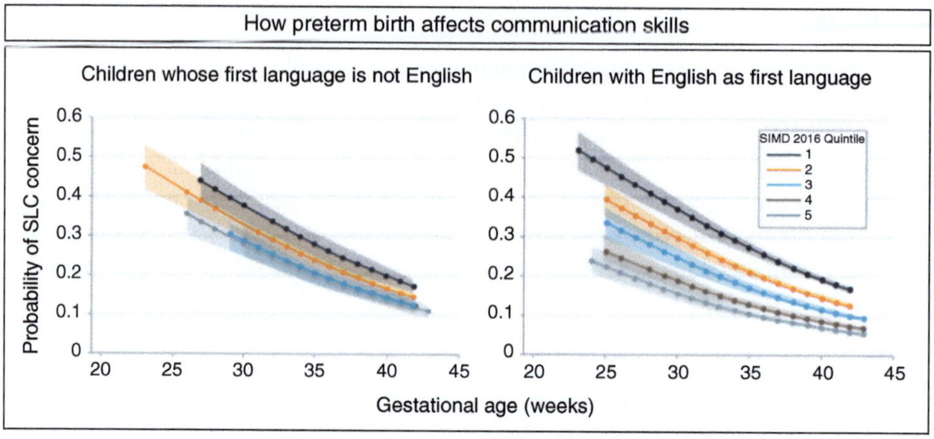

Figure 5.2 Effects of preterm birth on communication skills vary with socio-economic status. In this research, carried out in Scotland, UK, the medical records of children born at or before term were studied, to establish how many were recorded as having had speech, language and communication (SLC) problems at around 2.5 years of age. The children were assigned to five equally sized groups (quintiles) according to the socio-economic, health and educational status of their local environment, defined using a Scottish Index of Multiple Deprivation 2016 (SIMD 2016) score: quintile 1 is most deprived and quintile 5 is least deprived. In all groups, the more preterm the children had been born (that is, going to the left in each graph), the higher their SLC concerns. The ones that fared best had English as their first language and lived in the least deprived settings (quintiles 4 and 5 in the graph on the right; there were fewer concerns than for any of the groups in either graph). *Reproduced from Ene et al.*[127] */ JAMA Network / CC BY 4.0.*

heterogeneity of outcomes, which is probably due to the likelihood that other risk factors – that might be genetic or environmental – affect the susceptibility of different developing brains to the consequences of preterm birth.[126] Many genetic and/or environmental risk factors might contribute to make adverse outcomes more or less likely to emerge (see Figure 5.2 for one example of this). Some of these risk factors are complex and broadly defined, which leads me neatly on to the next section.

Complex Environmental Factors

The diverse nature of human brain development owes much to the highly complex, diverse environment in which we grow up. Unlike the specific, discrete factors I have discussed up to now in this chapter, much of the diversity in our environments arises from sources that we can only define in relatively general terms, largely because of their high degree of heterogeneity.

Some of this diversity is in aspects of our environment that we create. We make, maintain and update the numerous social and economic systems, from global to local, that impact our lives. The ways they do so vary greatly between populations, for example in different geographical locations, and the individuals within those populations.[128] Many studies indicate that our experiences within these systems early in life can contribute to the development of multifactorial disorder and neurodivergence, particularly in individuals whose genetics predisposes them to such outcomes.[129]

Other aspects of the diversity of our environments, both internal and external to our bodies, are not created by us, although we might influence them: examples are variation in the microorganisms in and around us (which we call our microbiome) or in the climate, which has an impact on pretty much everything.[130]

Socio-economic Disparities and Environmental Stressors

Variation in social and economic factors can contribute to differences in how our brains develop, and how our brains develop can affect our socio-economic status. This creates circularities of cause and effect that are liable to perpetuate the socio-economic status of individuals and their descendants. Families or individuals with low socio-economic status are more likely to experience hardship and stress due to a lack or poor quality of material resources, including money, and supportive social interactions, such as access to educational, healthcare and other social networks. Many studies have shown that children brought up in a low socio-economic setting are up to several times more likely to develop many types and degrees of emotional and behavioural difficulties than children reared in more advantaged families.

The reasons for this association are certain to be numerous and vary significantly in different parts of the world. Some possibilities include the following. Poor access to high-quality healthcare, including mental health resources, is liable to leave disorders undiagnosed or untreated. For example, infections that are potentially harmful to the developing brain, such as those described above, are more likely to occur and to be untreated or poorly treated in impoverished societies. Reduced access to nutritious food can result in malnutrition and the possible consequences exemplified earlier in this chapter. Socio-economic factors can determine access to quality education and the level of education of parents can influence the cognitive and emotional development of their children. Limited educational resources and opportunities can hinder cognitive development, potentially contributing to learning disabilities and neurodivergence. Economic demands on parents' time can limit their engagement in their children's lives.

A very large number of environmental factors, some of which are more likely to affect those in poor socio-economic circumstances, have the potential to affect brain development by raising the levels of psychological stress experienced by a pregnant mother or a growing child. The levels of stress a person feels are likely to depend on genetic factors as well as environmental factors including the following: levels of stress among parents and other family members; whether they live in an urban versus a rural environment, with individuals residing in urban environments likely to face higher levels of environmental stressors, including noise, crowds, crime and pollution; the frequency and extent of adverse encounters with stigmatization, discrimination, bullying, physical or sexual abuse and conflict. Many studies have suggested that both maternal stress during pregnancy and stress during childhood or adolescence may be associated with the development of neurodivergence, although this is a difficult association to establish definitively.[131]

Interpersonal Relationships and Lifestyle

As children grow up to become adolescents, potential sources of environmental stress increase greatly.[132,133] Initially centred around the child's immediate family and the local environment they create, they expand to include the individual's immediate peer group, teachers and numerous others that the person is required, chooses or is allowed to interact with. Nowadays, this includes interactions through the internet, for example on social media sites, and provides the potential for contact with a broad range of people and the ideas they espouse, be they valid or not.

Studies have revealed quite clearly the predominant importance of an adolescent's immediate peer group for their behaviour and development. At this stage of life, our drive to reduce our risk of isolation by conforming to the behaviours of our peer group is particularly strong, but can lead us to indulge in behaviours, such as alcohol or drug misuse, that place our mental health at risk for other reasons. There is less consensus on the effects of internet use, engagement with social media, playing computer games and so on.[134,135] For example, some research has found no evidence that the time children spend using computer screens impacts their brain function or mental health. On the other hand, a lack of evidence does not convince others that there is no association, which is a perfectly valid stance; in general, it is very difficult to prove that something does not exist. Another example is that violent video games are often suggested to trigger or enhance aggressive behaviour, but research on this has produced conflicting results, at least in part because different studies use different methods.[136,137] For now, the overall conclusion is that this is an extraordinarily difficult but extremely important area of research in which little is resolved.

The Microbiome

In the previous section, I discussed influences from the world we perceive consciously in our daily lives, the world of living and non-living things more or less similar in size to us. The microbiome is the complex community of many trillions of tiny microorganisms living in and on the body that we do not see. It is a crucial player in various aspects of health and disease, with the gut microbiome increasingly implicated in brain development. There is two-way communication between the gut microbiome and the brain through what we call the gut-brain axis.[138] Communication occurs in several ways including the following: via nerves that run between the gut and brain; via the immune and endocrine systems; via the production in the gut of important metabolites such as the brain neurotransmitter serotonin (or 5-hydroxytryptamine) (which we came across in Chapter 2). The composition of an individual's microbiome in early life is influenced by many factors, such as the way they were delivered (vaginal birth versus caesarean section), infection and use of antibiotics, nutritional intake (for example, whether breastfed or not), environmental stressors and, as ever, their genetics. The nature of an individual's gut microbiome is now being implicated in their likelihood of developing neurodivergence or neurodevelopmental disorders of various types and, later in life, conditions usually thought of as neurodegenerative (see Chapter 9).[139,140]

The Climate

Most scientists agree that our planet's climate is changing in a way that poses a major threat to human health. In the absence of sufficient ameliorating action by humankind, global warming is proceeding inexorably, progressively increasing average global temperatures and creating more local extremes of both hot and cold weather. The ability of babies and children to thermoregulate is poorer than that of adults, making them particularly susceptible to extremes of heat and cold, both of which can affect their mental health. For example, there is evidence that exposure to extremes of temperature early in life can exacerbate symptoms of anxiety, depression and aggression and can impair learning. The underlying mechanisms are unknown and are likely to be complex, involving the effects of extreme temperature on many developmental processes. These effects are likely to combine with other risk factors, including other environmental and genetic ones, to determine how the development of any individual person's brain is affected.[141]

Summary

- The developing brain's environment is made up of one part that is internal to the body and another that is external to the body.
- Aspects of both the internal and external environment can be influenced by genetic factors, originating either within the individual themselves of from others.
- Epidemiology uses studies of large populations to identify external environmental factors that are potentially harmful or protective.
- Longitudinal studies of defined cohorts over time have given useful insights into potential environmental risks for the developing brain.
- Mendelian randomization can help infer causation between risk factors and brain phenotypes.
- Specific potentially harmful external or internal environmental factors include some chemicals, infections, metabolic disorders, autoimmune conditions, malnutrition, trauma and preterm birth.
- Complex factors that can influence the developing brain include socio-economic status, maternal stress, interpersonal relationships, the world of invisible microorganisms around us (the microbiome) and the climate.

6

Ultimate Causes of Specific Neurodevelopmental Conditions

In this chapter, I will discuss the ultimate genetic and environmental causes – insofar as we know them – of specific named disorders and types of neurodivergence. For the most part, these disorders and neurodivergent conditions are defined – or diagnosed, to use the clinical term – on the basis of their characteristic symptoms (remember from Chapter 3, symptom is a term we can use to refer to an abnormal or divergent phenotype). I described many of the main types of symptoms, both structural and functional, that our brains can develop in Chapter 3; please refer back if you encounter unfamiliar terms in this chapter, or try the glossary. Many of the clinical diagnoses I will discuss are syndromes (meaning that a particular group of symptoms, some of which might be outside the brain, co-occur in the same individual).

For many people diagnosed with disordered brain development or considered as neurodivergent, the ultimate causes will be at best unclear and often simply unknown. This is because most result from combinations of factors only some of which are known to us, some of which are suspected to be involved and some of which we are completely unaware of. These factors, which range from mainly genetic to mainly environmental, combine and interact in complicated ways that can vary greatly from person to person. Even if we suspect that the causes of a person's symptoms are mainly genetic, mainly environmental or are a product of significant levels of both, it is often very difficult to pin down those origins in individual cases. We are most likely to be able to determine the ultimate causes of well-defined conditions that reliably associate with a clearly identifiable genetic abnormality, such as a large chromosomal or monogenic variant, or a specific environmental agent, such as a drug with undesirable side effects that has been administered to a cohort of individuals. Such conditions are often congenital, meaning that they are present at birth*.

* It is important to avoid misusing or misinterpreting the term congenital. The word is purely descriptive of the time of life at which a condition was *certainly* present. Some disorders are called congenital even though the symptoms do not emerge for weeks, months or even years after birth, if we can be certain that the underlying condition was indeed present, but lay undetected, at birth. We do not usually apply it to disorders that *might* have their origins in prenatal development. Most importantly of all, it does not imply anything about the cause of the disorder, although you might see it used as if it does. For example, you might sometimes see the terms genetic disorder and congenital disorder used interchangeably as if they mean the same thing. This is very misleading. There are many examples of congenital disorders whose causation is mainly or perhaps entirely non-genetic.

Although the ultimate causes of most types of neurodevelopmental disorder and neurodivergence are multifactorial, the fact that genetics plays an overwhelmingly predominant role in some makes it more straightforward to discuss these conditions before returning to consider others where environmental factors make, or are thought to make, a significant contribution.

Disorders Caused Mainly by Genetic Variation

Early Discoveries of Chromosomal Abnormalities

DNA and chromosomes were discovered in the 19th century. We knew that chromosomes are the carriers of hereditary information for many decades before the elucidation of DNA's structure and function in the 1950s revealed how they do it. The earliest descriptions of chromosome abnormalities in humans, which also came in the 1950s, were made possible by the development, over the preceding 100 years or so, of a technique called **karyotyping** for viewing and counting the chromosomes with a light microscope (Chapter 4, Figure 4.4). The first chromosomal abnormalities to be described were those causing Down syndrome, Turner syndrome and Klinefelter syndrome.

Down syndrome is named after John Langdon Down, who had described this syndrome about 90 years earlier. Most of us know people with Down syndrome and recognize that aspects of their overall appearance can seem unusual. The characteristic features of the brains of individuals with Down syndrome often include, to a variable degree, micrencephaly, neurodevelopmental delay, attentional, behavioural and emotional problems, learning delays and difficulties, intellectual disability, seizures and early dementia. About 95% of the time, Down syndrome is caused by full **trisomy** of chromosome 21, meaning the person has three copies of chromosome 21, instead of the usual two copies, in all cells. The other 5% have some type of partial trisomy 21. Translocation Down syndrome is due to a portion of chromosome 21 breaking off and becoming attached (translocated; Figure 4.6) onto another chromosome: there are the usual two copies of chromosome 21, but there is additional genetic material from chromosome 21 attached to another chromosome. Mosaic Down syndrome is a rare form in which only some of a person's cells have an extra copy of chromosome 21. This results when the abnormal cell division that produces a trisomic daughter cell occurs after the **zygote** has started to divide. The effects of partial trisomy 21 tend to be less severe.

During the 1950s, scientists also used karyotyping to identify abnormalities of the **sex chromosomes** as the causes of Turner syndrome in females and Klinefelter syndrome in males. People with one or other of these syndromes can show cognitive disabilities, although they are often relatively minor. Turner syndrome, first described early in the 20th century, is due to partial or complete absence of one X chromosome. It is often designated 45, XO, meaning there are 45 instead of 46 chromosomes and one of them is X while the other is absent (or largely so). Klinefelter syndrome, first described in the 1940s, is due to an extra X chromosome in males. It is often designated 47, XXY. Hot on the heels of these findings came the discovery that trisomy 13 and trisomy 18 cause Patau syndrome and Edwards syndrome respectively. Both syndromes involve severe developmental abnormalities, including **micrencephaly** and, in Patau syndrome, **holoprosencephaly**. Miscarriage, stillbirth or early neonatal death are by far the most likely outcomes. A small minority with Edwards syndrome survives childhood but with intellectual disability. Although named relatively recently after the scientists who found these trisomies, the first description of Patau syndrome was way back in 1657.

As karyotyping methods improved, most notably through the use of stains that detect reproducible banding patterns characteristic of each chromosome, scientists began to identify more subtle, although still relatively large (about 5 million bases or more), chromosomal abnormalities including deletions, duplications, translocations or inversions as the causes of clinical syndromes with abnormalities of brain development (see Chapter 4). For example, one early discovery made in the 1960s was that the deletion of a variably sized part of chromosome 5 causes Cri du chat syndrome, in which infants emit shrill cries sounding like those of a cat; other symptoms include micrencephaly, **hypotonia**, neurodevelopmental delay and intellectual disability.

More Recent Advances in Understanding Chromosomal Abnormalities

The explosive growth of molecular biology in the 1970s and 1980s allowed scientists to develop other ways of searching for chromosomal abnormalities. These involve probing for specific sequences on the chromosomes. One of the first techniques was fluorescence in situ hybridization (often shortened to FISH, which is an appropriate acronym for a method that is fishing for sequences in the genome!). How does it work? The name gives it away (Figure 6.1). Essentially, it uses probes made of two parts. The first part is designed to find the sequences we are looking for in intact chromosomes (they work in situ, or on site) and bind to them by forming stable chemical bonds (we call this hybridization). This is achieved by making the probe sequence complementary to the sequence we are looking for. The second part is a molecular tag that fluoresces a particular colour. By using multiple probes each designed to find a different sequence and fluoresce a different colour, an investigator can work out whether parts of chromosomes are present, absent, duplicated or inverted. One example of its use was the identification in the 1990s of deletions in chromosome 7 as a cause of Williams syndrome, whose symptoms can include structural defects and functional abnormalities such as neurodevelopmental delay and intellectual disability.

Subsequent developments allowed variants to be detected with greater resolution in a method known as array comparative genomic hybridization (or aCGH for short) (Figure 6.2). This technique allows **copy number variations** (**CNVs**: see Chapter 4) to be detected with a resolution up to 100 times that of FISH, depending on how many probes are on the array (the more there are, the higher the resolution). Examples of its use include the detection of copy number variations in chromosome 17 in Smith–Magenis syndrome (whose symptoms can include structural defects and functional abnormalities such as intellectual disability, behaviour and sleep problems) and in chromosome 4 in Wolf–Hirschhorn syndrome (whose symptoms can include structural defects and functional abnormalities such as intellectual disability and seizures).

Table 6.1 contains a list of relatively large chromosomal abnormalities causing syndromes that include structural and/or functional brain disorders and, collectively, affect a little under 1% of newborns. Some syndromes are named using the nomenclature for the

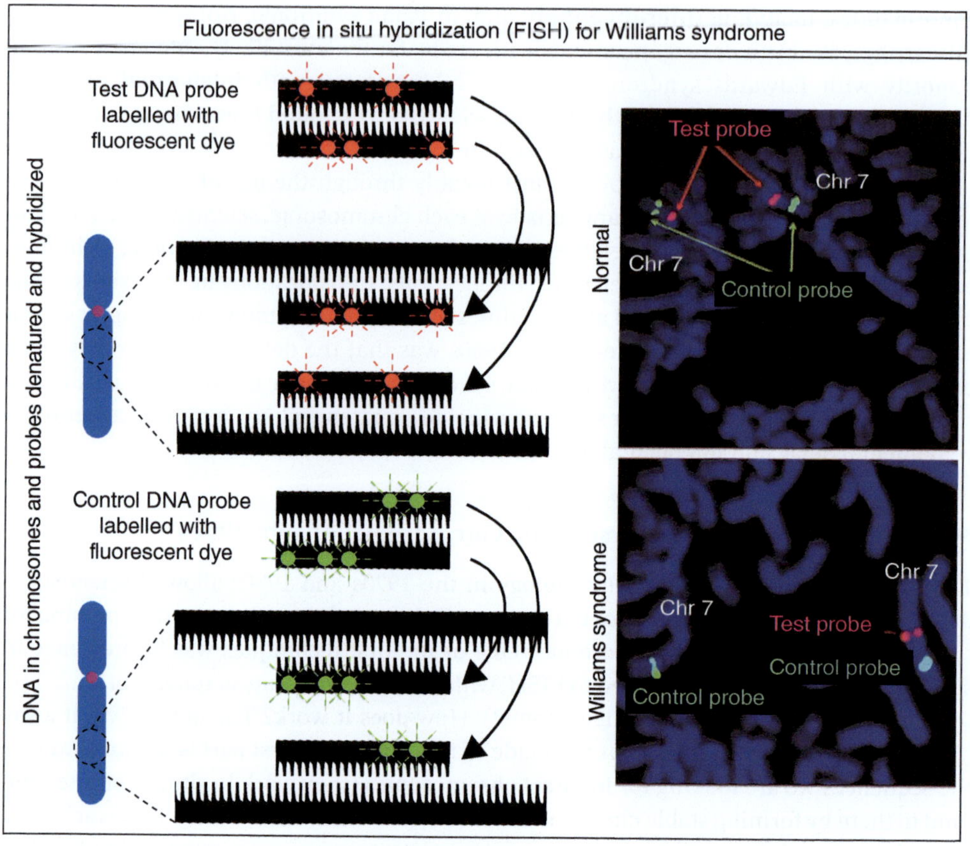

Figure 6.1 **Fluorescence in situ hybridization, detecting Williams syndrome.** Fluorescence in situ hybridization (FISH) is used in the laboratory to detect specific DNA sequences of interest on a chromosome. The researcher makes a fluorescently labelled probe that has the sequence of the chromosomal region of interest. They fix the full set of chromosomes from an individual to a glass slide and then expose them to the probe. The researcher first uses various biochemical methods to denature the DNA molecules in the chromosome and probe, which means they make their two strands come apart, and later allows them to get back together again. During this process, the probe will find and bind to its matching (complementary) sequence within the set of chromosomes, allowing the region of interest to be seen down the microscope. In the images, the test probe is to a region of chromosome 7 that is deleted in Williams syndrome and the control probe is to a different undeleted part. The images show both probes binding to both copies of normal chromosome 7 in a normal individual, but the test probe only binds to one copy of chromosome 7 in the person with Williams syndrome. *Hussein et al.*[142] */ Springer Nature / CC BY 4.0.*

affected chromosome or chromosomal region (explained in Figure 4.7 in Chapter 4). Others carry the name of the person or people who described it or discovered the underlying genetic cause. These syndromes are very likely to be polygenic in origin since the chromosomal abnormalities are so large that they affect multiple genes, although the phenotypic consequences of affecting each of these genes are likely to be greater for some than others. Working out which are the particularly crucial genes will require a huge amount of painstaking research involving gene manipulation in model organisms or cell culture systems to dissect the contributions of the affected genes individually and in combination.

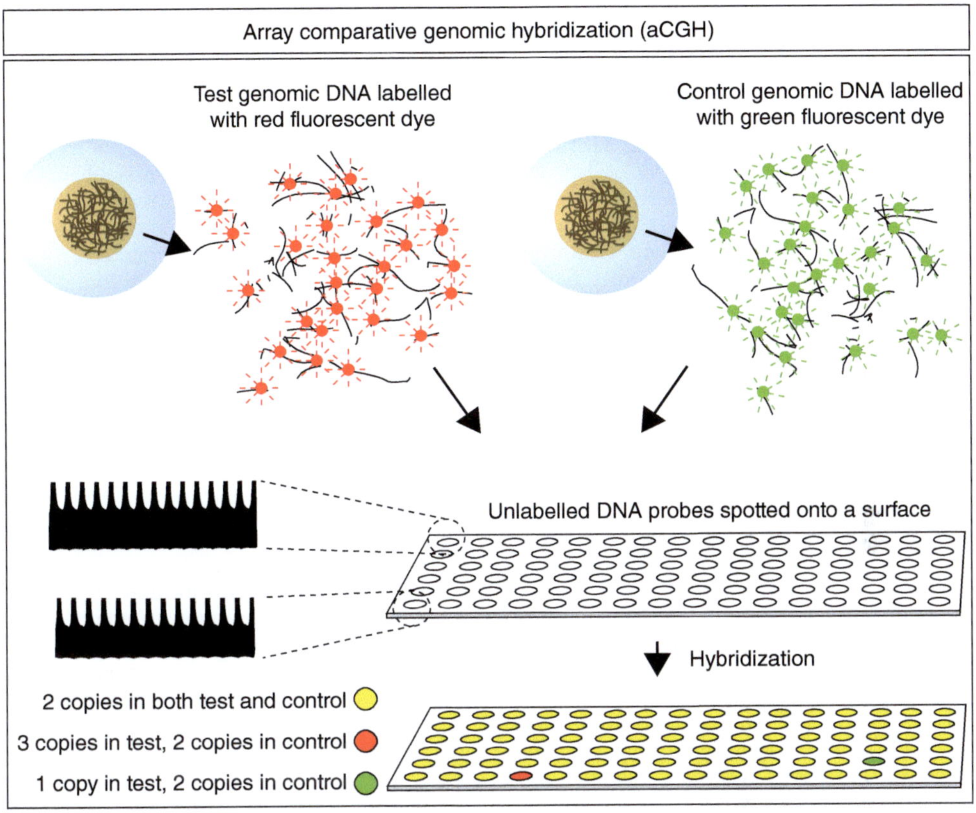

Figure 6.2 **Array comparative genomic hybridization.** In this method, large numbers of probes (which are unlabelled) that collectively cover a large number of genomic regions are stuck on glass slides in ordered microscopic arrays: each probe is a different tiny dot on the surface of the slide (in reality there would be many, many more microdots than shown here). In contrast to fluorescence in situ hybridization (Figure 6.1), it is the genomic DNA that is tagged fluorescently rather than the probes. DNA from a test subject, such as someone who we suspect has a variant affecting them (maybe someone with a neurodevelopmental disorder, for example), and DNA from someone who does not (a control subject, in other words) is added over the top of an array of stuck-down probes and allowed to hybridize. We know which DNA came from which person because we label the two samples with different fluorescent colours. By measuring the intensity of the two colours at each dot in the array (each dot is a probe representing a different, known part of the genome, remember), we can work out the relative amounts of each genomic region in our test versus our control sample. This allows us to deduce whether the test subject has a duplication, a deletion or neither of each region represented in the array. Examples of a duplication (red) and a deletion (green) are shown here; most regions are neither duplicated nor deleted (yellow).

Approaches to Understanding Monogenic Disorders

Now let us jump to the other end of the scale of genetic variants. Monogenic variation means variation in a single gene. Each individual monogenic disorder is rare but, since there are many thousands of them, collectively they are relatively common, affecting somewhere in the region of 5% of people. Usually, they have large effects on a person's phenotype, are highly **penetrant** and are inherited in a Mendelian fashion (these terms are

Table 6.1 Syndromes due to large chromosomal abnormalities.

SYNDROME AND/OR GENETIC ABNORMALITY	EFFECTS
1q21.1 deletion; 2p15p16.1 deletion; 2q23.1 deletion; 2q37 deletion; 3q29 deletion; 5q14.3 deletion; 7q11.23 duplication; 16p11.2 deletion; 16q24.3 deletion; 17q12 duplication; 17q23.1q23.2 deletion; 19p13.12 deletion; 20p12.3 deletion; 47, XYY; Cri du chat syndrome (deletions on 5p); Down syndrome (trisomy 21); Edwards syndrome (trisomy 18); Jacobsen syndrome (deletions on 11q); Miller–Dieker syndrome (deletions on 17p); Patau syndrome (trisomy 13); Prader–Willi syndrome (abnormalities of paternal chromosome 15); Smith–Magenis syndrome (deletions on 17p); Trisomy 9p; Wilms tumour–aniridia–genitourinary anomalies–mental disability syndrome (deletions on 11p); Williams syndrome (deletions on 7q); Wolf–Hirschhorn syndrome (deletions on 4p).	**Structural defects** include one or more of abnormal cortical gyration, cerebellar agenesis or dysgenesis, cerebral cortical atrophy, colpocephaly, corpus callosum agenesis or dysgenesis, heterotopias, macrencephaly, micrencephaly and ventriculomegaly. **Functional problems** include difficulty with one or more of attention, behaviour, cognition, communication, eating, intellect, learning, memory, mood, movement, muscle tone, rate of neurodevelopmental progress, seizures and sleep.
8p23.1 duplication; 8q12 duplication; 15q11.2 deletion; 15q13.3 deletion; 17q12 deletion; 22q11.2 deletion; 22q11.2 duplication; 22q13.3 deletion; 47, XXX; 48, XXXY; Xp11.22-p11.23 duplication; Klinefelter syndrome (47, XXY); Pallister–Killian mosaic syndrome (abnormalities of chromosome 12); Potocki–Lupski syndrome (duplications on 17p); Potocki–Shaffer syndrome (deletions on 11p); Turner syndrome (45, XO).	**Functional problems** include difficulty with one or more of attention, behaviour, cognition, communication, eating, intellect, learning, memory, mood, movement, muscle tone, rate of neurodevelopmental progress, seizures and sleep. **Structural Defects** are not usually apparent.

Many known syndromes are listed, but there will be others that I have missed. The nomenclature for chromosomal locations is in Figure 4.7, Chapter 4. Chromosomes' long arms are designated q and their short arms p.

explained in Chapter 4). Descriptions of them have been compiled in the Online Mendelian Inheritance in Man (OMIM) website.

During the 1980s and 1990s, research on monogenic conditions used **linkage** analysis and fine mapping. Linkage is a term used to describe a scenario in which two genes are close enough together on the same chromosome that their likelihood of being assorted and inherited together increases (remember, most **alleles** independently assort into the **germ cells** of the mother and father, with exceptions being those that show linkage: see discussion in the text around Figure 4.15 in Chapter 4). Linkage analysis aims to locate a gene that is causing a disorder by identifying co-inherited genetic markers whose chromosomal locations we know. A notable early success was the identification of the genetic basis of Huntington disease. In the 1970s and 1980s, researchers had defined the lines of ancestors (which we call the pedigrees) of large multigenerational families in the USA and Venezuela, some of whose members had inherited a genetic variant causing Huntington disease. They found that these individuals had co-inherited a **polymorphism** known to be on the short arm of chromosome 4, indicating to them that the Huntington disease-causing genetic variant was also somewhere in this region. Although close by relative to the overall size of a chromosome, 'somewhere in this region' was potentially up to 4 million bases away! Much painstaking work by many scientists was needed to narrow down its location.[143]

Things changed when, from 1990 to 2003, the Human Genome Project produced one of the greatest leaps forward in our understanding of human genetics.[144] By involving hundreds of scientists around the world in a large collaborative effort, it succeeded in giving us the first (almost complete) sequence of a human genome. It did not sequence the genome of one particular individual, but rather it amalgamated sequences from multiple individuals. Since then, the genomes of many individuals have been sequenced, giving us knowledge of the variation that exists between us. It is now possible to carry out whole genome sequencing on individuals in a fraction of the time and at a fraction of the cost expended on the Human Genome Project. This has been due to the development of so-called next-generation sequencing methods, summarized in Figure 6.3. Parallel advances in computer science have provided the means to analyse the vast amount of data produced by these projects.

The Human Genome Project and subsequent sequencing projects greatly accelerated progress in the search for monogenic variants causing phenotypic disorders by stimulating the invention of the new methods for DNA sequencing and new ways of handling the large amount of data produced. They provided lists of genes, the sequences of their exons and introns and their locations in the human genome. These lists created a reference allowing researchers to take the sequences they obtain, recognize them and identify their positions in the genome. This has made important new approaches possible, notably whole **exome** sequencing. The exome is the tiny portion (less than 2%) of the genome that contains all exons and, therefore, all its protein-coding parts (see Chapter 4). Since most known monogenic variants that cause disorders are concentrated in these regions, focusing next generation sequencing specifically on a person's exons is an efficient way of identifying variants likely to affect individuals. A powerful approach has been to apply this method to families in what is known as trio exome sequencing – the trio being the affected individual and their two parents – which can identify genetic variants that are likely to be the ultimate causes of a person's condition.[145] It can find **recessive** or **dominant** variants inherited from the parents as well as **de novo mutations** in the child. For example, a disorder in a child but not their parents might be suspected to be due to the inheritance of two harmful but recessive alleles if both parents are **heterozygous** and the child is **homozygous** for the allele, or due to a de novo mutation if neither parent carries it. A disorder in a child and one of their parents would suggest a dominant causative allele is likely to be identifiable in both of them but not in the other parent.

The complexity of analysing, comparing and interpreting the significance of the many differences between the exome sequences of even closely related individuals can result in many uncertainties about whether an exome variant is causative or not. The results from around two thirds of trios tested in this way fail to reach a strong conclusion about the likely ultimate cause of a person's condition. In some cases, this might happen because the cause is polygenic or multifactorial, which is too complex a scenario for trio exome sequencing to address and requires approaches such as **genome wide association studies (GWAS)**. In other cases, the causative variants might be outside exons; they might, for example, affect **non-coding RNAs** or **promoters** or **enhancers** (Chapter 4, Figures 4.9

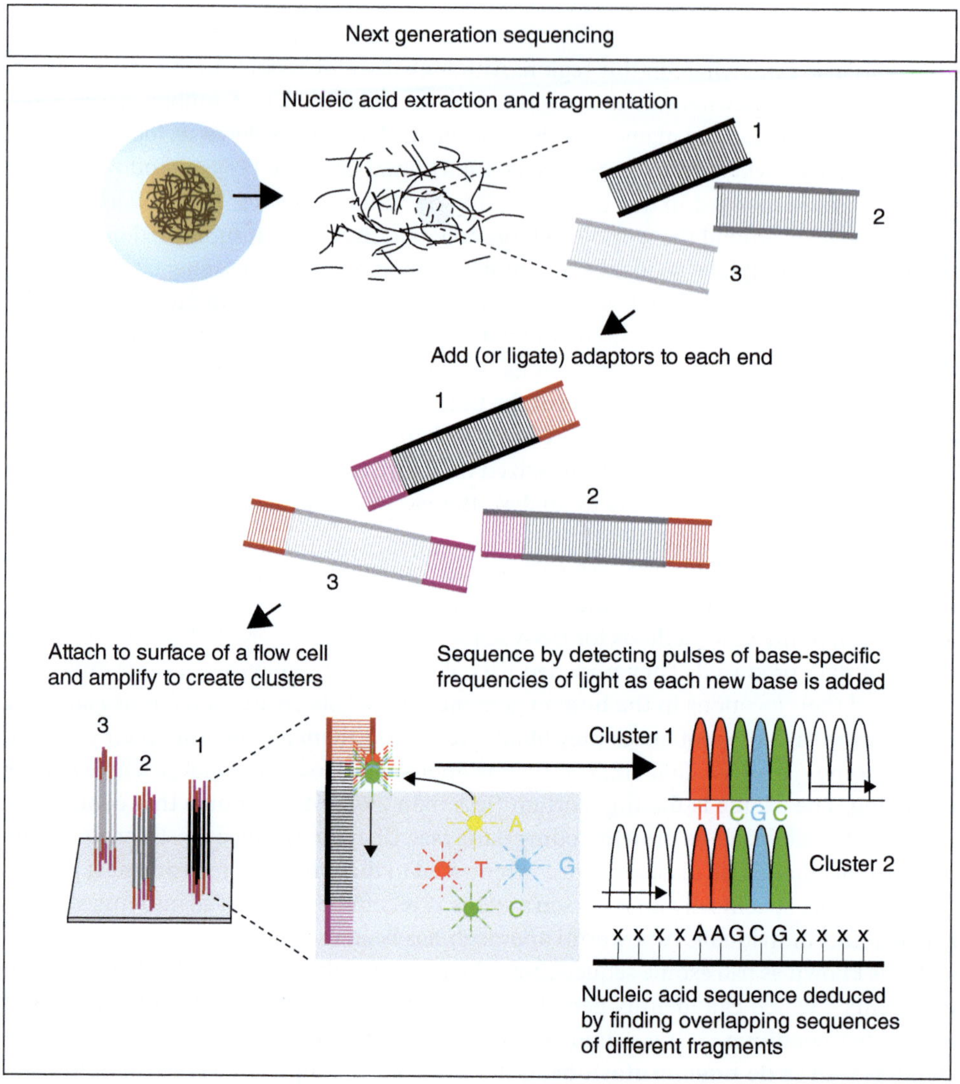

Figure 6.3 Next generation sequencing. Nucleic acids (DNA or RNA; in this illustration it is DNA) are extracted biochemically from cells and fragmented randomly into short sequences (just three of the millions of fragments that would be obtained are shown here). Adaptors (known sequences approximately 80 nucleotides long) are added to each end of each fragment. The adaptors prepare the fragments for replication and stick them to the surface of a flow cell. Through the flow cell are passed primers (which have been manufactured to bind to the adaptors), polymerase enzyme to catalyse nucleic acid synthesis starting at the primers and the nucleotides needed as the building blocks of new strands. Multiple rounds of replication create clusters, each containing many identical copies of the starting fragment plus its adaptors. The addition of nucleotides that fluoresce – each type a different colour – allows sequences of pulses of light of each of four wavelengths to be recorded as new strands are made. This allows us to deduce the sequences of the newly synthesized strands and, therefore, the sequences of their complementary strands. The discovery of the same unique sequences in more than one cluster allows partially overlapping sequences to be pieced together to work out the entire sequences of the very long nucleic acid molecules in our cells, from one end to the other (remember, fragmentation was random so there will be many overlaps between the fragments). Here, fragments 1 and 2 (which have become clusters 1 and 2) share a common sequence at a different place in each; in reality, unambiguous overlapping would be based on sequences much longer that 5 bases since any 5 base sequence will occur frequently in the genome, but I hope you follow the principle.

and 4.10), in which case sequencing the entire genome (whole genome sequencing) instead of just the exons can help. In practice, this is not only more expensive but generates much more potentially irrelevant data and, therefore, requires significantly more computing and statistical power. Whole genome sequencing really comes into its own for research involving information from the genomes of large numbers of people (discussed in Chapter 5).

One very important point to remember is that all of these methods *associate* genetic and phenotypic characteristics; they are not designed to prove that the former *causes* the latter. In some cases, it might be possible to use Mendelian randomization (Chapter 5) to gain insight into whether causation is likely. Ultimately, to test whether a particular variant causes a disorder and how it does so require controlled experiments, for example in model non-human species or in cultured cells, which might include human cells. As such approaches increase our understanding of the effects of monogenic variants, we can make better hypotheses as to whether any given variant is likely to be pathogenic. Meanwhile, many of the variants we have learned about still remain of uncertain significance.

Identified Monogenic Disorders

Tables 6.2–6.9 contain lists of known monogenic disorders. For some of these disorders (listed in Tables 6.2–6.8), there is evidence providing at least some level of certainty (even if the level is sometimes quite low) that variants of a particular gene (or sometimes one or other of a small set of genes) are causative and I have, therefore, named these genes in brackets after each condition. Tables 6.2–6.7 are organized, and internally subdivided, on the basis of key cellular processes whose disruption is likely to be a major proximate cause of the condition. For some, however, this is not possible (listed in Table 6.8). In addition, there are some disorders that are thought to be monogenic, but we are not yet confident of the genes involved (listed in Table 6.9); where there is evidence for the *types* of gene involved, I have indicated this in the table. Many such uncertainties are likely to be resolved through ongoing research. These tables provide a prelude to further discussion of how monogenic variants cause disorders in Part III.

Even Monogenic Disorders Might Be Polygenic

So far, in Chapter 4 and in this chapter, I have tended to draw a clear distinction between monogenic and polygenic causation. In fact, the distinction is likely to be more blurred than I have implied. Both polygenic and monogenic disorders, even those caused primarily by identical or similar variants that have indistinguishable effects on the molecular functions of the affected gene(s), can show diverse phenotypes with variable onset (for example, see Figure 6.4). Some individuals carrying variants known to be potentially deleterious can appear completely healthy (you might recall that I discussed these issues of penetrance and expressivity in Chapter 4).

There are several possible explanations for such phenotypic variation. One is based on the fact that each of us carry different variants of genes that interact with and *modify* the functions of the gene(s) primarily implicated in causing the phenotypes; in short, we say that the actions of genes can include *modifier* functions, and most genes are likely to have modifier functions in some circumstances. This explanation implies that multiple potentially unknown genes are involved in generating the phenotype even in so-called

Table 6.2 Monogenic conditions where variants affect DNA-related activities.

CELLULAR PROCESSES AND DISORDERS (AFFECTED GENES)	EFFECTS
DNA replication and repair Cockayne syndrome (*ERCC6* and *ERCC8*); Juberg–Hayward syndrome (*ESCO2*); Nijmegen breakage syndrome (*NBN*). **Chromatin remodelling/methylation/histone acetylation** Alpha-thalassemia X-linked intellectual disability syndrome (*ATRX*); Bainbridge–Ropers syndrome (*ASXL3*); Bohring–Opitz syndrome (*ASXL1*); GATAD2B-associated neurodevelopmental disorder (*GATAD2B*); Genitopatellar syndrome (*KAT6B*); Kabuki syndrome (*KMT2D* and *KDM6A*); Koolen de Vries syndrome (*KANSL1*); Nicolaides–Baraitser syndrome (*SMARCA2*); Rett syndrome (*MECP2*). **Transcription factors or co-factors acting on genomic regulatory elements** FOXG1 syndrome (*FOXG1*); Greig cephalopolysyndactyly syndrome (*GLI3*); Nonsyndromic holoprosencephaly (*TGIF1*); Pitt–Hopkins syndrome (*TCF4*); Proud syndrome (*ARX*); Syndromic microphthalmia, type 3 (*SOX2*); X-linked lissencephaly with abnormal genitalia (*ARX*).	**Structural defects** include one or more of abnormal cortical gyration, cerebellar agenesis or dysgenesis, cerebellar atrophy, cerebral cortical atrophy, colpocephaly, corpus callosum agenesis or dysgenesis, dysmyelination, hydrocephalus, intracranial calcifications, macrencephaly, micrencephaly and ventriculomegaly. **Functional problems** include difficulty with one or more of attention, behaviour, cognition, communication, eating, intellect, learning, memory, mood, movement, muscle tone, rate of neurodevelopmental progress, seizures and sleep.
DNA replication and repair Ataxia telangiectasia (*ATM*); Ataxia with oculomotor apraxia (*APTX*, *SETX*, *PNKP*) **Chromatin remodelling/methylation/histone acetylation** Autosomal dominant cerebellar ataxia deafness and narcolepsy (*DNMT1*); CHD2 myoclonic encephalopathy (*CHD2*); Floating–Harbor syndrome (*SRCAP*); SETBP1 disorder (*SETBP1*); MECP2 duplication syndrome (*MECP2*); PPM-X syndrome (*MECP2*); Sotos syndrome (*NSD1*); Wiedemann–Steiner syndrome (*KMT2A*). **Transcription factors or co-factors acting on genomic regulatory elements** Benign hereditary chorea (*NKX2-1*); Borjeson–Forssman–Lehmann syndrome (*PHF6*); Dentatorubral–pallidoluysian atrophy (*ATN1*); HIVEP2-related intellectual disability (*HIVEP2*); Intellectual disability-severe speech delay-mild dysmorphism syndrome (*FOXP1*); Marshall–Smith syndrome (*NFIX*); MED13L haploinsufficiency syndrome (*MED13L*); Myhre syndrome (*SMAD4*); Partington syndrome (*ARX*); SATB2-associated syndrome (*SATB2*).	**FUNCTIONAL PROBLEMS** include difficulty with one or more of attention, behaviour, cognition, communication, eating, intellect, learning, memory, mood, movement, muscle tone, rate of neurodevelopmental progress, seizures and sleep. **Structural defects** are not usually apparent.

DNA-related activities include DNA replication and repair, **chromatin** remodelling and the actions of **transcription factors** or co-factors on **regulatory elements** (see Chapter 4 for discussion of these activities). Note that the cellular processes under which disorders and genes are listed are prominent ones, but most variants are likely to affect multiple other processes as well.

Table 6.3 Monogenic conditions where variants affect RNA or protein synthesis or processing.

CELLULAR PROCESSES AND DISORDERS (AFFECTED GENES)	EFFECTS
RNA production, processing, transport or degradation Fragile X syndrome (*FMR1*); Mandibulofacial dysostosis with microcephaly (*EFTUD2*); Microcephalic osteodysplastic primordial dwarfism type 1 (*RNU4ATAC*); Pontocerebellar hypoplasia type 4 (*TSEN54*); Renpenning syndrome (*PQBP1*); RNAse T2-deficient leukoencephalopathy (*RNASET2*); *RNU4-2* syndromic neurodevelopmental disorder (*RNU4-2*); Syndromic X-linked mental retardation of the Snijders Blok type (*DDX3X*). **Protein synthesis or folding** Hypoparathyroidism–intellectual disability–dysmorphism syndrome (*TBCE*); Marinesco–Sjogren syndrome (*SIL1*); MEHMO syndrome (*EIF2S3*). **Protein modification by side group addition** Fukuyama type muscular dystrophy (*FKTN*); Multiple congenital anomalies–hypotonia–seizures syndrome (*PIGN*); Muscle–eye–brain disease (*POMGnT1*); Spastic paraplegia 26 (*B4GALNT1*); Other congenital disorders of glycosylation (*ALG1*, *ALG3*, *ALG9*, *ALG11*, *ALG12*, *ALG13*, *DPAGT1*, *MGAT2*, *PMM2*, *RFT1*, *SLC35A2*, *SLC35C1* and *SSR4*).	**Structural defects** include one or more of cerebellar agenesis or dysgenesis, cerebellar atrophy, cerebral cortical atrophy, cerebral dysplasia, dysmyelination, hydrocephalus, macrencephaly, micrencephaly and ventriculomegaly. **Functional problems** include difficulty with one or more of attention, behaviour, cognition, communication, eating, intellect, learning, memory, mood, movement, muscle tone, rate of neurodevelopmental progress, seizures and sleep.
Protein modification by side group addition Mucolipidosis III alpha/beta (*GNPTAB*); Other congenital disorders of glycosylation (*ALG6*, *ALG8*, *MAN1B1*, *MOGS* and *SRD5A3*).	**Functional Problems** include difficulty with one or more of attention, behaviour, cognition, communication, eating, intellect, learning, memory, mood, movement, muscle tone, rate of neurodevelopmental progress, seizures and sleep. **Structural defects** are not usually apparent.

RNA or protein synthesis or processing includes RNA production, processing, transport or degradation, protein synthesis or folding and protein modification by side group addition (see Chapter 4 and Part III for discussion of these processes). Note that the cellular processes under which disorders and genes are listed are prominent ones, but most variants are likely to affect multiple other processes as well.

Table 6.4 Monogenic conditions where variants affect structural or membrane proteins.

CELLULAR PROCESSES AND DISORDERS (AFFECTED GENES)	EFFECTS
Extracellular matrix proteins Familial porencephaly (*COL4A1*); Marshall syndrome (*COL11A1*). **Cytoskeletal protein** Alexander disease (*GFAP*); Autosomal dominant leukodystrophy with autonomic disease (*LMNB1*); Baraitser–Winter syndrome (*ACTB*); Spastic ataxia Charlevoix-Saguenay type (*SACS*); Subcortical band heterotopia (*DCX, PAFAH1B1* [*LIS1*]); X-linked periventricular heterotopia (*FNLA*). **Mitotic spindle protein** Filippi syndrome (*CKAP2L*); Lymphedema, microcephaly and chorioretinopathy syndrome (*KIF11*). **Myelin protein** Pelizaeus–Merzbacher disease (*PLP1*). **Membrane channel proteins** Gillespie syndrome (*ITPR1*). **Membrane transport proteins** Allan–Herndon–Dudley syndrome (*SLC16A2*); Andermann syndrome (*SLC12A6*); Christianson syndrome (*SLC9A6*); Glucose transporter type 1 deficiency syndrome (*SLC2A1*); Hypermanganesemia with dystonia polycythaemia and cirrhosis (*SLC30A10*); Occipital horn syndrome (*ATP7A*); Primary carnitine deficiency (*SLC22A5*); X-linked intellectual disability Najm type (*CASK*). **Cell adhesion protein** Bilateral frontoparietal polymicrogyria (*ADGRG1*); C syndrome (*CD96*); Graham–Cox syndrome (*IGBP1*); L1 syndrome (*L1CAM*).	**Structural Defects** include one or more of abnormal cortical gyration, cerebellar agenesis or dysgenesis, cerebral cortical atrophy, corpus callosum agenesis or dysgenesis, dysmyelination, heterotopias, intracranial calcifications, macrencephaly, micrencephaly and porencephaly. **Functional problems** include difficulty with one or more of behaviour, cognition, communication, eating, intellect, learning, mood, movement, muscle tone, rate of neurodevelopmental progress and sleep.

Cytoskeletal protein DNM1 epileptic encephalopathy (*DNM1*) **Myelin protein** Spastic paraplegia 2 (*PLP1*) **Membrane channel proteins** Birk-Barel syndrome (*KCNK9*); Convulsions benign familial neonatal dominant form (*KCNQ2*, *KCNQ3*); Dravet syndrome (*SCN1A*); *GRIN2B*-related neurodevelopmental disorder (*GRIN2B*); Hyperkalemic periodic paralysis (*SCN4A*); SeSAME syndrome (*KCNJ10*) **Membrane transport proteins** Alternating hemiplegia of childhood (*ATP1A3*); Dopamine transporter deficiency syndrome (*SLC6A3*); Hereditary folate malabsorption (*SLC46A1*); Rapid-onset dystonia-parkinsonism (*ATP1A3*); Riboflavin transporter deficiency (*SLC52A2*, *SLC52A3*); Spastic paraplegia 6 (*NIPA1*); X-linked creatine deficiency (*SLC6A8*) **Cell adhesion protein** PCDH19-related female-limited epilepsy (*PCDH19*); Zlotogora syndrome (*PVRL1*)	**FUNCTIONAL PROBLEMS** include difficulty with one or more of: behaviour, cognition, communication, eating, intellect, learning, mood, movement, muscle tone, rate of neurodevelopmental progress, sleep.

Structural or membrane proteins include those involved in the **extracellular matrix**, the **cytoskeleton**, **mitotic spindles**, **myelin**, membrane channels, membrane transport and cell adhesion (the functions of some of these types of protein, such as membrane channel proteins, were discussed in Chapter 2; I will cover others in Part III). Note that the cellular processes under which disorders and genes are listed are prominent ones, but most variants are likely to affect multiple other processes as well.

Table 6.5 Monogenic conditions where variants affect cell signalling.

CELLULAR PROCESSES AND DISORDERS (AFFECTED GENES)	EFFECTS
Cilia protein Acrocallosal syndrome Schinzel type (*KIF7*); COACH syndrome (*TMEM67*); Hydrolethalus syndrome (*HYLS1*). **Signalling protein** Achondroplasia (*FGF3*); Apert syndrome (*FGFR2*); CADASIL (*NOTCH3*); Crouzon syndrome (*FGFR2*); Curry Jones syndrome (*SMO*); Nonsyndromic holoprosencephaly (*SHH*, *FGF8*, *FGFR1*). **Second messenger system protein** Bannayan–Riley–Ruvalcaba syndrome (*PTEN*); Chudley–Mccullough syndrome (*GPSM2*); Costello syndrome (*HRAS*); IQSEC2 (*IQSEC2*); Martsolf syndrome (*RAB3GAP2*); OPHN1 syndrome (*OPHN1*); Tuberous sclerosis complex (*TSC1*, *TSC2*).	**Structural defects** include one or more of abnormal cortical gyration, cerebellar agenesis or dysgenesis, cerebral cortical atrophy, cerebral dysplasia, colpocephaly, corpus callosum agenesis or dysgenesis, dysmyelination, hydrocephalus, macrencephaly, micrencephaly and ventriculomegaly. **Functional problems** include difficulty with one or more of: attention, behaviour, cognition, communication, eating, intellect, mood, movement, muscle tone, rate of neurodevelopmental progress and seizures.
Signalling protein Neuronal intranuclear inclusion disease (*NOTCH2NLC*); Norrie disease (*NDP*). **Second messenger system protein** SYNGAP1-related intellectual disability (*SYNGAP1*).	**Functional problems** include difficulty with one or more of attention, behaviour, cognition, communication, eating, intellect, mood, movement, muscle tone, rate of neurodevelopmental progress and seizures. **Structural defects** are not usually apparent.

Cell signalling proteins include **signalling molecules** themselves, proteins involved in the structure and functions of cilia and those that operate in the cell's **second messenger** systems (I will be covering all these topics in Chapter 7). Note that the cellular processes under which disorders and genes are listed are prominent ones, but most variants are likely to affect multiple other processes as well.

Table 6.6 Monogenic conditions where variants affect cell metabolism.

CELLULAR PROCESSES AND DISORDERS (AFFECTED GENES)	EFFECTS
Protein degradation Angelman syndrome (*UBE3A*); Giant axonal neuropathy (*GAN*); ITCH E3 ubiquitin ligase deficiency (*ITCH*); Kaufman oculocerebrofacial syndrome (*UBE3B*); Spastic paraplegia 18 (*ERLIN2*). **Cellular trafficking** Acrodermatitis enteropathica (*SLC39A4*); Adrenomyeloneuropathy (*ABCD1*); Early-onset parkinsonism-intellectual disability syndrome (*RAB39B*); Griscelli syndrome type 1 (*MYO5A*); Mucolipidosis type 4 (*MCOLN1*); Northern epilepsy (*CLN8*); Troyer syndrome (*SPART*); X-linked cerebral adrenoleukodystrophy (*ABCD1*) **Autophagy** Autosomal recessive spastic paraplegia type 49 (*TECPR2*); Beta-Propeller Protein-Associated Neurodegeneration (*WDR45*); Spastic paraplegia 15 (*ZFYVE26*); Vici syndrome (*EPG5*). **Lysosomal degradation** GM1 gangliosidosis (*GLB1*); Hurler syndrome (or mucopolysaccharidosis type I; *IDUA*); Hunter syndrome (or mucopolysaccharidosis type II; *IDS*); Metachromatic leukodystrophy (*ARSA*). **Other metabolic processes** Adenylosuccinase deficiency (*ADSL*); Arts syndrome (*PRPS1*); Canavan disease (*ASPA*); D-glyceric acidemia (*GLYCTK*); Fatty acid hydroxylase-associated neurodegeneration (*FA2H*); Glutamine deficiency, congenital (*GLUL*); Lowe oculocerebrorenal syndrome (*OCRL*); Maternal hyperphenylalaninaemia (*PAH*); Multiple sulfatase deficiency (*SUMF1*); NBIA/DYT/PARK-PLA2G6 (*PLA2G6*); Pyridoxal 5'-phosphate-dependent epilepsy (*PNPO*); Pyridoxine-dependent epilepsy (*ALDH7A1*); Spastic paraplegia 5A (*CYP7B1*).	**Structural Defects** include one or more of abnormal cortical gyration, cerebellar agenesis or dysgenesis, cerebellar atrophy, cerebral cortical atrophy, cerebral dysplasia, corpus callosum agenesis or dysgenesis, dysmyelination, heterotopias, hydrocephalus, intracranial calcifications, macrencephaly, micrencephaly and ventriculomegaly. **Functional problems** include difficulty with one or more of attention, behaviour, cognition, communication, eating, intellect, learning, memory, mood, movement, muscle tone, rate of neurodevelopmental progress, seizures and sleep.

(*Continued*)

Table 6.6 (Continued)

CELLULAR PROCESSES AND DISORDERS (AFFECTED GENES)	EFFECTS
Protein degradation Schaaf–Yang syndrome (*MAGEL2*). **Cellular trafficking** Aceruloplasminaemia (*CP*); Ataxia with vitamin E deficiency (*TTPA*); Biotin–thiamine-responsive basal ganglia disease (*SLC19A3*); Free sialic acid storage disease (*SLC17A5*); GOSR2-related progressive myoclonus ataxia (*GORS2*); Niemann–Pick disease type C (*NPC1* and *NPC2*); PACS1-related syndrome (*PACS1*). **Lysosomal degradation** Alpha-mannosidosis (*MAN2B1*); Argininosuccinic aciduria (*ASL*); Aspartylglycosaminuria (*AGA*); Fucosidosis (*FUCA1*); Galactosialidosis (*CTSA*); Gaucher disease (*GBA*); Glucosidase acid-1,4-alpha deficiency (*GAA*); MPSIII (or mucopolysaccharidosis type III; *GNS*, *HGSNAT*, *NAGLU* and *SGSH*); Schindler disease type 1 (*NAGA*); Sialidosis, type II (*NEU1*); Spinal muscular atrophy–progressive myoclonic epilepsy syndrome (*ASAH1*); Tay–Sachs disease (*HEXA*); Unverricht–Lundborg disease (*CSTB*). **Neurotransmitter processing** Monoamine oxidase A deficiency (*MAOA*); Sepiapterin reductase deficiency (*SPR*); Tyrosine hydroxylase deficiency (*TH*). **Other metabolic processes** 6-pyruvoyl-tetrahydropterin synthase deficiency (*GCH1*, *PCBD1*, *PTS* and *QDPR*); ADCY5-related dyskinesia (*ADCY5*); Arginase deficiency (*ARG1*); Ataxia–hypogonadism–choroidal dystrophy or Boucher–Neuhäuser syndrome (*PNPLA6*); Crigler–Najjar syndrome (*UGT1A1*); GTPCH1-deficient DRD (*GCH1*); Guanidinoacetate methyltransferase deficiency (*GAMT*); Hydroxykynureninuria (*KYNU*); Laurence–Moon syndrome (*PNPLA6*); Lesch–Nyhan syndrome (*HPRT1*); MEGDEL syndrome (*SERAC1*); Orotic aciduria type 1 (*UMPS*); Phenylketonuria (*PAH*); Refsum disease, infantile form (*PHYH*); Sjogren–Larsson syndrome (*ALDH3A2*); Thyrotropin deficiency, isolated (*TSHB*); Trichomegaly with intellectual disability, dwarfism and pigmentary degeneration of retina (*PNPLA6*).	**Functional problems** include difficulty with one or more of attention, behaviour, cognition, communication, eating, intellect, learning, memory, mood, movement, muscle tone, rate of neurodevelopmental progress, seizures and sleep. **Structural defects are** not usually apparent.

Metabolic processes include protein degradation, cellular trafficking, **autophagy** (in which a cell breaks down old, damaged or abnormal proteins and other substances in its cytoplasm), lysosomal degradation (the lysosome is a cellular organelle involved in degrading unwanted molecules), neurotransmitter processing and others. I will be discussing these processes in Part III. Note that the cellular processes under which disorders and genes are listed are prominent ones, but most variants are likely to affect multiple other processes as well.

Table 6.7 Monogenic conditions in which variants affect mitochondria.

CELLULAR PROCESSES AND DISORDERS (AFFECTED GENES)	EFFECTS
Protein synthesis or degradation Spastic paraplegia 7 (*SPG7*). **Membrane transport** Mohr–Tranebjaerg syndrome (*TIMM8A*). **Other metabolic processes** Alpha-ketoglutarate dehydrogenase deficiency (*OGDH*); Combined D,L-2-hydroxyglutaric aciduria (*SLC25A1*); Ethylmalonic encephalopathy (*ETHE1*); Hyperprolinaemia (*ALDH4A1* and *PRODH*); Glycine encephalopathy (*GLDC*, *AMT*); Infantile cerebellar retinal degeneration (*ACO2*); L-2-hydroxyglutaric aciduria (*D2HGDH*, *L2HGDH*); Mitochondrial complex I deficiency (*FOXRED1*, *NDUFAF8*, *NDUFA6*, *NDUFAF2*, *NDUFS1*, *NDUFS2*, *NDUFS3*, *NDUFS4*, *NDUFS6*, *NDUFS7*, *NDUFS8*, *NDUFV1*, *NDUFV2*, *NDUFA1*, *NDUFAF4*, *NDUFAF5*, *NDUFA11*, *NDUFAF3*, *NDUFAF1*, *NDUFB9*, *NDUFB3*, *NDUFB10*, *NUBPL*, *TIMMDC1*, *TMEM126B*, *MT-ND1*, *MT-ND2* and *MT-ND3*); Mitochondrial DNA depletion syndrome, encephalomyopathic form with methylmalonic aciduria (*SUCLA2*); Mitochondrial DNA-associated Leigh syndrome (*MT-ATP6*, *MT-ND1*, *MT-ND2*, *MT-ND3*, *MT-ND4*, *MT-ND5*, *MT-ND6*, *MT-TL1*, *MT-TK*, *MT-TW* and *MT-TV*); Pyruvate decarboxylase deficiency (*PDHA1* and *LONP1*); Pyruvate dehydrogenase complex deficiency (*PDHA1*, *PDHB*, *DLAT*, *PDHX* and *PDP1*).	**Structural defects** include one or more of cerebellar agenesis or dysgenesis, cerebellar atrophy, cerebral cortical atrophy, cerebral dysplasia, corpus callosum agenesis or dysgenesis, dysmyelination, hydrocephalus, intracranial calcifications, macrencephaly, micrencephaly and ventriculomegaly. **Functional** problems include difficulty with one or more of: attention, behaviour, cognition, communication, eating, intellect, learning, mood, movement, muscle tone, rate of neurodevelopmental progress and seizures.
Mitochondrial DNA or RNA production Alpers–Huttenlocher syndrome (*POLG*); Ataxia neuropathy spectrum (*POLG*); Autosomal dominant progressive external ophthalmoplegia (*POLG*); Autosomal recessive progressive external ophthalmoplegia (*POLG*); Autosomal recessive spastic ataxia 4 (*MTPAP*); Childhood myocerebrohepatopathy spectrum (*POLG*); HSD10 disease (*HSD17B10*) ; Myoclonic epilepsy myopathy sensory ataxia (*POLG*). **Protein synthesis or degradation** 3-methylglutaconyl-CoA hydratase deficiency (*AUH*); HMG CoA lyase deficiency (*HMGCL*); Leukoencephalopathy with brain stem and spinal cord involvement and lactate elevation (*DARS2*); Myoclonic epilepsy with ragged red fibres (*MT-TK*). **Membrane transport** X-linked sideroblastic anaemia with ataxia (*ABCB7*).	**Functional problems** include difficulty with one or more of attention, behaviour, cognition, communication, eating, intellect, learning, mood, movement, muscle tone, rate of neurodevelopmental progress and seizures. **Structural defects** are not usually apparent.

(Continued)

Table 6.7 (Continued)

CELLULAR PROCESSES AND DISORDERS (AFFECTED GENES)	EFFECTS
Other metabolic processes 2-methylbutyryl-CoA dehydrogenase deficiency (*ACADSB*); Autosomal recessive spinocerebellar ataxia 9 (*SCA9*); Cerebelloparenchymal disorder (*PMPCA*); D-2-hydroxyglutaric aciduria (*D2HGDH* and *IDH2*); Friedreich ataxia (*FXN*); HIBCH deficiency (*HIBCH*); Fumarase deficiency (*FH*); Gamma aminobutyric acid transaminase deficiency (*ABAT*); Glutaric acidemia (*GCDH*); Leber hereditary optic neuropathy with dystonia (*MT-ND1*, *MT-ND4*, *MT-ND4L* and *MT-ND6*); Mitochondrial complex II deficiency (*SDHA*, *SDHB*, *SDHD* and *SDHAF1*); Mitochondrial complex III deficiency (*MT-CYB*, *MT-CYB*, *BCS1L*, *CYC1*, *LYRM7*, *TTC19*, *UQCRQ*, *UQCRB*, *UQCRC2*, *UQCC2*, *UQCC3* and *UQCRFS1*); Mitochondrial complex IV deficiency (*SURF1*, *APOPT1*, *COA3*, *COA5*, *COA6*, *COX4I1*, *COX5A*, *COX6A2*, *COX6B1*, *COX8A*, *COX10*, *COX14*, *COX15*, *COX20*, *COXFA4*, *LRPPRC*, *PET100*, *PET117*, *SCO1*, *SCO2*, *TACO1*, *MT-CO1*, *MT-CO2*, *MT-CO3*, *MT-TS1*, *MT-TL1* and *MT-TN*); Mitochondrial complex V deficiency (*ATP5F1E*, *ATP5F1A*, *ATP5F1D*, *ATP5MD*, *TMEM70*, *MT-ATP6* and *MT-ATP8*); N-acetylglutamate synthase deficiency (*NAGS*); Pantothenate kinase-associated neurodegeneration (*PANK2*); Pyruvate carboxylase deficiency (*PC*); Succinic semialdehyde dehydrogenase deficiency (*ALDH5A1*).	

Mitochondrial processes include mitochondrial DNA and RNA production, protein synthesis and degradation and membrane transport. Discussion of mitochondrial DNA was included in Chapter 4, and I will have more to say about mitochondria and their functions in Part III. Note that the cellular processes under which disorders and genes are listed are prominent ones, but most variants are likely to affect multiple other processes as well.

Table 6.8 Monogenic disorders in which the function of the affected gene is unknown.

DISORDERS (AFFECTED GENES)	EFFECTS
Feingold syndrome (*MYCN* or *MIR17HG*); Hypomyelination and congenital cataract (*FAM126A*); Megalencephalic leukoencephalopathy with subcortical cysts (*MLC1* or *HEPACAM*); Myelocerebellar disorder (*SAMD9L*); Spastic paraplegia 11 (*SPG11*); Spastic paraplegia 32 (locus on chromosome 14q12-q21); Temtamy syndrome (*C12orf57*).	**Structural defects** include one or more of cerebellar agenesis or dysgenesis, cerebellar atrophy, cerebral cortical atrophy, corpus callosum agenesis or dysgenesis, dysmyelination, macrencephaly and micrencephaly. **Functional problems** include difficulty with one or more of behaviour, cognition, communication, eating, intellect, learning, memory, mood, movement, muscle tone, perception, rate of neurodevelopmental progress and seizures.
CDKL5 deficiency disorder (*CDKL5*); Fragile XE syndrome (*AFF2*); Huntington disease (*HTT*); Paroxysomal nonkinesigenic dyskinesia (*PNKD*); Woodhouse Sakati syndrome (*DCAF17*).	**Functional problems** include difficulty with one or more of behaviour, cognition, communication, eating, intellect, learning, memory, mood, movement, muscle tone, perception, rate of neurodevelopmental progress and seizures. **Structural defects are** not usually apparent.

The nomenclature for chromosomal locations is in Figure 4.7, Chapter 4. Chromosomes' long arms are designated q and their short arms p.

monogenic disorders; in other words, they are actually polygenic. You might see the term oligogenic used instead of polygenic in situations where only a few modifying genes are known to be involved (*oligo-* is Greek for few), although whether this is particularly useful seems debatable since there might be many unknown modifying genes that are likely to be very hard to identify.

Do not forget that variability in other factors might contribute to the phenotypic variation under discussion here, including those arising from random variation in the developmental processes that create our brains or from the brain's environment. This brings me on to the next section, where I bring the environment into play.

Multifactorial Causes of Disorder and Divergence

The Origins of Most Disorder and Divergence Are Multifactorial

In the last section, I argued that most genetic disorders of brain development, even those generally thought of as monogenic, are probably to some degree polygenic. In fact, things are more complicated still because the ultimate causes of most developmental brain disorder and divergence are bound to be multifactorial, at least to some extent. In other words, they almost certainly involve a combination of both genetic and environmental factors, even if the contribution from one or other is tiny and undefined.

The relative contributions that genetic versus environmental factors make to each particular type of disorder or neurodivergence is hugely variable. At one end of a spectrum are

Table 6.9 Disorders that are probably monogenic, but evidence for the genes involved is weak.

DISORDERS	PHENOTYPES
Acquired acrodermatitis enteropathica; Aicardi syndrome; Aicardi–Goutieres syndrome; Autosomal recessive primary microcephaly; Cornelia de Lange syndrome[a]; Dandy–Walker syndrome; Deafness, dystonia and cerebral hypomyelination; Dubowitz syndrome; Dysgnathia complex; Early Infantile Epileptic Encephalopathy; FG syndrome; Fryns syndrome; Galloway–Mowat syndrome; Genoa syndrome; Hoyeraal–Hreidarsson syndrome; Joubert syndrome[b]; Juvenile amyotrophic lateral sclerosis; Leigh syndrome[c]; Megalencephaly–Polymicrogyria–Polydactyly–Hydrocephalus syndrome[d]; Micro syndrome[e]; Neu Laxova syndrome[f]; Oculocerebrocutaneous syndrome; Pachygyria–intellectual disability–epilepsy syndrome; Paine syndrome; Pontocerebellar hypoplasia spectrum; Pontocerebellar hypoplasia type 2; Septo-optic dysplasia spectrum; Spastic paraplegia 51[e]; Steinfeld syndrome; Walker–Warburg syndrome[g]; X-linked intellectual disability-dysmorphism-cerebral atrophy syndrome; Zellweger syndrome[h]	**Structural defects** include one or more of abnormal cortical gyration, cerebellar agenesis or dysgenesis, cerebellar atrophy, cerebral cortical atrophy, cerebral dysgenesis, corpus callosum agenesis or dysgenesis, dysmyelination, heterotopias, holoprosencephaly, hydrocephalus, macrencephaly, micrencephaly and ventriculomegaly. **Functional problems** include difficulty with one or more of: attention, behaviour, cognition, communication, eating, intellect, learning, memory, mood, movement, muscle tone, rate of neurodevelopmental progress, seizures and sleep.
Atypical Rett syndrome; Autosomal dominant nocturnal frontal lobe epilepsy; Autosomal dominant partial epilepsy with auditory features; Bardet–Biedl syndrome[b]; Benign rolandic epilepsy; Carpenter syndrome; Childhood apraxia of speech; Congenital hypothyroidism; Congenital mirror movement disorder; Developmental dysphasia familial; Disequilibrium syndrome; Fitzsimmons syndrome; Flynn Aird syndrome; Fountain syndrome; Harding ataxia; Hereditary hyperekplexia; Hereditary spastic paraplegia; Idiopathic hypersomnia; Juvenile myoclonic epilepsy; Landau–Kleffner syndrome; Lennox–Gastaut syndrome; Mitochondrial encephalomyopathy, lactic acidosis and stroke-like episodes[i]; Paroxysmal exertion-induced dyskinesia; Paroxysmal kinesigenic choreoathetosis; Spastic paraplegia 14; Tourette syndrome; West syndrome.	**Functional problems** include difficulty with one or more of attention, behaviour, cognition, communication, eating, intellect, learning, memory, mood, movement, muscle tone, rate of neurodevelopmental progress, seizures and sleep. **Structural defects** are not usually apparent.

In some cases (marked[a–i]), there is at least some limited evidence that the disorder might involve variants of genes involved in the following cellular processes:[a] encoding proteins that act on chromosomes (see Chapter 4);[b] encoding proteins that act in cilia (see Chapter 7);[c] encoding proteins that act in energy metabolism (see Chapter 8);[d] encoding proteins that act in PI3K-AKT-mTOR signalling pathway (see Chapter 7);[e] encoding proteins involved in cellular trafficking;[f] encoding proteins involved in serine biosynthesis (serine is an amino acid);[g] encoding proteins that act on glycosylation of alpha dystroglycan (which is involved in cell migration, see Chapter 7);[h] encoding proteins that act in peroxisomes (which are membrane-bound organelles involved in various metabolic processes);[i] mitochondrial function (see Chapter 4 and Part III).

Effects of *CHAMP1* variants

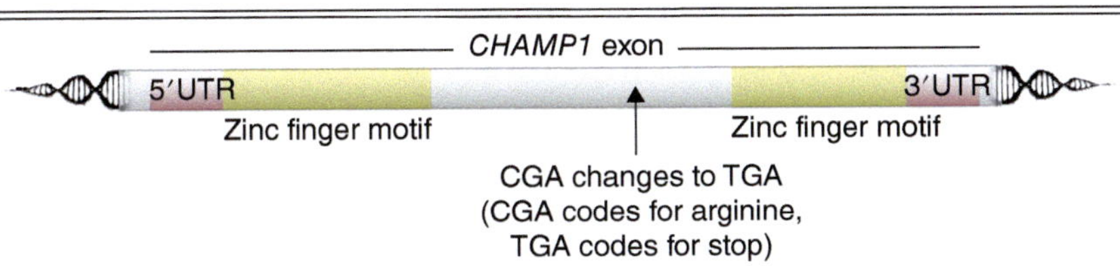

Biological sex:	Female	Male	Male	Female
Age (years):	2.2	3.3	10.3	13.4
Developmental delay or intellectual disability:	Yes	Yes	Yes	Yes
Hypotonia:	Yes	Yes	Yes	Yes
Microcephaly:	No	Yes	No	No
Seizures:	No	Yes	Yes	No

Figure 6.4 Monogenic disorders have variable phenotypes. People in whom the exact same genetic mutation has been identified as the cause of their disorder can have significantly different symptoms. Here is just one of many possible examples. These four children have a harmful variant of the *CHAMP1* gene, which encodes a protein whose normal function includes binding other molecules via its zinc finger **domains** (many of these molecules are proteins involved in chromosome segregation during cell division). All four children have exactly the same de novo single base change in the gene, which produces a premature stop codon (see Chapter 4). Although all four have major symptoms in common, such as developmental delay (in the younger pair) or intellectual disability (in the older pair) and hypotonia, only one has microcephaly and only two have seizures (one of the younger pair and one of the older pair). We can conclude that factors other than the single base change itself must contribute to causing symptoms. *Adapted from Garrity et al.*[146]

disorders such as those I have described up to now in this chapter, the phenotypes of which are determined predominantly by genetic factors. But even in these cases, environmental factors are usually involved in some way and to some extent: for example, they might influence the chances of a person acquiring a genetic variant in the first place and/or they might trigger or affect the progression of their disorder. The following examples will illustrate these points. The chances of an individual acquiring an additional chromosome 21, or part thereof, due to a problem with chromosome segregation during gamete formation (causing Down syndrome) is known to increase with maternal age and possibly other environmental factors such as low socio-economic status (incidentally, the term nondisjunction is used to describe a failure of sister chromatids to segregate properly, disjunction being the normal process of separation – horrible jargon in my opinion!). Environmental factors such as whether a person with Down syndrome is given speech and occupational therapy can influence whether they develop their full cognitive potential. As I mentioned above, environmental factors might contribute to phenotypic diversity even in monogenic disorders that are highly penetrant; for example, a person's lifestyle, diet and stress levels can modulate the onset and progression of the monogenic disorder, Huntington disease.

At the other extreme, some types of developmental brain disorder are caused primarily by environmental factors, although genetic factors probably play a role in most if not all. Examples of such disorders are given in Table 6.10. There is evidence that genetic factors

Table 6.10 Disorders with a strong environmental cause.

CAUSATIVE AGENTS AND DISORDERS	PHENOTYPES
Harmful exposure to substance (which is stated in brackets if required) Fetal alcohol spectrum disorders; Fetal aminopterin syndrome (methotrexate or aminopterin); Fetal hydantoin syndrome; Fetal retinoid syndrome; Lead poisoning; Manganese neurotoxicity. **Infection** Congenital cytomegalovirus; Congenital human immunodeficiency virus; Congenital mumps; Congenital rubella syndrome; Congenital syphilis; Congenital toxoplasmosis; Congenital varicella syndrome; Congenital Zika syndrome; Fetal parainfluenza virus type 3 syndrome; Neonatal chikungunya infection. **Lack of nutrient (which is stated in brackets if required)** Acquired acrodermatitis enteropathica (zinc); Folate deficiency.	**Structural defects** include one or more of abnormal cortical gyration, cerebellar agenesis or dysgenesis, cerebral cortical atrophy, cerebral dysplasia, corpus callosum agenesis or dysgenesis, hydrocephalus, intracranial calcifications; macrencephaly, micrencephaly and ventriculomegaly. **Functional problems** include difficulty with one or more of: attention, behaviour, cognition, communication, eating, intellect, learning, memory, mood, movement, muscle tone, rate of neurodevelopmental progress, seizures and sleep.
Harmful exposure to substance (which is stated in brackets if required) Arsenic poisoning; Cocaine antenatal exposure; Congenital Minamata disease (mercury); Fetal minoxidil syndrome; Fetal valproate syndrome; Kernicterus (bilirubin); Phenobarbital antenatal exposure. **Infection** Botulism; Congenital or neonatal malaria; Fetal parvovirus syndrome; Helminth infection of children; Nipah encephalitis; Pediatric autoimmune neuropsychiatric disorders associated with Streptococcus infections[a]; Poliomyelitis; Subacute sclerosing panencephalitis (associated with measles); Western equine encephalitis (mosquito-transmitted viral infections). **Lack of nutrient (which is stated in brackets)** Congenital hypothyroidism (iodine). **Autoimmune** Hashimoto encephalopathy; Rasmussen encephalitis; Susac syndrome. **Trauma** Neonatal stroke.	**Functional problems** include difficulty with one or more of attention, behaviour, cognition, communication, eating, intellect, learning, memory, mood, movement, muscle tone, rate of neurodevelopmental progress, seizures and sleep. **Structural defects** are not usually apparent.

[a] *Note:* Streptococcus bacteria have been implicated in triggering the long-term development of obsessive, compulsive and other neurodivergent behaviours in a condition known as paediatric autoimmune neuropsychiatric disorders associated with Streptococcus infections. Whether the underlying cause involves autoimmunity remains uncertain and controversy surrounds many aspects of the diagnosis.

can put some people at greater risk of these disorders than others. For example, the exact phenotypic features of a person with a fetal alcohol spectrum disorder depend on their own genetics, which can influence their brain's exposure to alcohol by affecting their ability to metabolize it, and their mother's genetics, which can influence her alcohol intake and metabolism.[147]

This section is about the ultimate causes of those numerous cases of neurodevelopmental disorder and most cases of neurodivergence that lie between the extremes and probably usually involve a balance of major contributions from both genetic and environmental factors. In many such cases, the genetic factors are clearly numerous but neither they nor the potentially complex environmental factors they combine with are likely to be fully worked out. These types of condition tend to be highly heterogeneous and are frequently defined based on how a person's brain is functioning, often in a context-specific way, although some conditions have detectable effects on brain structure. The genetic and environmental factors that need to combine to cause these conditions are often referred to as risk factors: on their own, they are insufficient to cause disorder or divergence, but they add to the chance of their development. I will be looking at examples of these conditions under their current diagnostic headings. It is worth bearing in mind that the large amount of heterogeneity within some of these diagnostic categories – particularly those based on functional traits – suggests that they might not adequately reflect the complexity of the conditions. It seems likely that some diagnostic categories will need to be revised as we learn more about them.

Prenatal Multifactorial Disorders

Many multifactorial disorders, and indeed many neurodivergent conditions, are likely to originate in prenatal events, but direct evidence is often lacking. I will focus here on examples of conditions that we know for sure emerge prenatally, namely **neural tube** defects. These are the commonest multifactorial brain disorders to emerge and be detectable prenatally. They include **anencephaly** and **exencephaly** (Chapter 3). We have evidence that their ultimate causation involves many genetic and environmental risk factors and that many of these factors need to combine for neural tube defects to occur. Most cases of neural tube defects occur sporadically rather than showing clear patterns of inheritance.

One of the best-known environmental risk factors for these neural tube defects is suboptimal blood levels of a vitamin, folate, which is causative in individuals whose genetics also puts them at risk. Other environmental risk factors are various therapeutic chemicals such as the anticonvulsant valproic acid, chemicals that occur naturally such as the fungal molecule fumonisin and maternal obesity and diabetes. Possible genetic risk factors are numerous, involving up to about 200 protein-coding genes. In most cases, their identification was based on studies of neural tube defects in non-human species, mainly mice, and on knowledge of genes involved in folate metabolism. Unlike for many other polygenic multifactorial conditions, GWAS have, so far, been relatively uninformative, probably because the risk factors are so numerous and vary greatly between individuals.

Multifactorial Conditions Emerging at Birth or in Early Childhood

Cerebral palsy is a relatively common type of multifactorial brain disorder affecting approximately one in every 500 young infants in many countries. Its symptoms include abnormalities of posture and movement attributable to brain dysfunction (palsy means muscle weakness or paralysis). The term cerebral palsy can be traced back to surgeon William Little's description, in 1843, of infants suffering from widespread muscle rigidity, resulting in contractures and deformities interfering with their ability to stand, walk or move normally in other ways.[148] In modern clinical usage, the term covers a wide range of permanent and relatively stable disorders of motor function, including movement and/or posture, attributable to problems in the developing brain. There are several types of cerebral palsy. Most common is spastic cerebral palsy, which – as William Little described – involves stiff or rigid muscles that have too much tone. Other types are dyskinetic cerebral palsy, which is characterized by uncontrolled movements, and ataxic cerebral palsy, in which a person has balance or coordination problems. Some people have more than one type. The nature and severity of the symptoms vary greatly from person to person. They are often associated with other functional symptoms, such as communication disorders and epilepsy.[149]

The ultimate causes of cerebral palsy are often unclear and are likely to be numerous. Until recently, the finger of blame has tended to point at environmental risk factors, such as infections or injury to the head during pregnancy, damage to the brain at birth, premature birth and low birthweight. Recently, however, evidence for a genetic contribution has been forthcoming, with many variants potentially involved.[150] Many of these variants have also been identified as potentially causative in other neurodevelopmental disorders, illustrating the point that many different phenotypes might share common ultimate causes.

Another relatively common condition, epilepsy, is diagnosed in around 0.5–1% of children, most commonly in the first year or so of life. In about 50% of people with epilepsy, no cause can be identified. In the other 50%, the causes can be genetic (rarely monogenic), environmental or, in most cases, both. Some epilepsy is hereditary and runs in families but some arises de novo. It is thought that, in most cases, genetic factors predispose a person to epilepsy by making their brains more sensitive to environmental factors that trigger seizures. Possible environmental triggers include direct trauma to the brain, poor nutrition or hypoxia, brain tumours, infections of the brain, substances of abuse such as prescription or recreational drugs and alcohol, hormonal changes during the menstrual cycle and stress.

Multifactorial Conditions Emerging Later in Childhood

Some disorders and neurodivergent behaviours start to appear as children's communication and cognitive skills develop, although it can take many years for some people to obtain an official diagnosis. One relatively common neurodivergent behavioural condition that receives a great deal of attention from clinicians, researchers and the general public is autism.

Autism occurs in about one in 100 people, many of whom show evidence of autism early in life but are not officially described as such until many years later (typically between the ages of 5 and 10 years in Europe and North America). The prevalence of reported cases of autism is increasing, with a major component of this increase likely due to an increase in diagnoses rather than an increase in the numbers of people with symptoms.[151] The term autism covers a very diverse range of phenotypes. Some prefer to use the term autism spectrum disorder (ASD) to reflect the condition's diversity, although this has the disadvantage for others of designating autism as a disorder. Many people diagnosed with autism, particularly those whose functional phenotypes are less divergent, are resistant to the notion that they have a disorder that needs treatment and prefer to consider they have a difference that society should accommodate. Awareness of this debate is important not only out of respect for autistic people but also because it highlights the fundamental fact that autism is a very diverse and imprecisely defined condition that is experienced and viewed in different ways by different people. An increased awareness of the condition coupled with a broadening of the diagnostic criteria likely accounts for much of the increased incidence of reported cases over the last few decades.

The phenotypes of autistic people usually include types and degrees of difficulty with social interaction and communication that most other people do not experience. Other characteristics can include atypical patterns of behaviour, such as repetitiveness of interests or actions (often called stimming, see Chapter 3), a greater degree of difficulty than most other people in changing from involvement in one activity to another, an unusually high attention to detail and unexpected reactions to sensations. Intellectual ability ranges from severe disability to above average intelligence. There is huge variation in the type and amount of support that autistic people need, both between individuals and as an individual grows up. Many autistic people live independent and very fulfilling lives, others require lifelong care and support.

Autistic people often have symptoms that lead to additional diagnoses; for example, a small majority (something like 50–70%) are also diagnosed with a condition known as attention deficit hyperactivity disorder (ADHD; although conventionally referred to as a disorder, it is better thought of as a divergence). This is often referred to as comorbidity (the simultaneous presence of more than one clinical condition in an individual). Epilepsy, depression and anxiety are also reported at higher rates in autistic people than in the general population. This raises the important, intriguing yet unresolved question of whether autism and other conditions such as ADHD are part of a single continuum with causative factors common to both.[152] I will say more about ADHD later in this section.

As you might imagine for a condition that is rather imprecisely defined and has a very variable range of phenotypes, researchers have found that determining autism's ultimate causative factors is a major challenge.[153] Difficulties are likely due to the number of potential causative factors and the somewhat nebulous nature of the condition itself, which creates methodological issues: for example, categorizing the condition so that methods such as GWAS can be applied is likely to be an inexact process. There is widespread agreement that autism has a strong basis in genetics[154,155]: its prevalence is higher in people with a family history of autism or whose siblings are autistic, especially if they are **monozygotic twins**. Its **heritability** is high, with estimates ranging from 40% to 80%.

A small proportion (about 5%) of people with symptoms of autism carry causative variants of a single gene. These include Angelman syndrome, fragile X syndrome, Rett syndrome and tuberous sclerosis, which are caused by variants of the *UBE3A*, *FMR1*, *MECP2* and the *TSC1* or *TSC2* genes respectively (I will say more about these and other genes in Part III). You will sometimes see autism that is associated with conditions such as these called syndromic autism, although not all researchers like this term.[156] To me, it seems unhelpful since we could consider all autism to be syndromic, meaning multiple phenotypes, or symptoms, co-occur in an individual (remember that autism is a clinical diagnosis made on the basis of multiple symptoms, it is not itself a symptom). Some authors say that these monogenic disorders result in a disorder that is autism-like or autism-related, although whether this is really useful seems debatable since there is significant overlap between the symptoms of people with these conditions and other autistic people. The ultimate causes of most cases of autism are unknown. You will sometimes see autism of unknown cause described as idio*pathic*, which seems undesirable given that the word idiopathic implies a state of disease or *path*ology.

Despite the challenges and uncertainties, researchers have made steady progress identifying genes whose variants are causative or are risk factors for developing symptoms that fall under the general heading of autism. Initial discoveries came from genetic mapping and linkage analyses. Whole-exome, whole-genome and GWAS added, and still continue to add, most of our current knowledge of inherited and de novo genetic causes and risk factors. Variants of hundreds of individual genes as well as many larger CNVs (see Chapter 4), which might be either deletions or duplications and might involve multiple genes, have been implicated in autism. The strength of the evidence supporting the involvement of these genetic variants is variable. If you want to find up-to-date curated lists of the large number (over a thousand) of probable or possible genetic causal and risk factors for autism, I suggest you look at the Simons Foundation Autism Research Initiative (SFARI) databases[†]. Bear in mind that the potential causes extend well beyond variants of discrete protein-coding genes. They are likely to include variants affecting genomic regions that are involved in gene regulation (for example, promoters, enhancers and so on), that produce non-coding RNAs (ncRNAs) or are responsible for alternative **splicing**, to name just a few (I discussed these regions in Chapter 4).[157] There remains a massive gulf to cross before we can understand the complex genetic factors contributing to conditions such as autism.[158]

Environmental factors are also implicated in causing autism, but they can be at least equally and often more difficult to identify unequivocally. Some environmental factors that are associated with autism might contribute to its causation by increasing the numbers of de novo genetic variants that are risk factors for the condition in the embryo. This might happen in the case of environmental factors that are mutagenic, such as exposure to high levels of mercury or cadmium. It might also help explain why older parents tend to have a greater risk of having autistic children, since the chance of de novo mutations arising in the germ cells increases with age, particularly in males (Chapter 4). It is still unclear whether other types of environmental factors might contribute to causation. Examples of possible contributory factors include the following, but remember that many are at best associations (and therefore might not be causative). Prenatal maternal stress has been associated with a higher risk of autism in the child. There is evidence for differences in the prevalence of autism between ethnic groups, although whether this is really because risk varies with ethnicity or whether it arises due to differences in detection rates is unknown.[159] Other environmental factors that have been associated with autism, and also with the frequently co-occurring ADHD, include the nature of the gut microbiome[160] (Chapter 5). Overall, our

† https://gene.sfari.org/.

understanding of the environmental contributions to the ultimate causes of autism is very limited. In the vacuum created by this lack of knowledge, many unsubstantiated claims have been made. These include high-profile claims of a causal association between autism and vaccination against measles, mumps and rubella. The fraudulent original report that suggested this link has been comprehensively debunked although, depressingly, the idea persists in the public psyche.[161]

One very consistent finding is that autism is between two to nine times more prevalent in males than in females. There is unlikely to be a straightforward genetic explanation for this, since most genetic variants that are risk factors for autism are found in the **autosomes** and not the sex chromosomes. A current widespread hypothesis is that the bias is due to protective factors of some sort in females, although what they might be is unknown.

There are many other neurodivergent conditions that start to appear as children's communication and cognitive skills develop. In at least those countries where health services record incidences, about two to three people in every 100 receive a diagnosis of ADHD. Many clinicians categorize symptoms of ADHD into two types: (i) inattentiveness; (ii) hyperactivity and impulsiveness. Many with ADHD show both types, some only one. For example, about 30% of people with ADHD have inattentiveness, but not hyperactivity or impulsiveness. I will not go into descriptions of the details and variability in how psychiatrists and related healthcare professionals categorize and subdivide these conditions. You will find many descriptions elsewhere, including ICD-11 and DSM-5[69,70] (also, see Chapter 3). Categories are frequently updated, as one might predict in a clinical field dealing with conditions that – relative to many disorders affecting other organs – are complex, hard to define precisely and relatively poorly understood. Another common condition that affects many children (estimated to be about ten in every 100 in the UK) is dyslexia. The symptoms of this condition include a variable degree of difficulty with reading, writing and spelling that interfere with a person's ability to learn and work effectively.

As with autism, the ultimate causes of these conditions in any given individual lie in a plethora of possible genetic and environmental factors that can affect many of the numerous processes that converge to control fundamental mechanisms of brain development. One of the many unresolved questions concerns the degree to which genetic risks for the development of conditions distinguished clinically are specific for each condition.[162] There is much evidence that similar genetic variants can contribute to multiple conditions, including epilepsy, the types of neurodivergence discussed in this section, as well as bipolar disorder, schizophrenia and Tourette syndrome. The last three of these conditions typically emerge in adolescence or early adulthood and I will discuss them further in the next section, where I will include further information to reinforce the general point about overlapping causation.

Multifactorial Conditions Emerging in Adolescence and Early Adulthood

Later in life, often in adolescence (by which I mean mainly the teenage years) or early adulthood (by which I mean mainly ages 20–30 years), some people experience types of disordered or divergent perceptions, thoughts and behaviours. They can be triggered by stress, trauma, drug misuse, drug side effects and so on. They might appear transiently in many people at this relatively turbulent stage of their life, but if symptoms are persistent and debilitating, they can lead psychiatrists to diagnose a disorder or a neurodivergent condition. I will first discuss three major disorders, often called psychoses, which affect around 2% of people: schizophrenia, bipolar disorder and severe depression.

Schizophrenia's characteristics are often described in terms of positive symptoms and negative symptoms. This can seem a bit confusing at first: the terms positive and negative do not imply that the symptoms are good or bad for the person. Positive symptoms include altered perceptions, thoughts or behaviours such as hallucinations and delusions (which I discussed in Chapter 3). Negative symptoms include a lack of interest or emotional response leading to withdrawal from the surrounding world and the social interactions on offer. Bipolar disorder is characterized by periods of hyperactivity or mania, during which a person might show a decreased need for sleep and increased talkativeness, might experience racing thoughts and be easily distracted and might show an increase in goal-directed activities. These periods are interspersed with severe depression. Severe depression, be it part of bipolar disorder or not, affects a person daily and is often associated with other symptoms such as significant weight loss, altered appetite, purposeless activity, chronic fatigue, feelings of worthlessness or guilt, an inability to concentrate, indecisiveness and thoughts of – or attempted – suicide.

Schizophrenia, bipolar disorder and severe depression are heritable, with genetic variation accounting for up to 80% of the phenotypic variation in those populations where studies have been done. The proportions of people that develop these conditions depend on whether they have close relatives who are affected. Incidences of all three conditions are 1–2% overall in most populations; about 10–20% if a parent, sibling or dizygotic twin is affected; about 40–50% if a monozygotic twin is affected. The fact that the incidence is 40–50%, rather than 100%, when a monozygotic twin is affected is clear evidence that the conditions are not caused by a person's genetic make-up alone. Rather, a person's genetic make-up somehow affects the likelihood of them developing one of these conditions, but exposure and responses to other factors are also critical.

Identified risk factors for these conditions are numerous.[163] Aside from a family history, they include being male, being born in an urban environment, complications during embryonic and fetal life such as maternal illness, stress, infection or malnutrition, physical damage to the brain, hypoxia or low body weight at birth, being born in winter, being brought up in a city, childhood adversity and the use of drugs such as cannabis. The extent to which such factors are likely to contribute across the population varies, and we almost certainly do not know about all the possibilities. As I have mentioned before, you should remember that much of what we know is based on statistical associations, leaving us with a poor understanding of whether, and if so how, these factors are causative.

GWAS have implicated variants of many hundreds of coding and non-coding genomic regions in the aetiology of schizophrenia, bipolar disorder and severe depression.[164–167] Some researchers have combined the results of many GWAS, carrying out what we call meta-analyses. Meta-analysis is a general term for analysing a combination of data from multiple studies of a similar type, the idea being that using data from multiple sources gives a more robust understanding than using data from just one source. This approach has identified variants consistently associated with schizophrenia, bipolar disorder and severe depression, as well as with conditions that often co-occur, such as autism and ADHD. The considerable overlap in the genetics of all these disorders probably goes a long way to explaining why combinations of them are often diagnosed in the same person.[168] The genes that share implication in these conditions encode a variety of proteins involved in a very wide range of tasks, for example, regulating the expression of other genes; guiding the growth of axonal connections between nerve cells; allowing nerve cells and their connections to function properly; modulating **synaptic plasticity**; controlling immune responses. As you will probably anticipate from comments I have made up to now in this book, it is no surprise to find that variants of genomic regions that do not code for proteins have also been implicated in these conditions.[169]

Some researchers argue that schizophrenia, bipolar disorder, severe depression and possibly other conditions (such as autism, anorexia nervosa, Tourette syndrome and OCD, which I discuss below) are probably best thought of as overlapping syndromes whose causes are drawn from a common pool of genetic and environmental factors that have the potential to affect brain development, even though clinicians currently subdivide them into categories based on differences in observable symptoms.[170–172] By way of backing up this point, I include Table 6.11, which lists a number of genes whose variants have been associated with several of eight cognitive conditions, including schizophrenia, bipolar disorder and severe depression (in one case, all eight: this is the gene *DCC*, which encodes a receptor involved in intercellular signalling, a process I will discuss further in Part III). The products of these genes are involved in many types of cellular processes, as summarized in Table 6.11. There are many more genes not listed in Table 6.11 that are associated with just one or two of these conditions; again, their gene products are involved in many types of cellular processes.

Other major psychiatric conditions that usually manifest during adolescence and whose ultimate causes include both genetic variants and environmental factors include anorexia nervosa, extreme anxiety, obsessive-compulsive disorder (often abbreviated to OCD; although conventionally described as a disorder, it is better thought of as a divergence) and Tourette syndrome.

Anorexia nervosa affects about 4% of females and about 0.3% of males during their lifetime. People with anorexia nervosa have a disturbed image of their body giving them a powerful urge to lose weight because they think, incorrectly, that they are overweight. They fear gaining weight, even though they are very likely to be undernourished and consequently unhealthily thin. They limit the amount or type of food they eat so as to prevent weight gain or ensure continued weight loss. Anorexia nervosa has a higher mortality than any other psychiatric condition. There is significant comorbidity with other conditions. For example, most people with anorexia nervosa have major depression, many have severe chronic anxiety and the incidence of OCD is much higher than in the general population. Anorexia nervosa tends to run in families with estimates of heritability ranging between 25% and 75%. Early environmental risk factors include adverse prenatal, perinatal and neonatal events, such as preterm birth. It is most common in young women and its average age of onset is about 15 years, which coincides with profound changes in a teenager's internal and external environment. These include hormonal changes as the body undergoes puberty and socio-cultural pressures that promote ideas of ideal or fashionable body size and shape. It is possible that factors such as these can trigger anorexia nervosa in genetically susceptible individuals.

OCD affects 1–2% of people worldwide. Its defining characteristics are persistently intrusive thoughts, or obsessions, and patterns of repetitive behaviour, which we call compulsions, that a person initiates to help alleviate the distressing obsessions they experience. The symptoms cause mental torment, consume a great deal of a person's time and prevent

Table 6.11 Genes whose variants are associated with multiple clinically defined conditions.

CELLULAR PROCESSES AND GENES					CONDITIONS							
DNA-related activity	**RNA/ protein synthesis/ processing**	**Membrane/ECM protein**	**Mitochondrial protein**	**Other**	**ADHD**	**AN**	**Autism**	**Bipolar**	**Depression**	**OCD**	**Schizophrenia**	**Tourette**
		DCC			■	■	■	■	■	■	■	■
	RBFOX1				■		■	■	■	■	■	■
PHOX2B				*DCAF4L1*	■	■	■		■		■	■
		OR5AK2, OR5B21, SLC43A1			■		■		■	■	■	
	TRIM49D1, TRIM64B, TRIM77	*GRM5*				■	■	■	■		■	
		NEGR1				■	■		■		■	■
		SORCS3			■		■		■		■	■
L3MBTL2, TEF	*ST13*	*CHADL*							■	■	■	■
KDM7A, WEE2	*MKRN1*	*TAS2R3*	*NDUFB2*				■	■		■	■	
BCL11A, SOX5							■	■	■		■	
PHF7, NEK4, BNC1, ZNF592, ZSCAN2, PBX1, DCLK3, TFAP2B, TFAP2D		*ABHD14A, GLT8D1, RFT1*		*POC1A, ADAMTSL3, GOLGA6L4, WDR73*			■	■			■	

CELLULAR PROCESSES AND GENES					CONDITIONS							
DNA-related activity	RNA/ protein synthesis/ processing	Membrane/ECM protein	Mitochondrial protein	Other	ADHD	AN	Autism	Bipolar	Depression	OCD	Schizophrenia	Tourette
ZNF687, MAZ, ZNF689, ZNF771, TAOK2	*PRPF3, SCNM1, SF3B4*	*FCGR1A, ASPHD1, C16orf92, CDIPT, NPIPB11, SEZ6L2, TMEM219*		*ADAMTSL4*				■		■	■	
	DHX38, RSRC1						■	■	■			
RCOR2, MARK2	*STIP1*	*DNAJC4, FKBP2, GPR137, SLC22A10, FLRT1, TMPRSS5, GRIK3, PXDNL*	*COX8A*	*PPP1R14B, TTC12*				■	■		■	
		UNC5D			■			■			■	
		GPR20				■				■	■	
		KIZ			■		■				■	
NPAS3							■	■				■
				PPP2R2B				■			■	■

A selection of genes with known or strongly suspected functions whose variants have been associated with three or more conditions from ADHD, anorexia nervosa (AN), autism, bipolar disorder, major depression, OCD, schizophrenia or Tourette syndrome. Genes are listed under several broad headings indicating the types of cellular processes they are involved in or where in the cell they likely act (ECM: extracellular matrix). Black shading under each condition indicates a statistically probable association. The table is based on data extracted from Cross-Disorder Group of the Psychiatric Genomics Consortium.
Adapted from Cross-Disorder Group of the Psychiatric Genomics Consortium. Electronic address PMHE, Cross-Disorder Group of the Psychiatric Genomics C.[171]

them from functioning effectively. Commonly, there is comorbidity with other psychiatric conditions such as anorexia nervosa and severe anxiety. Estimates of heritability range from about 35–50%. Specific ultimate causes have proven hard to define. There is evidence that environmental risk factors include prenatal exposure to stress or infection.

Tourette syndrome affects about 1% of people. People with Tourette syndrome make involuntary sounds and movements. These are called tics; they are not the same as seizures which, unlike tics, result from temporary uncontrolled bursts of activity within the brain. Tourette syndrome affects about 1% of individuals worldwide and has high heritability of around 70–80%. It usually starts during childhood and has significant levels of comorbidity with conditions including ADHD, OCD and autism. Genetic risk factors have proven hard to identify; there is evidence they might include variants of genes whose proteins are involved in intercellular signalling or **neurotransmission** using dopamine and serotonin (or 5-hydroxytryptamine).[173] Knowledge of environmental risk factors is also limited: there is evidence that some infections and disorders with an allergic basis might trigger the symptoms of Tourette syndrome by inducing abnormal **autoimmune** responses, and factors such as psychological and social stress are also implicated in its causation, possibly affecting its severity.

On the face of it, anorexia nervosa, extreme anxiety, OCD and Tourette syndrome are defined by rather distinct phenotypes. Nevertheless, we might ask: what is the degree of overlap between the genetic factors contributing to their causation and that of schizophrenia, bipolar disorder, depression and autism? This is not an easy question to answer due to the diverse nature of all these conditions and their causes and the fact that, for now, our understanding of what their ultimate genetic and environmental causes actually are remains limited. What evidence there is suggests that the degree of overlap among them all is variable, with some outlier conditions such as OCD and Tourette showing less overlap with the others and with each other (Table 6.11). It is also more than likely that conditions affecting other parts of the body share genetic foundations with conditions that affect the brain.[174] Evidence supporting this is now emerging and further strengthens the need to look at many neurodevelopmental disorders from the perspective of their ultimate causation rather than their often highly variable phenotypes, which are currently used to define them in clinical practice. One notable example is a very recent study examining the associations between genetic variants and 14 major psychiatric conditions, including all those discussed here, in over a million people.[175] The results have given some of the strongest evidence to date that the genetic variants responsible for causing these conditions fall into categories that cut across current diagnostic boundaries. Findings such as these bolster the belief that changes in the way we define psychiatric conditions in research and clinical practice are inevitable.

Summary

- Large chromosomal abnormalities affecting numerous genes often cause disordered brain development; examples include Down, Turner, Klinefelter and Patau syndromes.
- Many individually rare monogenic disorders caused by variants of single genes can also cause disordered brain development; their identification has been accelerated greatly by our ability to sequence human genomes.

- The phenotypes of most genetic disorders, even monogenic ones, are probably affected at least to some extent by environmental and additional undetected genetic factors (such as other variants carried by an individual).
- Many of the commonest neurodevelopmental disorders and most types of neurodivergence are caused by the combined effects of multiple genetic variants (polygenic) and environmental factors (multifactorial).
- Among the commonest prenatal defects are those in which the early embryonic neural tube fails to close properly; the causes are multifactorial and variable and our understanding of them is limited.
- Conditions that often emerge in early childhood include cerebral palsy and epilepsy; they are multifactorial, with many – often unknown – genetic and environmental factors contributing to their ultimate causation.
- Multifactorial conditions that often emerge later in childhood include autism and difficulties with attention and learning; we have a relatively poor understanding of their causation.
- Multifactorial disorders that often appear in adolescence or early adulthood include schizophrenia, bipolar disorder and major depression.
- Research is showing much overlap in the genetic and environmental factors that contribute to the causation of many distinct, phenotypically defined neurodevelopmental disorders and types of neurodivergence.

Part III

Proximate Causes of Neurodevelopmental Disorder and Neurodiversity – Linking Ultimate Causes to Phenotypes

7

Proximate Causes of Prenatal Disorders

In this and the following two chapters, I will attempt to bridge the gap between our knowledge of ultimate causes (discussed in Part II) and specific clinical conditions (discussed and listed in Chapter 6). As you will see, our understanding of the steps that link ultimate causes to the detectable symptoms or phenotypes that they produce is far from complete.

One reason for this is that the routes from ultimate causes to detectable phenotypes are usually very complex. Figure 7.1 gives an extremely simple overview of some types of direct and indirect routes (look ahead to Figure 7.16, discussed later in the chapter, to see a more realistic view of them, but even it is restricted to only a tiny fraction of the total number). In some cases, different ultimate causes might bring about similar phenotypes by converging on the same proximate causative mechanisms. Usually, ultimate causes operate through multiple relatively independent direct and indirect routes to have many phenotypic effects, creating syndromes. Many syndromes whose phenotypes include brain disorders or types of neurodivergence also include symptoms in other parts of the body, which can have indirect, knock-on effects on the developing brain. Discovering all these routes for any given condition can be very challenging.[176]

Another reason is that the phenotypes we observe are the result of effects on the vast number of *interactions* between molecules (looking ahead again to Figure 7.16, you can see how complex these interactions can be for even a small collection of molecules). This is important when we consider a question relevant to many conditions caused by genetic variation: why are the effects of many variants more severe in, or limited to, only some regions of the brain in which they are expressed? There are many possible reasons for this, but one important one is that the effects of a variant might depend on what other molecules each cell contains. As I will discuss below (in the section on how the specification of cell types can be disrupted), the molecular constitution of cells varies considerably between different cell types. For example, **transcription factors** operate in combinations to specify cell types, with different cell types expressing different cocktails of transcription factors, meaning that the effects of a variant of any particular transcription factor will be context-dependent and possibly tissue-specific.

Yet another issue is that our basic understanding of the mechanisms that control brain development is limited. For example, the fact that we rarely, if ever, know what all of the functions of any given region of the genome are, and we sometimes know nothing about them at all, hampers our ability to predict the effects of variants of those regions. Even if the regions are protein-coding, we face problems such as the fact that most proteins have multiple functions (a property we refer to as pleiotropism), many of which we might not

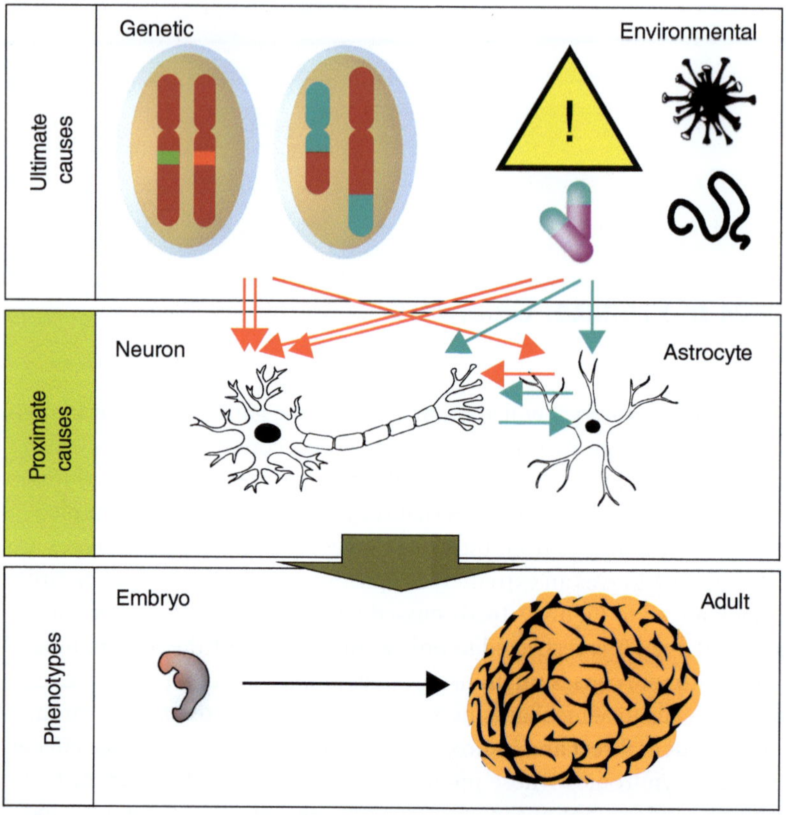

Figure 7.1 The routes from ultimate causes to phenotypes are complex. The red arrows illustrate multiple genetic and environmental factors converging on neurons to generate the proximate causes of the diversity in our brain phenotypes. Effects on neuronal development and function can be either direct or indirect (the latter is illustrated here by a route via astrocytes, which is just one of many possibilities). The blue arrows indicate factor(s) of one main type, here environmental, having proximate effects on multiple interacting cell types. Remember that further diversity is introduced by the random variability of developmental mechanisms involved in generating the brain, which is not illustrated here. This picture gives only a very, very simplified impression; if you look forward to Figure 7.16, you will see a more complex representation of the possible interactions between a relatively small number of key molecules that regulate development, but even that picture is only a small fraction of all the interactions needed to make a brain.

know about and which might vary between different developmental stages or in cells of different types. When it comes to actually identifying rather than merely predicting the effects of genetic variants or environmental factors, a practical difficulty rears its head, especially if changes arise prenatally: how do we spot them when they are likely to be too subtle to be detected with current technologies?

These limitations create a major uncertainty: do some types of brain disorder and neurodivergence that we first perceive after birth actually originate prenatally? There are circumstantial reasons for thinking that this is likely in many cases. First, the chances of genetic variants having a prenatal impact of some sort are increased by the fact that most (at least three quarters) of our protein-coding genes are expressed in our prenatal brains.[177] Second, our prenatal brains are very sensitive to potentially harmful environmental factors such as radiation, toxins, infective agents and so on, since they are in a state of rapid growth and their protective blood–brain barrier (Chapter 1, Figure 1.11) is immature and not as effective as it will become later in life.

In this chapter, I will focus mainly on brain disorders that we know for sure originate prenatally for the very simple reason that they are detectable at birth. In subsequent chapters, I will focus mainly on brain disorders and divergence that are usually detected postnatally. There will be grey areas. For example, conditions first detected at ages around the time of birth might be discussed here or the next chapter, depending on how they are usually caused; conditions that are first detected postnatally might be included in this chapter if there is overwhelming evidence supporting their prenatal causation. It would not surprise me if, in years to come, some – maybe even most – of the conditions I have included in later chapters need to be moved into this one.

The overall structure of the main body of this chapter follows that of Chapter 1, in which I discuss the key cellular processes that create the brain prenatally: proliferation, migration and so on. I will describe how disruption of these processes can cause disorders, illustrating the points I make by selecting some of the better-understood examples of conditions listed in the tables in Chapter 6.

But, before starting on that, I need to explain some essential details of the molecular mechanisms by which the cells of the brain collaborate together. To work effectively as a growing ensemble, they need to communicate with each other. Up to now in this book, I have alluded to the process of intercellular signalling without giving details of this complex, multifaceted process. In this and subsequent chapters, the details of how it is achieved become important.

The molecules involved in intercellular signalling contribute to the *micro*environment surrounding and impacting each and every cell in our developing brain and allow each and every cell to respond in particular ways to the specific signals around it. The following section fills a gap between my descriptions in Chapter 4 of how each cell's genome is regulated and those in Chapter 5 of how external and internal *macro*environmental factors can impact brain development. In what follows, I will explain how the microenvironment of cells – which is created by the complex interplay between both genetic and macroenvironmental factors – is configured, how it is detected by cells and how it influences the operation of their genomes. This will give you information important for understanding how variation in these processes might contribute to types of disorder and divergence.

The Microenvironment of Developing Brain Cells

Some molecules in the microenvironment of developing and mature brain cells are less free to move than others, either because they are very large or they are firmly tethered to cell surfaces. Others, including many **signalling molecules**, are freer to move among cells and often have distribution patterns that are important for the ways in which they function (we saw an example of this in the gradient model in Chapter 1, Figure 1.5).

Some Extracellular Components Are Relatively Static

In all multicellular organisms, cells need to stick to each other. Intimate associations with their neighbours allow them to act collectively to maintain the structures of tissues, to communicate effectively with each other and so on. Cells use molecules located either entirely or partially in the space around them to do this. We call this cell adhesion, and many proteins are involved. These proteins are often referred to as cell adhesion molecules, but be sure to remember that, as for most proteins, they have multiple functions, and while cell adhesion might be a prominent role for some, for others it might be less so. Also, cell adhesion molecules – and it is often the very same ones – have important roles to play throughout life, not just during early development.

Three large families of cell adhesion molecules that are particularly important during brain development are the cadherins, the immunoglobulin superfamily of cell adhesion molecules (or IgSFs for short), and the integrins (Figure 7.2). They are components of the cell membrane and interact with molecules in the membranes of other cells or with the complex meshwork of molecules surrounding cells, the **extracellular matrix**. The cadherins are distinguished from other cell adhesion molecules by their need for calcium for their activity. The part of them that is outside the cell (which we call their extracellular **domain**) binds to other cadherin molecules on the surface of neighbouring cells. For some cadherins, the part that is inside the cell (the cytoplasmic domain) links through other proteins (we often call them, collectively, adaptor proteins) to the **cytoskeleton**, which provides intracellular mechanical support. Other cadherins, which we call protocadherins, do not link with the cytoskeleton. The immunoglobulin superfamily of cell adhesion molecules has immunoglobulin-like extracellular domains (in other words, they are like those in antibodies), and they operate in similar ways to cadherins. Some of the best-known members of this very diverse group include neural cell adhesion molecule 1 (NCAM1), L1 cell adhesion molecule (L1CAM), Down syndrome cell adhesion molecule (DSCAM) and NECTIN1 (also known as PVRL1). The integrins mostly connect cells to the extracellular matrix, with many then linking it to the cytoskeleton.

Cell Signalling Molecules and Second Messenger Pathways Are More Dynamic

Intercellular signalling molecules provide the means for populations of cells to coordinate their activities. They act, effectively, as messengers. The receiving cell gets information from parameters such as the type of signalling molecule it encounters, its concentration and what form it is in (for example, whether it is an entire, cleaved or modified version of a signalling protein). The receiving cell has receptors, usually on its surface, and their activation kicks off a series of internal biochemical reactions in the cell's **second messenger** pathways; we refer to this process in its entirety as **signal transduction** (remembering that transduction means the process of converting something into another form, in this case it refers to the conversion of the signal into a cascade of biochemical events within the cell). Figure 7.3 shows key examples of signalling pathways most relevant to brain development.

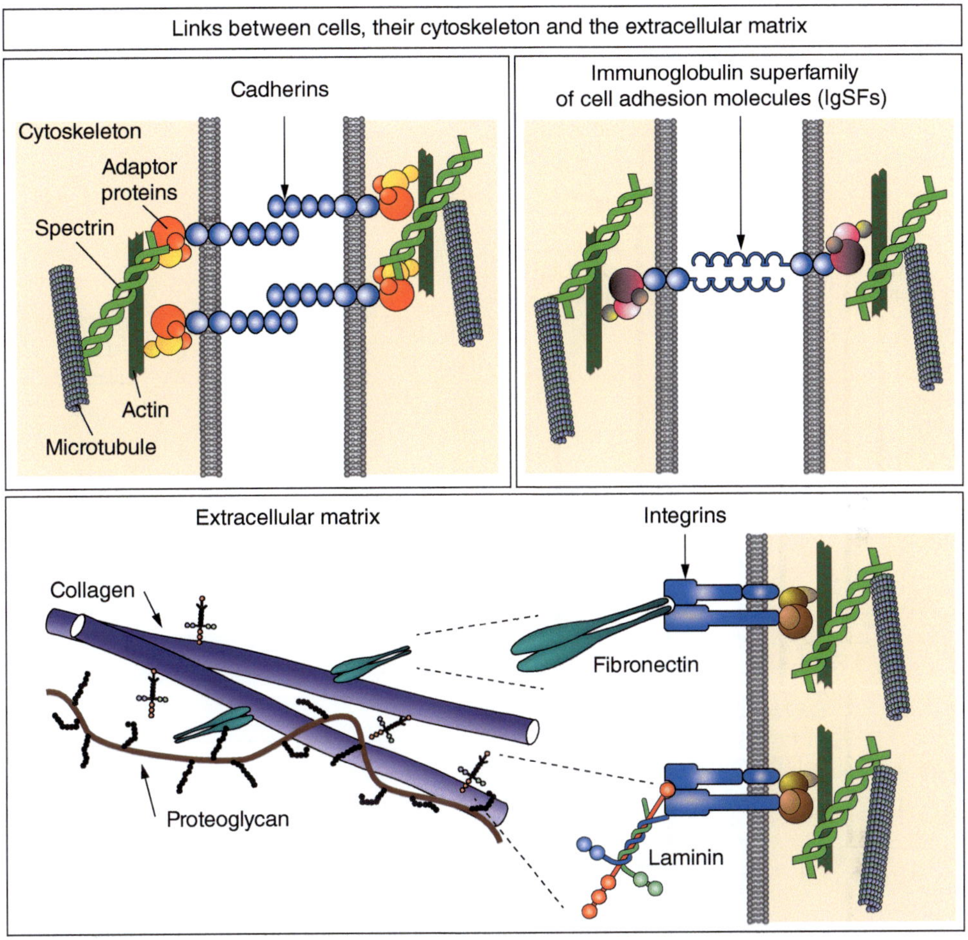

Figure 7.2 Cell adhesion molecules and the extracellular matrix. Cadherins and members of the immunoglobulin superfamily of cell adhesion molecules (IgSFs) bind cells together. Intracellularly, they connect via adaptor proteins to the cell's cytoskeleton, which is a complex arrangement of microtubules, actin, spectrin (long heterodimers composed of two protein subunits with actin-binding domains) and other molecules. The cytoskeleton gives cells their shape and internal organization, providing mechanical support for functions such as cell division and movement. Integrins link the cell's cytoskeleton with the extracellular matrix, which is a complex gel-like material that provides external support for cells, provides signals for them and transmits signals between them. The extracellular matrix comprises long fibrous molecules secreted by cells, mainly glycoproteins (that is, proteins covered by a variety of monosaccharides ... or simple sugar molecules, in other words) and proteoglycans (that is, proteins covered with chains of repeating disaccharides, which are pairs of simple sugars joined together). Major glycoproteins include collagen, fibronectin and laminin; major proteoglycans include heparan sulphate proteoglycans (HSPGs) and chondroitin sulphate proteoglycans (CSPGs). Laminin (which consists of three different subunits) and fibronectin (which consists of two similar subunits) bind to collagens and to integrins.

The diagram in Figure 7.4 expands a bit (although it is still very simplified) on the second messenger pathways through which some of the ligands listed in Figure 7.3 operate. These are examples of pathways activated when ligands bind to G-protein coupled receptors

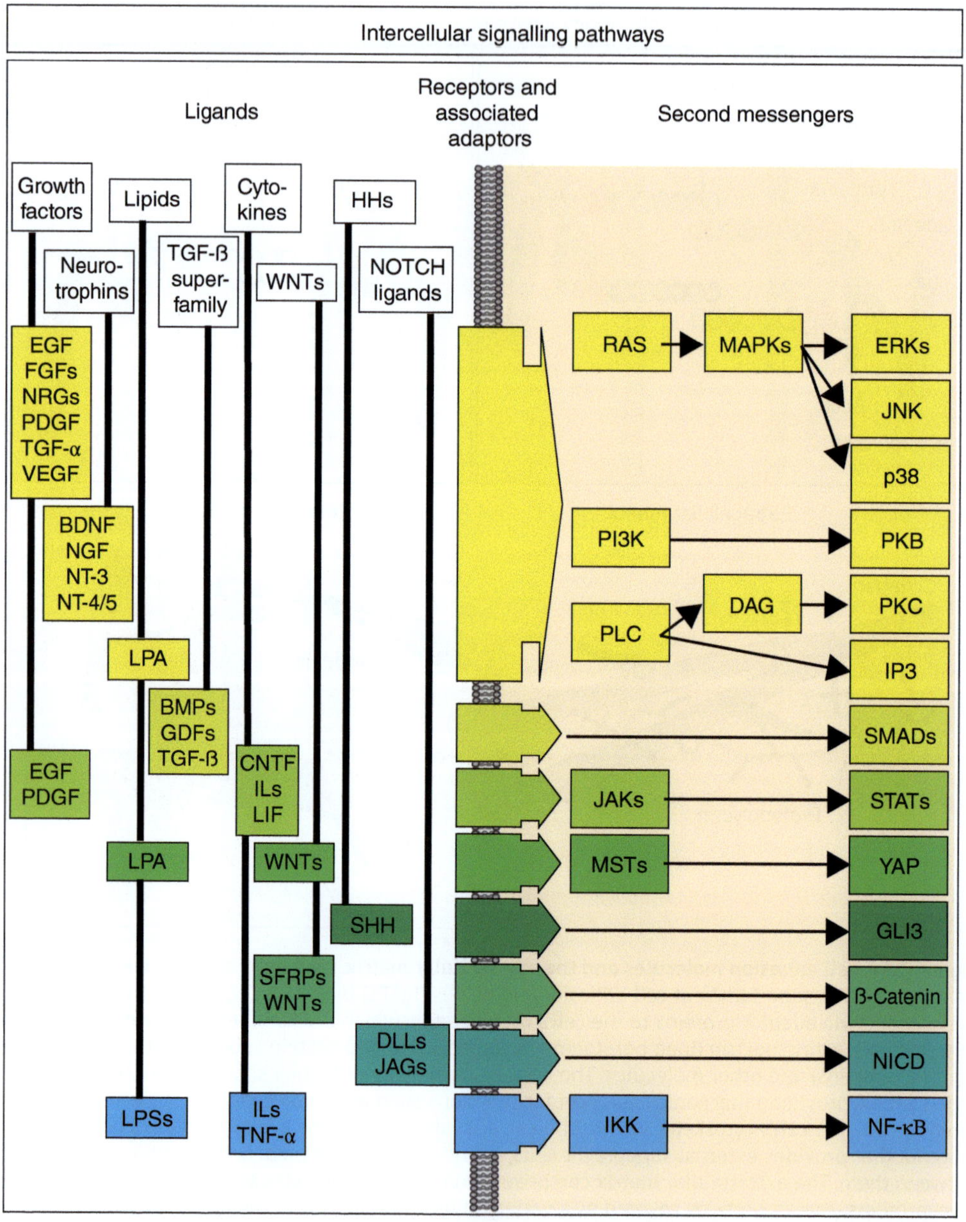

Figure 7.3 **Major intercellular signals and their second messengers.** Cells have many signalling pathways allowing them to respond in a wide variety of ways to molecules they encounter outside themselves, mediating the coordination of cellular activities across populations of cells. Some important signalling molecules used by developing brain cells are listed on the left, under the heading of ligands; in biochemistry, ligands are molecules that bind with other molecules to form complexes, in this case with specific cell surface receptors. I have organized the ligands into families based on shared characteristics and the types of receptors and second messenger pathways they act through; some ligands can act through more than one pathway. Colour-coding links ligands to the second messenger pathways they activate. The arrows in the cellular pathways

Figure 7.3 (Continued) represent cascades of biochemical reactions involving many intermediate molecules that are not included here. Activation of the pathways can have many effects on numerous cellular processes. Abbreviations of ligands: BDNF, brain-derived neurotrophic factor; BMP, bone morphogenetic protein; CNTF, ciliary neurotrophic factor; DLL, delta-like; EGF, epidermal growth factor; FGF, fibroblast growth factor; GDF, growth differentiation factor; HH, hedgehog; IL, interleukin; JAG, jagged; LIF, leukaemia inhibitory factor; LPA, lysophosphatidic acid; NGF, nerve growth factor; NRG, neuregulin; NT, neurotrophin; PDGF, platelet-derived growth factor; SFRP, secreted frizzled-related protein; SHH, sonic hedgehog; TGF, transforming growth factor; VEGF, vascular endothelial growth factor. Abbreviations of second messengers: DAG, diacylglycerol; ERK, extracellular signal-regulated kinase; GLI, glioma-associated oncogene; IKK, IκB kinase; IP3, inositol 1,4,5-trisphosphate; JAK, Janus kinases; JNK, c-Jun N-terminal kinase; MAPK, mitogen-activated protein kinases; MST, mammalian sterile-20-like; NF, nuclear factor; NICD, notch intracellular domain; PI3K, phosphatidylinositol 3-kinase; PK, protein kinase; PL, phospholipase; STAT, signal transducer and activator of transcription; YAP, yes-associated protein. Notes: many of these molecules' names, such as RAS, stem from their original discovery and, for many, the original abbreviations have now effectively become their names; WNT and SMAD are not straightforward abbreviations but combinations of other ones; p38 is not an abbreviation but is a specific set of stress-activated protein kinases; PKB is also known as AKT.

(GPCRs) and receptor tyrosine kinases (RTKs). GPCRs are a large family of receptors with a particular structure: along their length, they weave backwards and forwards across the cell membrane seven times. Their structure changes when a ligand binds, and this activates intracellular G-proteins which then trigger second messengers. G-proteins, one example being RAS,* are a family of proteins that are, effectively, molecular switches whose activity is regulated by molecules that control their ability to hydrolyse guanosine triphosphate (GTP) to guanosine diphosphate (GDP): when bound to GTP, they are active (or on), and when bound to GDP, they are inactive (or off). RTKs are simpler than GPCRs, with only one domain crossing the membrane, an external ligand-binding domain and an internal tyrosine-kinase domain. When a ligand binds, two RTKs come together (we say that they dimerize), which activates their tyrosine kinase domains to add phosphate groups to specific intracellular proteins (we call this phosphorylation), which in turn triggers second messenger signalling cascades. These two pathways are closely integrated with much cross-talk.[178] They exert control over many cellular processes including cell division and **apoptosis** by affecting underlying molecular events such as the **transcription** and/or **translation** (in the case of protein-coding genes) of molecules involved in those processes and subsequent changes in their 3-dimensional conformation and/or their biochemical modification (for example, through protein phosphorylation). The pathways are controlled by proteins that activate and others that inhibit (yellow and red respectively in Figure 7.4). In many cases, activation can increase proliferation and reduce apoptosis, but it can have the opposite effect, depending on what else is happening in the cell; in other words, the cellular responses to a signal can be dependent on context (indicated by open circles beside division and apoptosis in the top diagram in Figure 7.4).

* RAS comes from rat sarcoma virus, which gives a clue as to how RAS was discovered in cancer biology (a sarcoma is a type of cancer), but RAS is the full name of the RAS protein and is not considered an abbreviation. This is the case for some other proteins too.

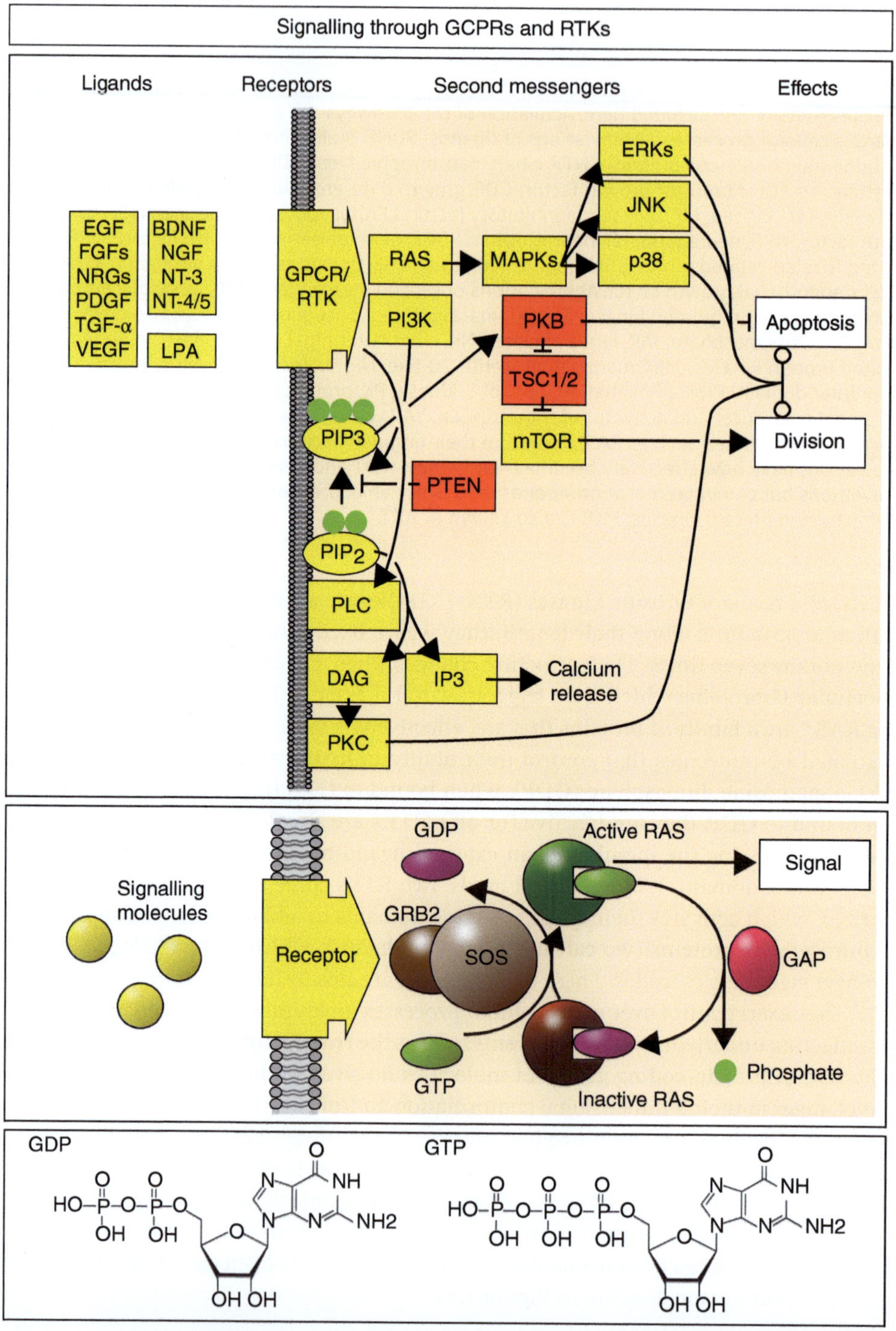

Figure 7.4 A simplified summary of signalling through G-protein coupled receptors (GCPRs) and receptor tyrosine kinases (RTKs). The schematic at the top shows some of the major consequences of activation of GPCRs and/or RTKs by some of the signalling molecules (ligands) listed in Figure 7.3. Receptor binding can activate RAS, which in turn can activate MAPKs, then ERK, JNK and p38 to have context-dependent effects on, among other things, cell division and apoptosis. Receptor

Figure 7.4 (Continued) binding can also activate PI3K, which produces PIP_3 (phosphatidylinositol triphosphate, which is in cell membranes) by phosphorylating PIP_2 (phosphatidylinositol bisphosphate, also in cell membranes); this is indicated by the change from two to three green circles. PIP_3 can then activate PKB (also known as AKT) to, among other things, inhibit apoptosis. PKB phosphorylates TSC2, which forms a functional complex with TSC1. This complex inhibits mTOR (mammalian target of rapamycin), thereby counteracting its effect, which includes the promotion of cell division. PTEN (phosphatase and tensin homolog) is a crucial negative regulator of this generally pro-proliferative pathway: it limits the conversion of PIP_2 to PIP_3 by promoting the reverse reaction (it might come as no surprise that mutations that result in PTEN losing its inhibitory function are often associated with cancer). These pathways branch out in other ways, including through other membrane-associated proteins such as PLC and PKC. PLC is activated by stimulation of RTKs or GPCRs to break down PIP_2 into smaller molecules, the lipid DAG (diacyl glycerol) and IP3, which themselves serve as second messengers. DAG can activate PKC to affect many cellular processes including apoptosis and cell division. IP3 regulates release of intracellular calcium (Ca^{2+}) from stores in cellular organelles such as the endoplasmic reticulum, which can affect a huge number of enzymes, proteins and, therefore, molecular and cellular processes. The schematic in the middle and at the bottom illustrates signalling through G proteins such as RAS. Binding to the transmembrane receptor activates guanine nucleotide exchange enzymes such as SOS (son of sevenless; the obscure name comes from a gene in fruit fly) that are bound into complexes by proteins such as GRB2 (growth factor receptor-bound protein 2). The guanine nucleotide exchange enzymes activate RAS by catalysing its transition from an inactive to an active form. This is achieved by removing GDP and allowing the binding of GTP, for which RAS has a higher affinity. To inactivate RAS, GAPs (GTPase-activating proteins) remove a phosphate, returning RAS to its inactive GDP-bound state. Abbreviations not defined here are defined in Figure 7.3.

Primary Cilia Mediate Much Intercellular Signalling

Most cells rely on small solitary structures that project from their surfaces, called primary cilia, for at least some of their ability to respond to extracellular signals.[179] Primary cilia are often described as the cells' antennae.[180–182] There is another type of cilia, called motile cilia, that you might be more familiar with since defects of them can result in disorders such as cystic fibrosis. Motile cilia move in a coordinated way. Their synchronized movements allow them to carry out specialized functions such as moving secretions and foreign particles from the lungs up towards the mouth so they can be expelled (which is a problem for those with cystic fibrosis). Primary cilia share some similarities with motile cilia, but they lack the ability to move as motile cilia do. They are rich in receptors and act as signalling hubs mediating the ability of cells to carry out signal transduction, including through SHH, GPCR and probably WNT and other pathways. I have summarized key aspects of their structure and function in Figure 7.5.

How the Earliest Steps in Brain Development Can Be Derailed

Derailment of Neural Induction

In Chapter 1, I described how, in the early embryo, **neural induction** converts some **ectodermal** cells to more specialized ***neuro*****ectodermal** cells, which then go on to make the nervous system, including the brain. This process is so fundamental that significant

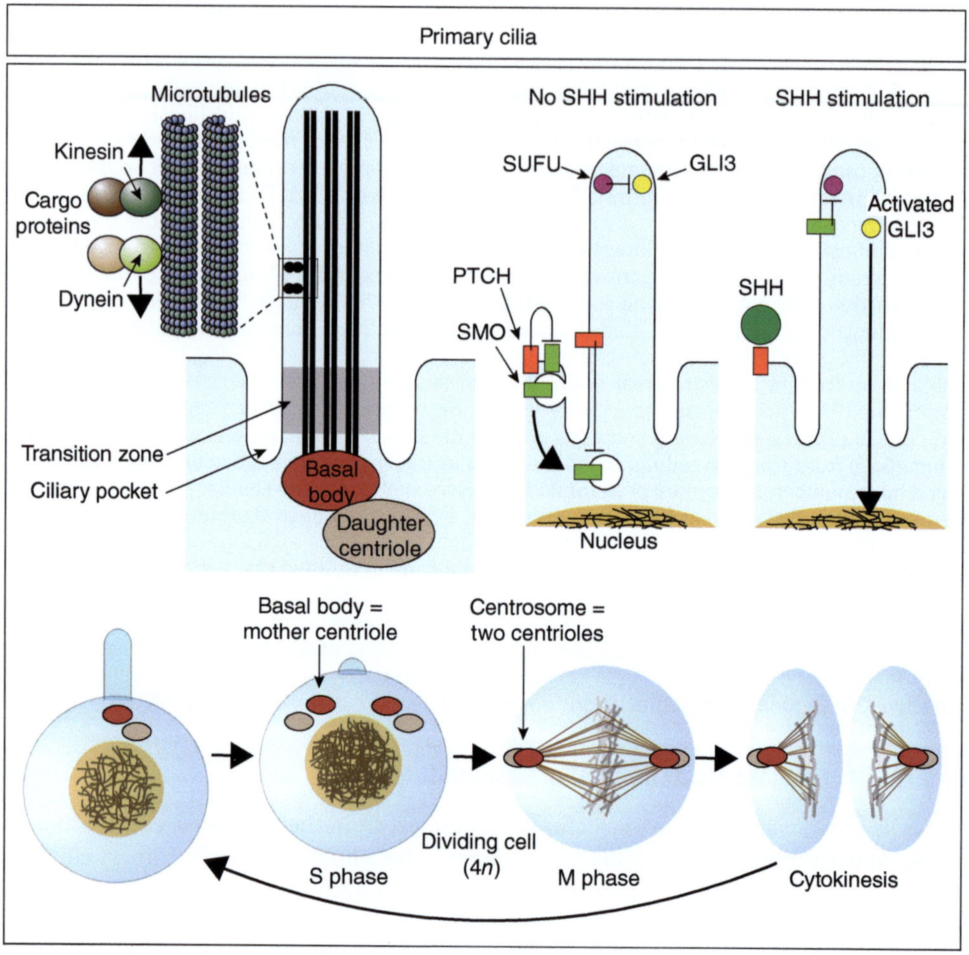

Figure 7.5 **Cilia are cells' antennae.** Cilia contain the machinery to transport molecules required for cell–cell signalling within themselves and between themselves and the rest of the cell. This is called intraflagellar transport, and it is needed for assembling and maintaining the cilia and for them to carry out their roles in signalling. Intraflagellar transport is achieved by transport proteins, called kinesins and dyneins, that use microtubules running the length of the cilia (a bit like railway tracks) to carry cargo proteins towards the tip of the cilia or towards the cytoplasm. A specialized domain at the base of the cilium, the transition zone, acts as a gate controlling protein entry and exit. Microtubules are anchored in a protein-based structure called the basal body, which is in fact one of the cell's **centrioles** (see Figure 7.10) and beside which sits the cell's other centriole. When a cell is getting ready to divide, it retracts its cilium and duplicates its centrioles (as well as its DNA of course). The newly produced centrioles are called daughter centrioles (shown here in light brown). Each of the two new cells gets one daughter. The other is called the mother centriole, and it is that one that forms the basal body of the cilium as it forms in each new cell. If you follow this through you will see that, at each cell division, the mother centriole stays as a mother and the daughter becomes a new mother. A relatively well-studied example of signalling through the cilium is shown here in a simplified form. In the absence of SHH, the SHH receptor patched-1 (PTCH1), which in its unbound state has the property of inhibiting a GPCR called smoothened (SMO), is enriched in the ciliary membrane. From there, it prevents SMO integrated into vesicles from entering into the cilium. This allows a pathway involving SUFU (suppressor of fused) to inhibit transcription factor GLI3's ability to act as an activator of gene expression by promoting its proteolytic cleavage to a repressor form. When SHH comes along and binds PTCH1, its inhibition of SMO is lifted. PTCH1 leaves the cilium allowing SMO in, where it inhibits the pathway through SUFU, thereby stopping proteolytic cleavage of GLI3. Full length GLI3 can now act as an activator of gene expression.

disruption by genetic or non-genetic factors is likely to have catastrophic outcomes that would probably compromise the viability of the early embryo, resulting in early miscarriage. For practical and ethical reasons, finding out whether this is indeed the case would be extremely difficult in humans. Less extreme variation of neural induction might be tolerated by embryos, allowing them to survive until later prenatal or postnatal ages, when the individual might show some type of disorder or neurodivergence. It would be hard to get unequivocal evidence of this because any conditions caused this way would be detected long after neural induction, making it difficult to rule out other causes arising at later ages.

Despite these difficulties, we do have evidence that some neurodevelopmental disorders *probably* have their origins during neural induction. This evidence comes from research on human stem cells (see Chapter 1 and Figure 1.1) and on development in other species. The research addresses how the early embryo gets its genome ready to receive and respond to the signals that specify cell fates during events such as neural induction. Developmental biologists often describe the ability of a cell to respond to such signals as its competence and refer to the process of establishing a competent state – in which a cell primes itself for action – as **epigenetic** priming. Why *epigenetic* priming? Remember from Chapter 4 that DNA's intimate association with, and regulation by, proteins whose molecular states and positions control gene expression is described as epigenetic. Changes in this association have a major impact on a cell's competence. The best understood example of epigenetic priming is the creation of bivalent **chromatin**, which is chromatin that is effectively poised for action by being in an intermediate ambiguous state (Figure 7.6).[183–185]

Kabuki syndrome is an example of a condition whose proximate causation might include a failure of epigenetic priming. This failure probably affects events as early as neural induction. Kabuki syndrome, which is characterized by abnormalities throughout the body and varying degrees of intellectual disability, is usually caused by variants of either the *KMT2D* or the *KDM6A* gene. What do these genes normally do? Both code for enzymes that affect **histone** methylation (note, to avoid confusion, that the methylation here is not of DNA) and KMT2D also enhances histone acetylation. KMT2D adds methyl groups to histones called H3K4 and helps acetylate histones called H3K27, and both of these actions loosen chromatin to allow gene activation. KDM6A is a demethylase that helps remove methyl groups from H3K27 histones, and since these methyl groups on H3K27 tighten chromatin and cause gene repression, their removal by KDM6A also allows gene activation. These actions will move chromatin from a bivalent, poised state to a more open state. These actions along with other histone modifications are required for the epigenetic regulation of genes including those encoding products required for neural induction.

In Kabuki syndrome, variants causing either KMT2D or KDM6A to lose function will result in too little gene activation (Figure 7.6). Which genes might be affected by the failure of the genome to open properly? We know very little about this. It seems likely that genes encoding transcription factors that regulate the development of specific cell types would be affected. One example is the transcription factor SOX2, whose major (although not exclusive) actions are to act internally within multipotent ectodermal cells to create a neuroectodermal identity in response to inductive signals. How SOX2 does this remains unclear. Variants of *SOX2* can cause numerous symptoms of brain disorder including delayed development, seizures and intellectual disability, and it is certainly possible that the origins of these phenotypes might eventually be traced back to disrupted neural induction.

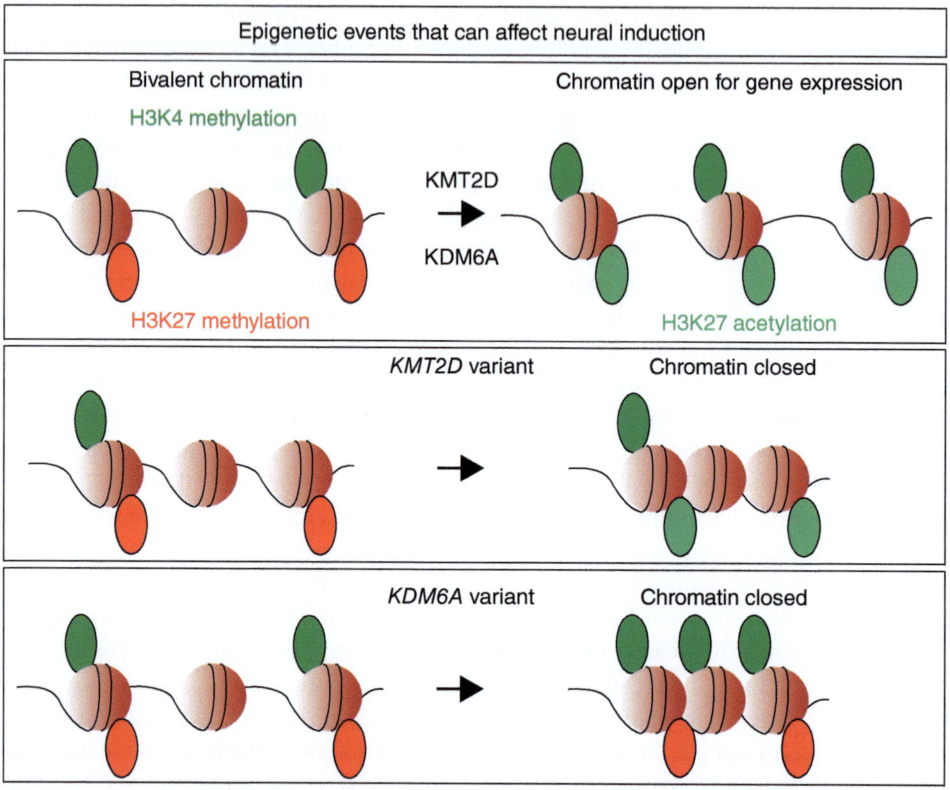

Figure 7.6 Bivalent chromatin and the effects of KMT2D and KDM6A variants in Kabuki syndrome. Chromatin is the complex of DNA and proteins in chromosomes. For a region of the genome to express its genes effectively that region needs to be in a loose or open state (see Chapter 4). Bivalent chromatin is in an intermediate, poised state: its DNA is wrapped around histones many of which are methylated so as to simultaneously enhance gene expression (H3K4 methylation) and repress it (H3K27 methylation). Normally, KMT2D and KDM6A would, respectively, add further H3K4 methylation and replace H3K27 methylation by acetylation, both of which loosen the chromatin for gene expression. Loss of KMT2D function in *KMT2D* variants will limit H3K4 methylation and prevent proper chromatin opening. Loss of KDM6A function in *KDM6A* variants will limit H3K27 acetylation, again preventing proper chromatin opening.

Derailment of Neurulation

Things are a bit clearer when we come to neurulation which, as I described in Chapter 1, follows neural induction and is accomplished by the folding of the initially relatively flat **neural plate** to become the **neural tube**. We know more about this event than we do about neural induction in humans. Disorders of neural tube formation are well recognized and studied. We can deduce when neural tube defects first arose based on the nature of the structural abnormalities combined with our knowledge of the process of neurulation in humans and other species. In Chapter 3, I described the main types of neural tube defects affecting the brain, of which **anencephaly** and **exencephaly** are prominent examples. The question for now is: how do the numerous genetic and non-genetic risk factors that ultimately cause these and other neural tube defects (which I discussed in Chapter 6) bring about failure of anterior neural tube closure?

The fact that the ultimate causes of neural tube defects are variable and multifactorial makes it hard to know what exactly has triggered the problem in an individual case. Nevertheless, in general terms, we know enough about many of the relevant risk factors to generate reasonable hypotheses as to the likely proximate causes of these sorts of defects. We know that the products of many of the genes whose variants might be involved play important roles in shaping the neural plate. They make it fold and zipper together to form the neural tube and ensure that the numbers of cells in the rapidly enlarging structure increase at rates that allow it to complete these manoeuvres.

Essentially, to turn a flat sheet of neural plate cells into a tube, the sheet must distort in such a way as to bring its lateral edges together. The first major step in this process is the bending of the neural plate down the middle of its anterior-to-posterior length. Researchers call this the median hinge point. They remain uncertain as to where the biomechanical forces that cause bending originate and how they operate. They probably come from numerous sources (Figure 7.7). One possibility is that cells at the median hinge point, by becoming wedge shaped with their narrow end towards the inside of the fold, pull cells on either side towards each other. These cells become wedge shaped at least in part due to the contraction of long thread-like intracellular structures made up of actin and myosin molecules, which are located at the cells' apex and act a bit like a tightening drawstring around the top of an old-fashioned purse (more about them in the next paragraph). Another possibility is that the lateral edges of the neural plate are pushed towards each other by the growth of the non-neural ectoderm (or surface ectoderm, which forms the skin). It is likely that this pushing force also contributes to the formation of additional bends in the neural plate on each side of the midline, the dorsolateral hinge points, which allow the lateral edges of the folding neural plate to approach each other. Once these edges get close enough, they contact each other initially through cellular protrusions that look like lamellipodia and filopodia. Lamellipodia are sheet-like structures (*lamella* is Latin for sheet) and filopodia are long and thin (*filum* is Latin for thread). Both are packed with actin molecules and they probably help pull the two edges together so they can fuse. Subsequent remodelling produces two discrete epithelial structures: neural tube overlaid by intact surface ectoderm (Figure 1.4 in Chapter 1).

I have mentioned actin twice in the previous paragraph. Its importance for neurulation cannot be overstated. Actin is the most abundant protein in almost all our cells. It is essential for creating, maintaining and modifying their external shapes and their internal organization. It does this by combining with other accessory and regulatory proteins to create the actin cytoskeleton (Figure 7.2). Individual actin molecules (or actin monomers) assemble into long chains (which we call actin polymers or actin filaments). The actin cytoskeleton is the cell's main force-generating machinery: when actin molecules polymerize to make longer filaments, they push; when they interact with myosin molecules, they slide along them and pull (which, as I mentioned above, occurs at the apex of the median hinge cells). Please note that although I am describing these functions of actin in the context of neurulation, it is essential in a wide array of other cellular processes too. For example, we shall see its role in cell division very soon.

The proteins that actin filaments combine with are critical for their function; for example, they regulate the length and stability of the actin filaments. We call them actin-binding

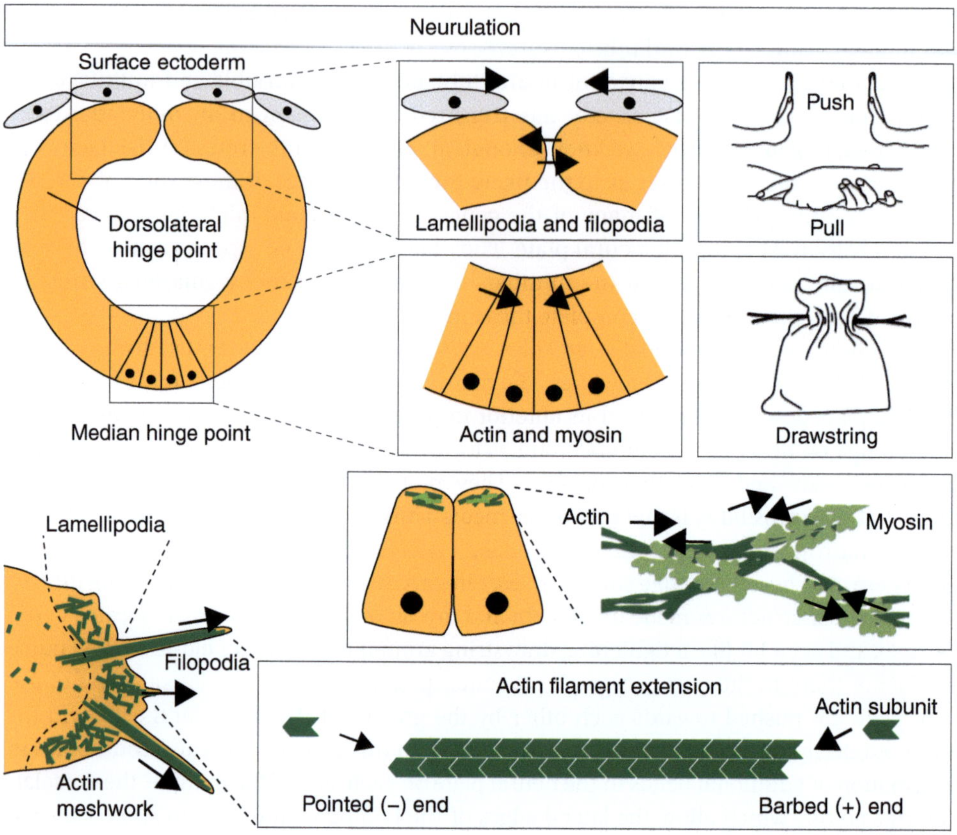

Figure 7.7 Probable mechanisms of neural tube closure. Forces that cause hinging of the neural plate and push and pull its two edges together are likely to be critical for proper closure. Many of these forces can be generated from within the neural plate by long filamentous molecules such as actin, which generates force through elongation or shortening and by interaction with other molecules such as myosin. Actin generates force by changing its length through the addition or removal of actin subunits at its ends (called, respectively, polymerization or depolymerization). Essentially, polymerization pushes and depolymerization pulls. Cellular processes called lamellipodia and filopodia protrude from the cells to reach across the narrowing gap between the two sides of the closing neural tube. Their reach is due to the elongation of actin filaments as they polymerize, more rapidly at their so-called barbed (or +) ends. Once contact is established, pulling the sides together would involve depolymerization to shorten the actin filaments, and this occurs mainly from their pointed (or −) ends (which is, essentially, the reverse of what is shown here). Some forces pushing the two sides of the closing neural tube together might be generated from outside, as cells in the surface ectoderm move together, dragging neuroectodermal cells with them.

proteins. The gene *SHROOM3*, variants of which have been particularly strongly implicated in neural tube defects including anencephaly, encodes one of these actin-binding proteins. It seems likely that variants of other actin-binding protein encoding genes will also turn out to be risk factors. In addition, variants of *MYO1E*, which encodes a type of myosin that interacts with actin in processes such as those described above, are considered likely risk factors. More research will be needed to identify how the dynamic events that contribute to neural tube formation are affected by these variants.

Front-line molecules such as actin and myosin must be directed and supported by many other cellular mechanisms that provide guidance, energy and so on. Neural tube closure is a highly choreographed process in which cells need to undergo complex movements and shape changes reliably and stereotypically. To complicate matters further, as they go through all of this, they are dividing to double their numbers many times over. Just imagine how precisely a ballet in which the number of ballet dancers keeps doubling throughout the performance would need to be choreographed! It is probably little wonder that numerous genetic and non-genetic risk factors with the potential to compromise processes fundamental to the inner workings of neuroectodermal cells can cause neural tube formation to go awry. Just as in a ballet, suboptimal performances for whatever reasons are likely to bring things crashing down. This probably explains why the risk factors for neural tube defects are so many and varied. For example, one well-known risk factor is low folate levels due to inadequate intake in the diet or resulting from recognized genetic risk factors such as **polymorphisms** of *MTHFR*, which encodes an enzyme required for the conversion of folate into its active form. Low folate can affect a cell's ability to proliferate or produce the energy it needs from its mitochondria. If the efficiency of either of these processes were to be undermined, the elaborate choreography of neural tube closure would surely be put at risk.

Turning to disruptions affecting the mechanisms that *direct* or *guide* the events of neural tube formation – in my analogy of a ballet, this would include the choreographers – there is evidence implicating variants of genes such as *CELSR1*, *DACT1*, *FUZ*, *FZD6*, *PRICKLE1*, *SCRIB*, *VANGL1* and *VANGL2*, whose protein products are involved in regulating what is known as planar cell polarity. These molecules coordinate the polarity of individual cells within the plane of a sheet of cells, such as the neural plate. Cells have polarity because their shapes and the distributions of their internal components are asymmetric (the exact nature of the asymmetry depends on what type of cell they are). Cell polarities need to be coordinated within a sheet so that the sheet as a whole can undergo appropriate shape changes. For example, the neural plate undergoes a process known as convergent extension, in which it gets longer and thinner shortly before – and in preparation for – folding to become a tube (Figure 7.8). If the actin and myosin that makes neural plate cells do this were to start pushing and pulling in an uncoordinated way, a mess would ensue and proper neural tube closure would be jeopardized. Variants of genes whose protein products are involved in intercellular signalling, including many of those I discussed in the previous section, have also been implicated in the generation of neural tube defects. Exactly how these signals help coordinate the choreography of neural tube formation and why their disruption derails it remain unclear.

As well as genetic variants, environmental factors other than low folate can increase the risk of neural tube defects. For example, exposure of the embryo to environmental toxins, a prominent example being fumonisin (see Chapter 5), are likely to cause defects because they kill cells, many of which will be those attempting to create the neural tube. Fumonisin has this effect by acting on biochemical pathways normally involved in intercellular signalling (Figure 7.9).[186]

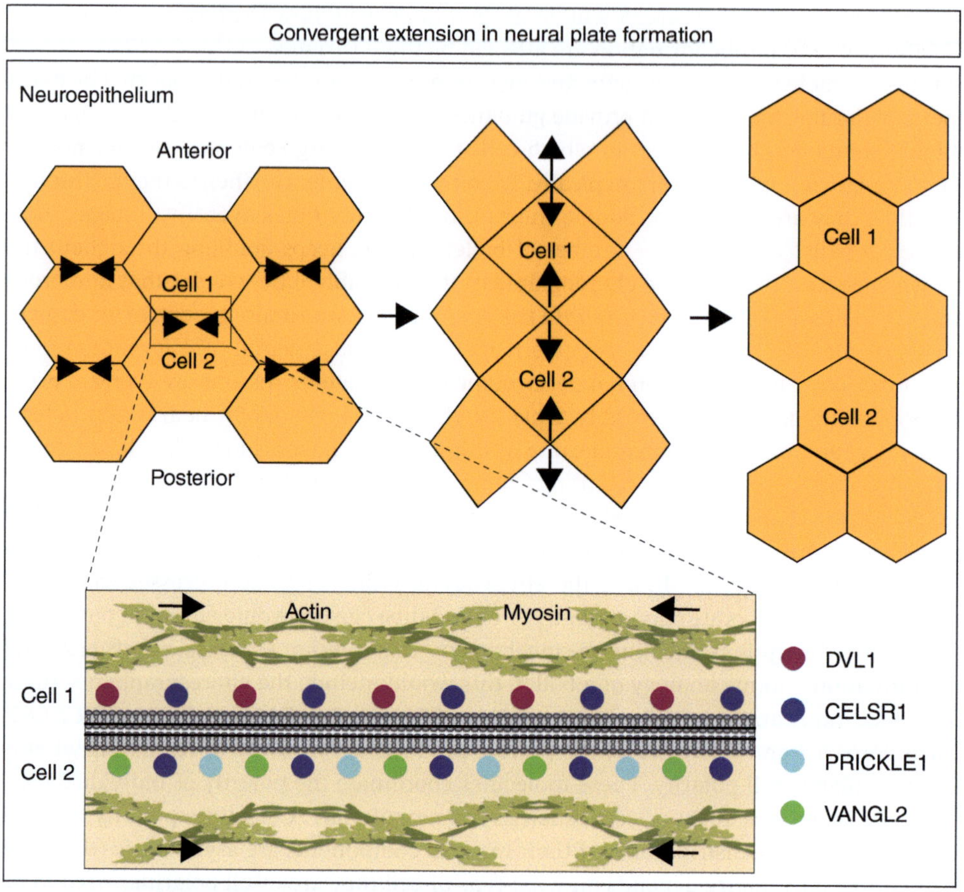

Figure 7.8 Convergent extension. The shaping of the neuroepithelium into the relatively elongated structure of the neural plate involves convergent extension. Individual cells are polarized with different sets of molecules beneath their anterior and posterior membranes. It is thought that these molecules regulate the contractility of actin and myosin complexes beneath them in the cell. Contraction of these complexes shortens the faces of the cell membranes where each cell contacts the one in front of it, as shown by the arrowheads in the diagram at the top left or by the arrows in the enlargement. This contraction reduces contact between cells along their anterior and posterior faces, but creates new points of contact between cells that were initially separated along the medial-to-lateral axis. Note how contact between Cells 1 and 2 in the middle diagram has shrunk to a point, but the cells to the right and left of Cells 1 and 2 are now in contact. These new contacts then expand, as shown by the arrows, thereby separating Cells 1 and 2, as seen at the top right. The overall result is elongation along the anterior-to-posterior axis.

How Cell Proliferation Can Be Disrupted

How Genetic Variants Might Affect Cell Proliferation

Many genes have direct or indirect effects on the process of cell division. Variants of these genes are likely to cause disorders of brain size, particularly **micrencephaly** and

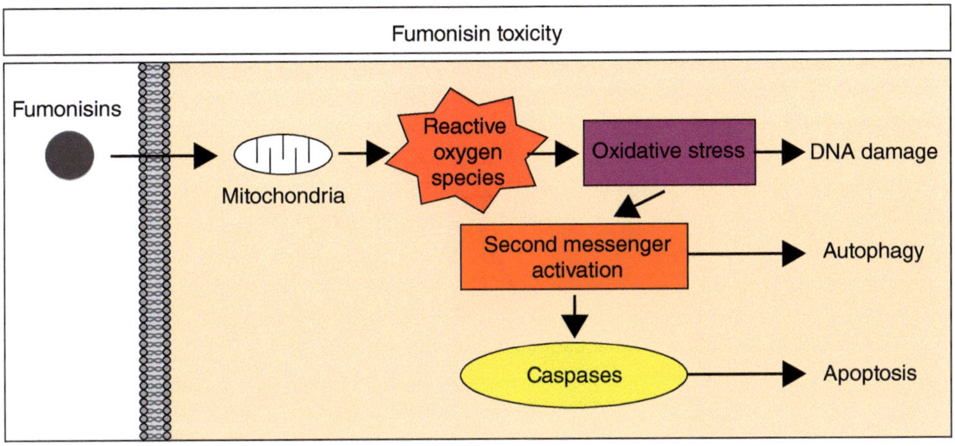

Figure 7.9 Mechanism of fumonisin toxicity. Fumonisins operate on mitochondria to induce oxidative stress and the production of **reactive oxygen species** (oxidative stress is an imbalance between reactive forms of oxygen and antioxidants). This affects the second messenger pathways that have evolved to respond to intercellular signalling, in such a way as to activate caspase enzymes responsible for apoptosis (programmed cell death) and autophagy (the breaking down of cellular components).

megalencephaly (Chapter 3), as well as many other abnormalities. It is worth bearing in mind that the actions of these genes are rarely restricted to dividing brain cells, and their variants are likely to affect the development of other parts of the body too, which might then impact brain development indirectly.

Sometimes disorders of cell proliferation occur because variants *directly* affect the functions of proteins with the most immediate involvement in the mechanics of cell division – proteins that, in other words, act at the frontline as part of the machinery that makes cells divide. Some variants affect proliferation rather more *indirectly*, affecting the mechanisms that control the machinery of cell division via direct effects primarily on other cellular processes. For example, they might affect the types, levels and activities of frontline proteins by acting on the expression of the genes that encode them or they might affect cell signalling molecules that control frontline protein types, levels and activities.

Genetic Variants Can Directly Affect the Machinery of Dividing Cells

Some of our proteins are needed by every dividing cell, be it a brain cell or not, to carry out the organization, condensation and separation of its chromosomes during the **cell cycle** (Figure 7.10). These proteins create the molecular machinery required for the very act of splitting a cell into two new ones. Variants of genes such as *ASPM*, *CDK5RAP2*, *KIF11* and *MCPH1* affect **centrosomes** and **mitotic spindles**, which pull replicated chromosomes apart into the two new daughter cells[187]; variants of the *ESCO2* gene affect mechanisms holding replicated chromosomes together (we call this **sister chromatid** cohesion) until the time comes for them to separate. Variants of these genes can cause micrencephaly, which might develop as a part of a broader syndrome, such as Juberg–Hayward syndrome in the case of *ESCO2* variants.

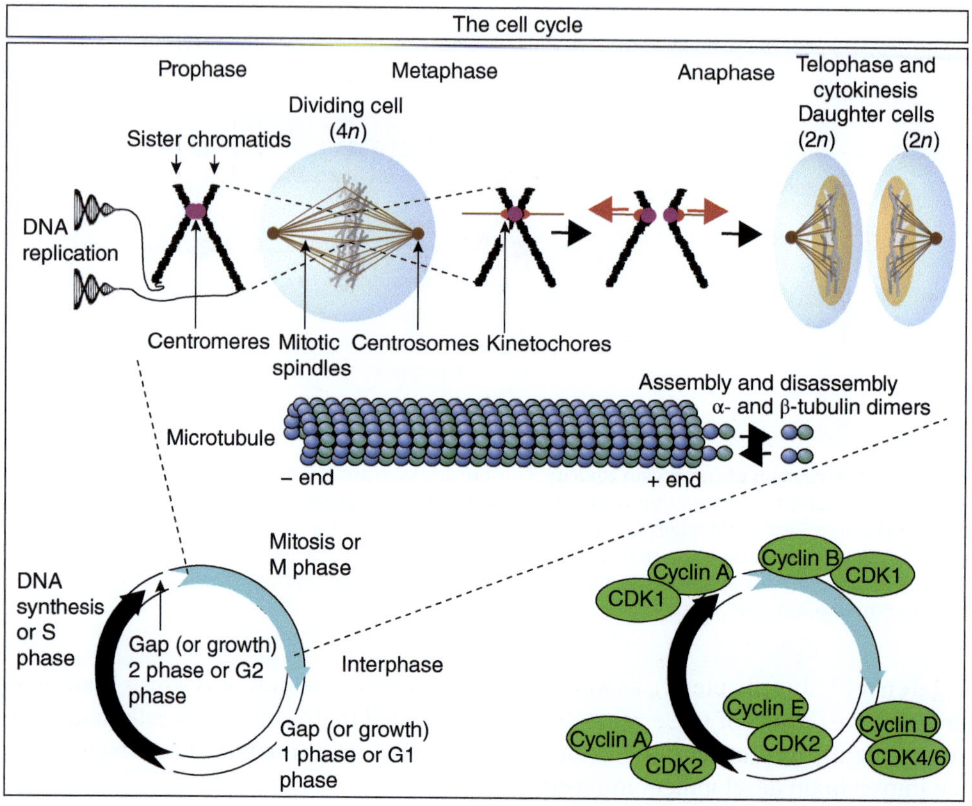

Figure 7.10 The cell cycle. Cell division (or cytokinesis, from the Greek *cyto* meaning cell and *kinesis* meaning movement) occurs during M (standing for mitosis) phase of the cell cycle. M phase is split into prophase, metaphase, anaphase and telophase. During prophase, the replicated DNA (4n, meaning four copies of each chromosome) condenses into sister chromatids linked at their **centromeres**. During metaphase, sister chromatids align along the middle of the cell, with their centromeres attached via kinetochores to mitotic spindles, which are made of microtubules. The mitotic spindles tether each of a pair of sister chromatids to one or other of the cell's centrosomes (centrosomes are the main organizers of microtubules in animal cells; they are often referred to as microtubule organizing centres). It can be hard to remember the difference between the words we use: centromeres (meaning the central part; mere means part) of sister chromatids are attached to centrosomes (meaning the central body; soma means body), and centrosomes are made of centrioles (meaning small centres). Microtubules are made of a protein called tubulin; at one end, their + end, they grow through the assembly of tubulin subunits, and they contract by their disassembly. Having extended to link centromeres and centrosomes, they contract during anaphase to pull pairs of sister chromatids apart. During subsequent telophase, a nuclear membrane reforms around each set of chromosomes (now back down to 2*n* per cell again). A dividing cell that is *not* in M phase is in a state we call interphase. During interphase, a cell that is going to divide must replicate its DNA, which it does during S (standing for synthesis) phase. Before and after S phase, there are two gap (or G) phases, G1 and G2, during which the cell grows and is influenced by signals received from its surroundings (if a cell is not going to divide again, we say it is in G0). Influences such as signals from a cell's surroundings affect the cell cycle by acting on its regulatory proteins, prominent among which are cyclins and cyclin-dependent kinases (CDKs). They, together with other proteins not shown here, form complexes: different combinations of regulatory proteins are responsible for controlling progression at different stages of the cell cycle.

Variants of genes encoding cyclins and cyclin-dependent kinases (CDKs), the evolutionarily highly conserved direct regulators of a cell's progression through the cell cycle, are commonly associated with cancer (Figure 7.10; these genes are sometimes collectively referred to – rather vaguely – as cell cycle genes). Some have been associated with rare neurodevelopmental disorders: for example, variants of several of these genes are implicated as the causes of poor brain growth, manifested as micrencephaly, **lissencephaly**, cerebellar **hypoplasia** and intellectual disability.[188] Whether these phenotypes result from effects on only – or even mainly – cell cycle progression remains unclear since, as is so often the case, many of these cell cycle regulators can have effects on other cellular processes too.

Transcription Factor Variants Can Have Indirect Effects on the Machinery of Dividing Cells

Some proteins that control cell proliferation, as well as many other processes I will discuss below, are transcription factors. Genes whose variants can affect the proliferation of brain **progenitors** include *ARX* (which is also involved in cell migration, see below), *FOXG1* and *PAX6* (both of which are also involved in the specification of cell types, see below) and *FOXP1* and *NFIX* (which are also involved in progenitor cell differentiation). Variants of these genes cause various conditions including Proud syndrome, FOXG1 syndrome, congenital aniridia, intellectual disability-severe speech delay-mild dysmorphism syndrome and Marshall–Smith syndrome. Deletions of part of chromosome 11 causing loss of both the *PAX6* and the *WT1* genes (WT1 is also a transcription factor) are thought to cause Wilms tumour–aniridia–genitourinary anomalies–mental disability syndrome, in which an increased risk of kidney cancer (Wilms tumour) is associated with eye problems (aniridia) and abnormalities of the genitourinary system, developmental delay, intellectual disability and other psychiatric symptoms.

Individual transcription factors have the potential to control the expression of many hundreds of other genes (referred to as downstream genes). This is one reason why the syndromes associated with their variants include such widespread effects, as illustrated by examples in the previous paragraph. Exactly which ones are controlled by a particular transcription factor depends on variables such as the cell's stage of development, its type, the nature of its extracellular environment and so on. This makes it challenging to work out how such transcription factors influence cell proliferation and, therefore, how exactly the variants produce symptoms. There is evidence that some of them can directly affect the expression of proteins that are part of the machinery of cell division: for example, the proteins whose expression *NFIX* variants affect include mitotic spindle components. Transcription factors might also, or instead, exert indirect effects by altering the expression of cyclins and cyclin-dependent kinases (CDKs) (Figure 7.10). Effects on the production of these sorts of proteins is likely to affect parameters such as the time it takes cells to divide. For example, there is evidence that the PAX6 transcription factor regulates cell cycle length by controlling the levels of expression of genes including *CDK6*. Another way in which transcription factors might indirectly affect cell divisions is by influencing the expression of genes encoding intercellular signalling molecules involved in regulating cell divisions.

For example, there is evidence that variants of the FOXP1 transcription factor do this in causing intellectual disability–severe speech delay–mild dysmorphism syndrome.

Signalling Molecule Variants Can Have Indirect Effects on the Machinery of Dividing Cells

Variants of the genes encoding either signalling molecules themselves or their receptors and second messenger pathways can have profound effects on the numbers of cells the brain generates. We can see how such variants might cause structural malformation through a catastrophic breakdown in the coordination of brain cell proliferation by considering possible causes of **holoprosencephaly**.

If you remember from Chapter 3, holoprosencephaly originates from problems occurring as the early single-chambered **forebrain** forms into the two distinct **telencephalic** vesicles. In extreme cases, it can leave the forebrain as just one chamber. Normally, the conversion of a single chamber to two chambers is achieved by ensuring that the midline of the forebrain expands less than the right and left cerebral hemispheres. Imagine blowing up a balloon with a rubber band around the middle of it and you have the idea. It is thought that one reason for the midline expanding less is that its cells proliferate less and more of them undergo apoptosis (or programmed cell death; Chapter 1). Scientists think that these differences between the midline cells and those in the surrounding, more rapidly expanding cerebral hemispheres are controlled by signalling proteins that are concentrated in particular parts of the midline (Figure 7.11). These proteins include many of those listed (as ligands) in Figure 7.3: SHH, BMP, WNTs and FGFs. Variants that disrupt the functions of genes encoding some of these proteins – including *SHH*, one of the *FGFs* (*FGF8*), and a gene encoding one of the FGF

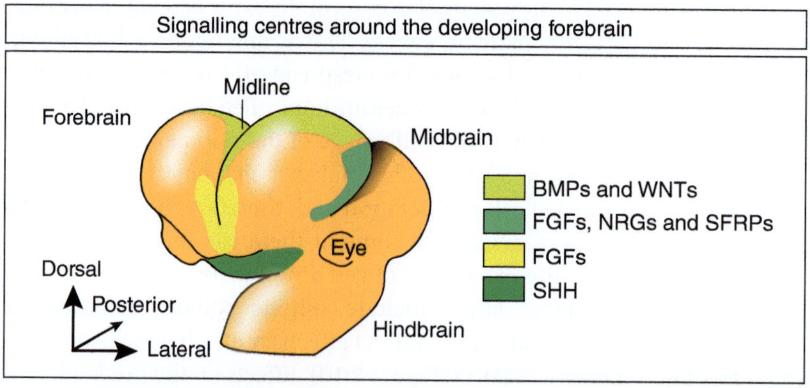

Figure 7.11 Regions of the embryonic forebrain expressing high levels of signalling molecules. A number of important signalling molecules are produced at high levels around the edges of the early embryonic forebrain. One of their likely roles is ensuring that rates of cell proliferation in different regions are regulated to create structures of specific sizes. For example, they might keep proliferation lower in the midline than in surrounding cerebral cortex, thereby allowing the latter to expand more than the former to form two distinct cerebral hemispheres (for abbreviations, see Figure 7.3).

receptors (*FGFR1*) – are likely to cause some cases of holoprosencephaly.† It remains unclear how these variants disturb the coordination and balance of cell proliferation between different parts of the developing brain. In general terms, they might reduce differences in proliferation between different regions by reducing the activity of signalling molecules normally enriched in specific regions. By levelling out the overall effects of signalling on proliferation across the developing brain, they might hamper the emergence of regions of very different size, such as the midline and the cerebral hemispheres. For overall context and balance, I should say that many other genetic and environmental factors have also been implicated as causes of holoprosencephaly, although how they have their effects is not well understood.

Moving on to other conditions, we find many instances of genetic variants affecting development by disrupting key components of second messenger pathways. These pathways are very complicated, and I will focus on some illustrative examples here.

Genetic variants that render some of the proteins in second messenger pathways persistently active can cause disorders and increase the likelihood of tumour development. These variant forms are often described as oncogenes (*onco-* is Greek for tumour); the non-persistently activated form of the gene is known as a proto-oncogene (*proto-* is Greek for 'earliest form of'). Costello syndrome, for example, in which neurodevelopmental delay and intellectual disability are associated with abnormalities of cell proliferation, is caused by persistently active variants of the *HRAS* gene, which encodes a member of the RAS family of proto-oncogenes (Figure 7.4).

Molecules that normally inhibit these second messenger pathways include a protein called PTEN, which keeps growth under control by limiting proliferation and triggering apoptosis. It does this by removing the phosphate groups from other proteins, such as phosphatidylinositol triphosphate (PIP_3; Figure 7.4). *PTEN* variants that interfere with this function can cause conditions such as Bannayan–Riley–Ruvalcaba syndrome, in which megalencephaly frequently occurs, probably as a result of abnormally high rates of cell proliferation and reduced rates of apoptosis. Since proteins like PTEN suppress the uncontrolled cell proliferation that can sometimes result in tumour formation, they are referred to collectively as tumour suppressors. Other examples of tumour suppressor proteins include TSC1 and TSC2, which join together (or we can say they *hetero*dimerize, which means two different molecules combine) to inhibit mTOR (which stands for mammalian target of rapamycin), resulting in inhibition of cell proliferation among other processes (Figure 7.4). TSC1 and TSC2 integrate signals from numerous signalling molecules in addition to those in Figure 7.4, including WNT signalling through β-catenin (Figure 7.3). Genetic variants of *TSC1* or *TSC2* that interfere with their function cause a condition known as tuberous sclerosis (which is where TSC comes from), in which benign tumours appear in the developing brain (see Chapter 3, Figure 3.2), associated with a range of symptoms of variable severity, including intellectual disability and epilepsy.

† Variants of genes encoding the receptors of SHH (which are called patched and smoothened) can have the opposite effect of *increasing* the activity of SHH's second messenger pathway, in which case uncontrolled proliferation and childhood cancers such as medulloblastoma are the result.

Variants Affecting Primary Cilia Can Interfere with Signalling and Affect Proliferation

Progress on understanding the roles of primary cilia in intercellular signalling, which I discussed earlier in this chapter, has shed light on how some genetic neurodevelopmental disorders are caused. Genetic variants that interfere with the construction, structure or function of primary cilia interfere with signalling to cause a range of disorders known collectively as ciliopathies.[179,189,190] Some of these variants are likely to have their primary effects on ligands, receptors and second messengers whose actions are required for the construction of cilia. If disruption of these signalling pathway components results in the production of faulty cilia with reduced ability to detect those very same signals as well as other signals, a vicious circle is likely to be entered in which the initial failure of signalling is compounded by a resulting failure of cilia to detect that and other signals, thereby exacerbating the problem. This might underlie conditions such as Curry–Jones syndrome, in which variants of the gene *SMO* (which encodes the SHH receptor smoothened, which is a GPCR; see Figures 7.4 and 7.5) interfere with SHH signalling, leading to **agenesis** of the **corpus callosum** and intellectual disability. Because SHH signalling seems to be involved in initiating the construction of cilia, the proximate causes of this condition might include not only SHH signalling disruption due directly to defective SMO function but also disordered ciliogenesis. The latter will both worsen the problems with SHH signalling and cause collateral damage to other signalling pathways that use cilia. It is important to keep an open mind given the myriad of possible interactions that can be affected in situations such as these.

More than thirty ciliopathies have been identified so far, and many of their phenotypes are likely to result from an effect on cell proliferation. For example, Joubert syndrome, Coach syndrome or Meckel syndrome are ciliopathies caused by variants of genes including *ARL13B*, *INPP5E* or *TMEM67*. These genes encode proteins that play multiple roles in cilia, including in regulating their size (*ARL13B*), membrane composition (*INPP5E*) and the gating function of their transition zones (*TMEM67*). The hypoplasia (underdevelopment) of parts of the brain seen in these conditions probably occurs at least in part because these variants, by interfering with the assembly, disassembly and proper function of cilia in signalling, affect progression through the cell cycle.

Polygenic conditions that disrupt primary cilia formation and function include Down syndrome (Trisomy 21).[191] Evidence indicates that this is due in part to raised cellular levels of a protein called pericentrin, which is a centrosome protein encoded on chromosome 21. Pericentrin is known to play a role in ciliogenesis – not surprisingly, since centrosomes contain **centrioles** and centrioles are essential components of cilia (Figure 7.5) – but exactly how increased levels of pericentrin disrupt ciliogenesis is unclear.

How Environmental Variation Can Affect Cell Proliferation

At the start of Chapter 5, I explained that I take environmental factors to include not only those that arise outside the body but also those that affect the developing brain from outside it but from within the rest of the body, and I pointed out that some of the latter might originate in an individual's own genetics. We can see examples of this when we look at the causes of the brain abnormalities in conditions such as Apert syndrome or Crouzon syndrome, which are caused by *FGFR2* variants. The main primary effects of these variants are not on the brain directly but on the bones of the skull, causing them to fuse abnormally early, before birth. This condition, known as craniosynostosis, prevents the normal expansion of the growing brain, which increases the pressure inside the **ventricles** (**hydrocephalus**) causing their expansion (**ventriculomegaly**), thereby reducing brain cell proliferation.

Other environmental factors that impact the developing brain indirectly, by having their primary effects on structures outside the brain, include cerebral hypoxia arising for reasons such as an interruption of the supply of blood to the brain due to trauma or infection. Cerebral hypoxia in the fetus can impair growth of the cerebral cortex, leading to conditions such as **porencephaly** (Chapter 3).

A number of chemicals can affect brain development in a range of ways that include reduced brain cell proliferation and excessive cell death. Notable examples include mercury or its organic form methylmercury (which is produced by microbes in water contaminated with mercury) and alcohol. These chemicals likely have their detrimental effects on proliferation – and on other developmental processes – through fairly ubiquitous molecular effects. For instance, mercury has a high affinity for thiol groups (also called sulfhydryl groups) that link parts of a protein together to give them their three-dimensional structures (their tertiary and quaternary structures: Figure 7.12). By interfering with this process, mercury can inhibit DNA synthesis, alter the ability of proteins involved in the cell cycle to function, disturb intercellular signalling and so on. Alcohol has been shown to have a broad effect on many aspects of the intercellular signalling pathways I described in the previous section.

There is strong evidence that some infections can have direct effects on brain cell proliferation. For example, Zika virus, spread to pregnant women by mosquitoes, can impair brain cell proliferation and increase brain cell apoptosis, causing micrencephaly. The evidence suggests that infection of the fetus has a direct effect disrupting particularly the divisions of cerebral cortical **radial glial progenitors** (Figure 7.13).

There are probably countless other types of environmental factors that affect early brain cell proliferation, many of which we do not know much if anything about. Factors that are being studied increasingly include those that fall under the heading of maternal mental health, such as prenatal anxiety and stress. For example, evidence is accumulating that maternal anxiety and stress can have effects on the developing brain by increasing maternal levels of stress hormones, called glucocorticoids, which circulate in the blood and can stimulate the proliferation of cerebral cortical cells.[192] Remember that problems can arise during development due to both over- and underproliferation, since both will interfere with the choreography of brain development and, for any of a large number of possible reasons, leave it more vulnerable than it has evolved to be.

How Cell Migration Can Be Disrupted

The migration of cells from the place of their birth to their eventual site of differentiation and function is a major activity in the embryonic and fetal brain (Chapter 1). Many neurons and glia migrate, radially and/or tangentially, entirely within the brain. Other cells, such as

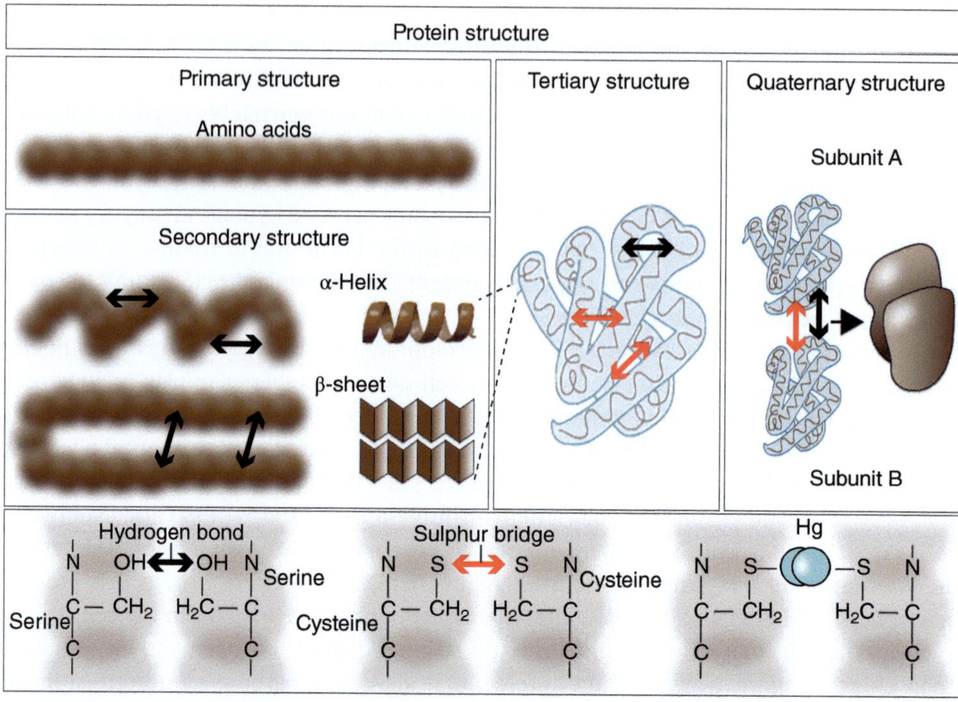

Figure 7.12 Protein structure and the effect of mercury. A protein's primary structure is the sequence of amino acids along its length. A protein's secondary structure is the way in which its primary structure folds locally to form, most commonly, either an α-helix or a β-sheet. Secondary structure is held in place by hydrogen bonds between the oxygen atom of one amino acid and the hydrogen of another (the example here is of bonding between two serine amino acids). Tertiary structure is the overall three-dimensional structure of a protein that is folded into its secondary structure. Many types of bonds can hold the tertiary structure in place, including hydrogen bonds, but one type, the disulphide bond or sulphur bridge, is stronger than others. Disulphide bonds form between the sulphur atoms of cysteines, which are in their sulfhydryl (or thiol) groups, which consist of a sulphur atom bonded to a hydrogen atom (-SH). They stabilize protein folding and overall structure. Many proteins join specifically with others (in which case we call each of them a protein subunit) to make a larger conglomerate protein; this is their quaternary structure, and it is held together by the same types of bonds that hold tertiary structure in place. Mercury (Hg) has a high affinity for sulphur atoms, and it disrupts the strong disulphide bonds that hold tertiary and quaternary structures in place.

microglia, migrate into the brain from elsewhere. Before discussing how cell migration can be disrupted, I need to explain some of the key cellular processes involved in cell migration and introduce some important molecules, of which there are many. Interference with any of them is likely to affect cell migration. Any genetic or non-genetic factor that disrupts the ability of cells to migrate, by affecting either the migrating cells themselves or the environment in which they are migrating, is likely to impact the development of the prenatal brain. Some of those factors, probably many of them, are likely to be ones also affecting other processes discussed in the previous and future sections.

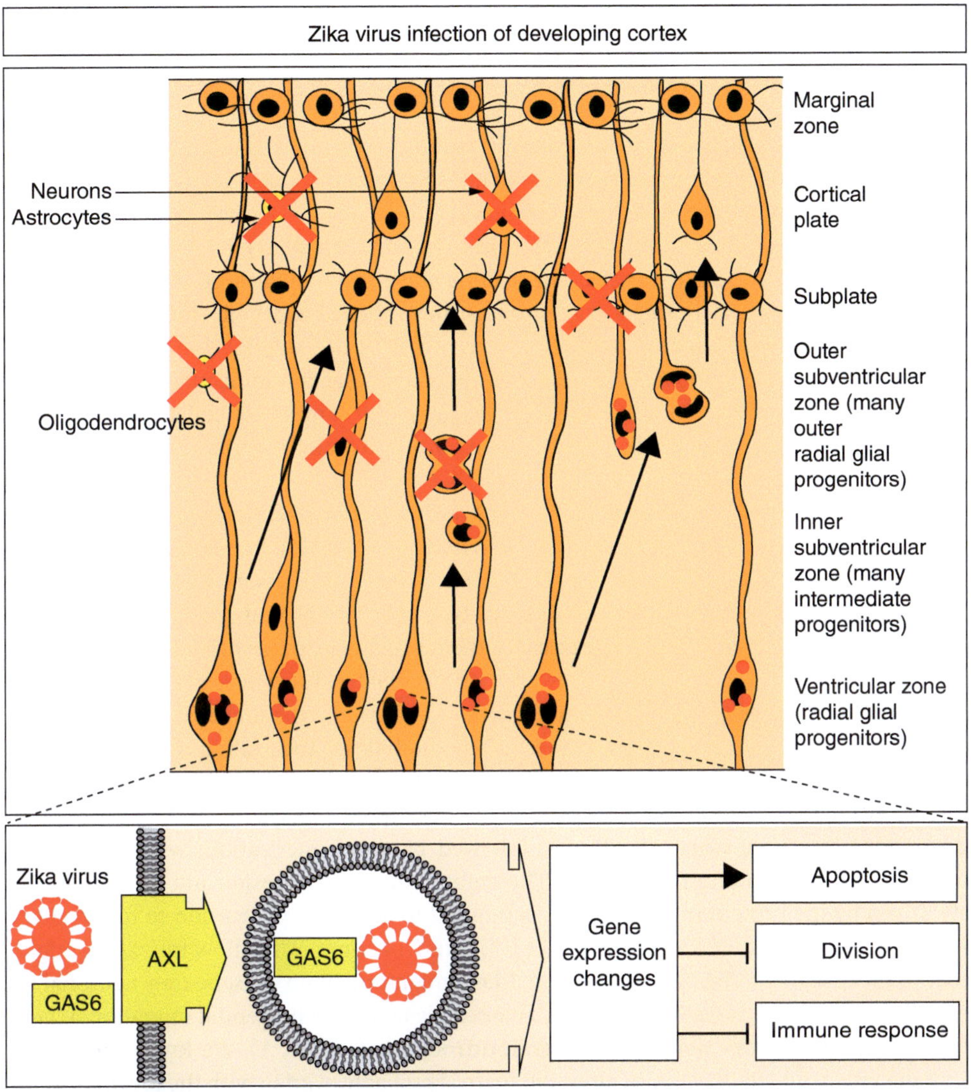

Figure 7.13 **The brain cells targeted by Zika virus.** Zika virus can cross the placental barrier of infected pregnant women to reach the fetal brain. Once it has reached the brain, its main targets include radial glial progenitors and intermediate progenitors (red dots). This reduces the numbers of those cells and their output of other cell types, including neurons, astrocytes and oligodendrocytes as a consequence of apoptosis, cell cycle dysregulation and suppressed immune responses, leading to micrencephaly and other symptoms. Zika virus gains entry to cells that express a tyrosine kinase receptor called AXL, which brain progenitors do. AXL is activated by a ligand called GAS6 (growth arrest-specific 6) which also happens to bind to a molecule (phosphatidylserine) exposed on the surface of the virus and bridges the virus to the AXL receptor. GAS6 binding to AXL causes a process called endocytosis in which a vesicle is pinched off from the membrane containing both ligand and receptor; if Zika virus is bound to GAS6, it comes along too. Once in the cell, it can act on internal pathways leading to gene expression changes that have detrimental cellular consequences. This is an example where the biochemistry of some cell types renders them particularly vulnerable.

Mechanisms of Cell Migration in the Developing Brain

Cell migration involves intimate interactions between cells that must stick to others only at the appropriate times. The roles of cell adhesion molecules during cell migration are likely to be particularly complex and dynamic because migrating cells need to change their adhesion to other cells in a highly coordinated way in order to move. Our understanding of how cells control these dynamics is still quite limited.

As I discussed in Chapter 1, the long fibre-like extensions of radial glial progenitors guide many migrating neurons in the brain. The leading edges of the migrating cells wrap themselves around and adhere to these extensions as they move along them (Figure 7.14). The results of experiments with cells in culture or in other species suggest that cadherins (see previous section and Figure 7.2) together with two cell adhesion molecules, astrotactin 1 and astrotactin 2, which do not belong to any of the major families of cell adhesion molecules I described earlier, are likely to be essential for this process in human cerebral cortical and cerebellar development.

Changes in the cytoskeleton drive movement of the cell, but here the main forces come from the cytoskeleton's microtubules rather than its actin filaments. Remember, from earlier, that microtubules are polymers of tubulin (Figure 7.10). They are very dynamic and their behaviour is regulated by a number of different types of proteins that, for example, bind to them to make them more or less stable or cause them to elongate or to shrink and, as a consequence, power cell movements. They grow from that specialized structure we met before, the centrosome. In these cells, the centrosome comes to lie between the nucleus and the leading edge of the migrating cell and extends one set of microtubules to the leading edge and another to surround the nucleus. Existing evidence suggests that movement of the cell occurs in a series of short repeated bursts, each comprising microtubule polymerization pushing the leading-edge forwards, where it adheres, followed by depolymerization, which pulls the nucleus and the cell's trailing edge after it. The trailing edge loses adhesion and contacts, probably driven by the protein myosin, which is thought to add extra propulsion to the nucleus (like squeezing the end of a toothpaste tube). Figure 7.14 summarizes how migration might occur in the case of guided radial migration. Many migrating brain cells, be they neuronal or non-neuronal, do not use radial guidance, an example being the tangential migration of cerebral cortical **inhibitory** neurons or **oligodendrocytes** (Chapter 1). We know much less about the way in which they move, although there is probably overlap with the mechanisms I have described in this section.

As was the case for genetic variants affecting cell proliferation, cell migration can be affected either directly, as a result of variants directly impacting the molecules that actually move cells, or indirectly, as a result of variants altering the control mechanisms that regulate, guide or modulate how cell movements occur in different cell types, at different locations and at different times.

Genetic Variants Directly Affecting the Mechanics of Migrating Cells

Variants of genes that encode proteins directly involved in the mechanisms that cells use to move themselves, as outlined in the previous section, can cause migration defects. These variants are also liable to affect many additional cellular processes that rely on cell

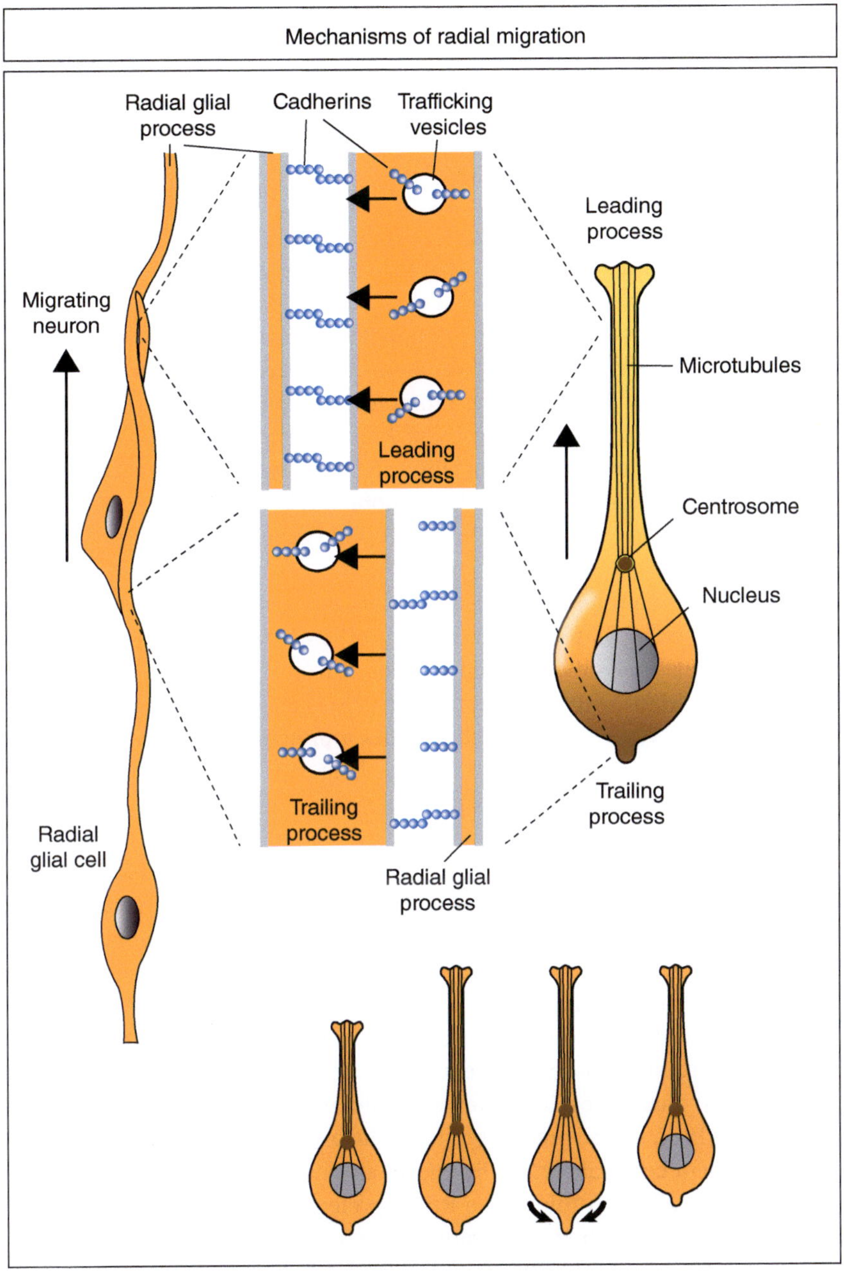

Figure 7.14 Current model of how radial neuronal migration occurs. Localized trafficking of cadherins to and from the plasma membrane of a migrating neuron allows its leading process to stick to its guiding radial glial process and its trailing process to detach. Cadherins are added to or removed from the plasma membrane by intracellular vesicles. In migrating neurons, a network of microtubules emanating from that busy body, the centrosome, extends forward into the leading process and backward to enmesh the nucleus. The sequence of diagrams at the bottom summarizes the key stages of cell migration: first, the leading process extends, likely driven by microtubule extension; next, the centrosome moves forward into the leading process and, at the same time, the trailing end of the cell contracts (arrows), driving the rear of the cell forward; then the nucleus moves forward towards the centrosome and the trailing process retracts. The cell repeats these steps until it reaches its destination.

adhesion, the cytoskeleton and so on, including cell proliferation (see above) and **dendrite** and axon growth (see later).

Some variants affect migration by interfering with the cytoskeleton. Variants of the *ACTB* or *ACTG1* genes that encode actin cytoskeletal proteins (Figure 7.2) can cause **pachygyria** (in which the cerebral cortex has abnormally few sulci and gyri). Lissencephaly can be caused by variants of *TUBA1A* and *TUBB2B*, which encode tubulin proteins; *PAFAH1B1* (also known as *LIS1*), whose protein product is a regulator of a so-called motor protein, dynein, which walks along microtubules transporting molecular cargo and generating force (we encountered this molecule in the section above on cilia; Figure 7.5)[193]; *DCX*, which encodes a protein that interacts with dynein and promotes the organization and stability of microtubules.[194] It is thought that loss of one copy of *PAFAH1B1* plays a major role in the development of lissencephaly, intellectual disability and seizures in Miller–Dieker syndrome, which is caused by a deletion of part of chromosome 17. Variants of *PAFAH1B1* and *DCX* have also been implicated in causing brain **heterotopias** (misplaced collections of cells that have failed to migrate normally; see Chapter 3, Figure 3.2). So too have variants of *FNLA*, which encodes a filamin protein whose functions include attaching to actin to help it create the branching network of filaments that make up the cytoskeleton. It does this by binding to two actin molecules and holding them at an angle to each other, creating a meshwork. Other genes whose variants cause structural brain disorders through their effects, at least in part, on migration include *TBCE*, which encodes a protein that is involved in the correct folding of tubulin proteins (we call proteins that help with folding chaperone proteins) and *MYO5A*, which encodes a type of myosin (remember, myosin 'squeezes the toothpaste tube at the end', Figure 7.14).

Some variants might affect migration by interfering with cells' ability to adhere. This can result in abnormalities on the lissencephaly–pachygyria spectrum (Chapter 3). For example, variants of *L1CAM* have been implicated as a common cause of congenital hydrocephalus. Variants of *FKTN* that modify a protein called α-dystroglycan, which helps moving cells grip their surroundings, cause a brain malformation called cobblestone lissencephaly in which the cerebral cortex has reduced gyri and sulci, looks bumpy and has numerous other defects. Alpha-dystroglycan is an example of an extracellular glycoprotein (that is, a protein with carbohydrate side chains) that helps link the extracellular matrix to the cytoskeleton (Figure 7.2). Variants of *RELN*, whose protein product is also a large extracellular glycoprotein, can also cause a form of cerebral cortical lissencephaly and cerebellar hypoplasia. This is possibly because the cell adhesion role of the RELN protein helps instruct immigrating cells where to stop. Interference with cell adhesion is also known to cause polymicrogyria (too many small gyri; see Chapter 3, Figure 3.2). For example, variants of the *ADGRG1* gene have been implicated in this disorder. *ADGRG1* encodes a type of cell surface GPCR (receptors discussed earlier in this Chapter) involved in cell adhesion, among other processes, by virtue of their especially extensive extracellular regions containing a variety of adhesion domains. Examples of polygenic disorders whose phenotypes might be caused by disruption of processes including cell adhesion include Down syndrome (trisomy 21); the *DSCAM* gene is located on chromosome 21, and there is evidence implicating alterations of DSCAM function in some of the neurological features of Down syndrome. The details of how the adhesion defects caused by these variants result in, or contribute to, their associated disordered phenotypes remain unclear.

Some Genetic Variants Affect Migrating Cells Less Directly

Unlike the variants affecting cell migration by creating variation in the nature of molecules that contribute to the machinery of cell movement itself, some variants operate at least one step removed from the machinery of cell movement. They might, for example, influence how well the molecules that produce movement can work, not by directly affecting their structure, but by altering their production and hence their levels. For example, they might be variants of transcription factors that regulate the expression of the genes that encode the molecules that actually move cells. An example is the gene *ARX*, which codes for a transcription factor that can influence, indirectly, the migration (as well as the proliferation, see above) of cerebral cortical **excitatory** and inhibitory neurons.[195] One reason (probably of many) that variants of *ARX* are likely to have this effect is because some ARX-regulated genes code for microtubule proteins. If ARX function was to decrease, this would affect the levels of microtubule proteins and might hamper the ability of cells to move. This might help explain why variants of *ARX* can cause structural abnormalities of the cerebral cortex, including incomplete cortical formation resulting in lissencephaly. These variants also cause functional symptoms such as seizures. Seizures might result from reduced migration into the cortex of cortical inhibitory neurons – remember from Chapter 1 that tangential migration from the **ganglionic eminences** (future **basal ganglia**) provides the cortex with at least many of its inhibitory neurons (Figure 1.6). A lack of inhibitory neurons is likely to tip the cortex towards over-excitation, rendering it prone to seizures (it is worth pointing out here that *ARX* is on the X chromosome and so effects are more severe in males, who have only one X chromosome; see Chapter 4 for further explanation).

Migration can also be affected indirectly by variants whose direct effects are on molecules that allow the proper functioning of the cell movement machinery. For example, I said earlier that some cell adhesion molecules require calcium to work: in fact, many cellular processes, including interactions between actin and myosin, require a correct level of calcium. It is no surprise, therefore, that variants that alter the transfer of ions such as Ca^{2+} across cell membranes can create problems with migration. A severe disorder known as Timothy syndrome is caused by variants in a gene encoding a subunit of a calcium channel (*CACNA1C*).[196] People with this condition have major problems with their heart – whose electrical activity depends on calcium – and develop brain phenotypes including developmental delay, intellectual disability and seizures. There is evidence suggesting that these brain symptoms might arise at least in part because of defects in the migration of inhibitory neurons into the cerebral cortex.

There are certain to be many other ways in which variants affecting processes one step removed from the molecular machinery that moves cells can influence their migration. For example, it is likely that migration is affected by variants that disrupt intercellular signalling. In non-human species, there are countless examples of intercellular signalling being important for guiding migrating cells, either towards or away from a source of signalling molecules. Biologists refer to this as chemotaxis. At present, however, evidence supporting a role for this in human brain development is limited, which is probably due mainly to the practical difficulty of searching for it.

How Environmental Variation Can Affect Cell Migration

Many of the environmental factors that affect proliferation (see above) probably also have an impact on cell migration, in part because these two processes are intimately related during fetal brain development and in part because the effects of many environmental agents on the biochemistry of cells are rather ubiquitous. For example, as I pointed out above, mercury can interfere with the three-dimensional structure of most proteins and this is likely to have widespread effects on cellular biochemistry (Figure 7.12). More specifically, there is evidence that the effects of exposure to methylmercury include impaired polymerization of microtubules, which is critical for both cell migration and formation of the mitotic spindle during cell division (Figure 7.10).

How the Specification of Cell Types Can Be Disrupted

Multicellular organisms comprise many different, often highly specialized, cell types. These cell types develop because different sets of genes are active in different cells. The genes that any given cell expresses play a major role in determining how that cell develops, what it looks like and how it functions. When examined at the finest of all molecular level resolutions, no two cells are likely to be identical, but they will share many more characteristics with some cells than others. This allows them to be categorized into different cell types, such as those commonly recognized and discussed throughout this book. Examples of major categories include neurons, with subcategories being excitatory or inhibitory neurons, or glia, with subcategories being **astrocytes** or oligodendrocytes. Subcategories can be further subdivided: for example, excitatory neurons might be divided according to their location, such as cerebral cortical layer, connection patterns and so on. A question is where should we stop subdividing categories or, if you like, how many different cell types does the developing and mature brain contain? This question is difficult to answer. It depends on how and with what degree of resolution you define each cell type, which depends on what it is that you want to know. For example, if you are interested in comparing neurons with glia, you need to only identify two categories; if you are interested in comparing layer 2 and 3 cortical neurons with layer 5 and 6 cortical neurons, you need greater resolution; if you are interested in the possibility that some rare cell type or particular cell state exists, then you will need even higher resolution giving more categories. Using modern molecular biological and computational methods to examine the molecular make-up of individual cells, we can select a level of resolution appropriate for the issue we are trying to resolve. I have included some details of how this can be done in Figure 7.15.[197–203]

In Chapter 1 (Figure 1.5), I outlined the principles of cell type specification and early brain **patterning**, in which signalling molecules confer different molecular identities on cells in different regions of the early brain. One of the most important aspects of the molecular identity of early brain cells is what transcription factors they express, since transcription factors are key regulators of hundreds or thousands of other genes whose activities steer cells towards their eventual fates. Any variants that impact the ability of these transcription factors to function are likely to create disorder (see Figure 7.15 for an example).

What the disorder might be can be hard to predict because rarely, if ever, does any particular transcription factor uniquely define a cell type. Rather, it is unique *combinations* of transcription factors that specify cell types. In this way, individual transcription factors can potentially contribute to a vast number of different cellular identities, and they can be involved in multiple functions in different locations or even at different times, depending what other factors they are co-expressed with.

How Transcription Factor Disruption Can Derail Cell Specification

Disorders of brain development caused by variants that disrupt the function of the transcription factors that pattern the embryonic brain include benign hereditary chorea (*NKX2-1* variants); FOXG1 syndrome (*FOXG1* variants); Partington syndrome, Proud syndrome and **X-linked** lissencephaly with abnormal genitalia (*ARX* variants); SATB2-associated syndrome (*SATB2* variants); **schizencephaly** (*EMX2* variants). The genes whose variants give rise to these types of disorder are fundamental for creating the major cell types and subregions in the brain. They generate the molecular identities that guide region-specific developmental programmes leading to the emergence of specific structural and functional attributes. Losing the function of both copies (that is, **homozygosity**) usually leads to a catastrophic failure of brain development and prenatal or perinatal death. In most cases, survival after birth is restricted to individuals who are **heterozygous** for such disruptive variants. Each of these types of transcription factor regulates the expression of many hundreds or even thousands of downstream genes, the details of which scientists are still working out. Although we do not understand the exact workings of these complex networks of downstream interactions, which we often call gene regulatory networks (Figure 7.16), we know many of the effects they have.

For example, ARX, FOXG1 and NKX2-1 play critical roles in specifying the development of the ganglionic eminences and their production of particular types of inhibitory neurons that migrate into the cerebral cortex. The disordered phenotypes caused by *ARX*, *FOXG1* and *NKX2-1* variants include some symptoms, such as **chorea** and epilepsy, that might be attributable to the compromised development of the basal ganglia and of cortical inhibitory neurons. The syndromes caused by these variants comprise many additional symptoms both inside and outside the brain, reflecting the involvement of these genes in many other processes (for example, I have already discussed the effects of *ARX* and *FOXG1* variants on cell proliferation and migration; see earlier in this chapter).

SATB2 and EMX2 are examples of transcription factors that play critical roles in developing cerebral cortex and its production of cortical excitatory neurons. SATB2 regulates the development of specific types of cerebral cortical excitatory neurons that either project axons across the corpus callosum to the other side of the brain or to regions beneath the cortex on the same side of the brain. Disruption of these processes might help explain how variants of SATB2 can cause intellectual and cognitive disabilities, although later disruptions probably contribute too since it appears that SATB2 also regulates synaptic functions in adulthood. Variants that disrupt EMX2 can cause schizencephaly, in which large portions of the cerebral hemispheres may be absent; exactly how this occurs is unclear.

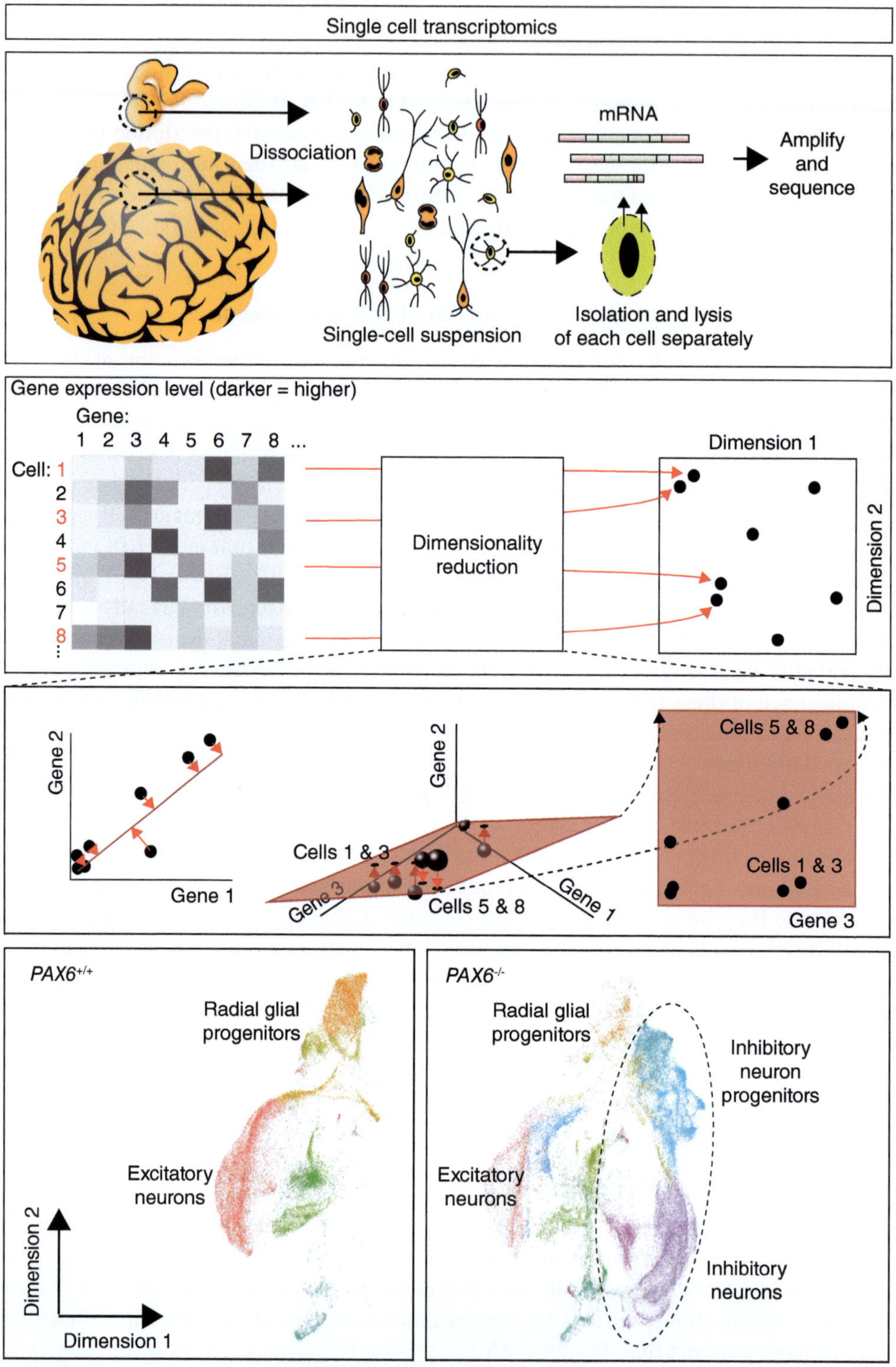

Figure 7.15 **Using single-cell transcriptomics to explore cell types.** The careful treatment of brain tissue of any age with a mixture of chemicals and physical force can produce suspensions of intact single cells of all sorts, a process we call dissociation. While we have known how to do this for

Figure 7.15 (Continued) many decades, it is only relatively recently that we have been able to isolate and process large numbers of these cells individually. This has been achieved through advances in microfluidics, an area of technology where fluids are manipulated and processed through ultrathin channels with diameters on the microscale (typically tens to hundreds of micrometres). Individually isolated cells can be processed in a variety of ways. Typically, they are lysed to release their mRNAs, which can then be amplified and sequenced to tell us which genes are expressed at what level in each cell (note: other types of processing are possible, depending on the purpose of the work; for example, the nuclei might be isolated and the DNA and/or RNA they contain might be sequenced or the DNA might be studied to discover which parts are open or closed for transcription; Figure 4.11 in Chapter 4). These procedures generate vast datasets detailing the expression of all genes in many thousands of cells; the matrix here shows just the top left corner of a fictitious dataset, containing data on the mRNA levels from only eight genes in only eight cells as shades of grey. We say, in the jargon of the field, that the data for each cell have eight dimensions. The complex computational analysis required for us to understand the vast amounts of data produced by single-cell analyses invariably involves dimensionality reduction to extract essential elements from the data and allows us to see patterns, such as the fact that – based on the expression of these eight genes – Cells 1 and 3 in the matrix have rather similar patterns of expression that are different to those of the other six cells; the same is true for Cells 5 and 8. You can see that here because we are only looking at eight dimensions in the data for each cell, but if we added more data per cell it would start to get much harder to see in the matrix. Moreover, of course, these similarities and differences might or might not be retained if we included data on all genes in every cell (but we are going to stick with just eight genes per cell for the purposes of explaining the principle here). In the second row, I have reduced each set of eight values to a single point for each cell in the two-dimensional graph; in other words, I have reduced eight dimensions to two. You can see that Cells 1 and 3 are close together and Cells 5 and 8 are close together. How did I do this? The principles of dimensionality reduction are shown in the third row. If we are dealing with just two dimensions – in this case, the expression levels of Genes 1 and 2 – we can plot them on a standard graph with two axes. Alternatively, we could project each point onto a *single* axis drawn through the datapoints in such a way as to maximize their variance (note that the manipulation shown has spread the values over a greater distance along the red line than their values were spread along either of the two original axes). Clearly, this loses us information, since we now have just one value (or dimension) for each cell instead of two; it is the case that dimensionality reduction sacrifices some information. If we move on to add information on Gene 3, we can plot a 3-dimensional graph and we can calculate mathematically a two-dimensional plane onto which the points can be projected so as to maximize variation. We can see that cells are starting to separate and cluster together across this plane, and it is much easier to see this at a glance than it was in the original matrix. We can keep going, adding data from more and more genes per cell and projecting all the dimensions onto a two-dimensional plane that maximizes overall variance (this cannot be shown visually once we get above three dimensions). Going from two dimensions to one, or from three or more dimensions to two, as in these illustrations, is what we mean by dimensionality reduction. Including data on all genes in thousands of cells yields plots looking like those at the bottom, in which each cell is a tiny dot, and the closer the dots are, the more similar are their gene expression patterns. The cells are split into groups and given different colours; how many groups we split them into depends on what level of resolution we are interested in (the maximum we can use is the total number of cells, one per group, but this would be useless when searching for patterns in the data!). Often, we can use our pre-existing knowledge of key genes expressed by different major cell types to identify them: for example, the ochre dots represent radial glial progenitors and the red dots represent excitatory cortical cells. These data are from human cortical cells that carry either two fully functional *PAX6* genes or two non-functional ones (PAX6 is a transcription factor highly expressed in developing cerebral cortex). In the absence of functional *PAX6*, there is an abnormal production of many inhibitory progenitors and neurons (blue and purple cells) instead of excitatory ones. This shift in fundamental cell type production might help explain the severely disordered phenotype of the brains of humans lacking functional PAX6. *Adapted from Chan et al.*[204]

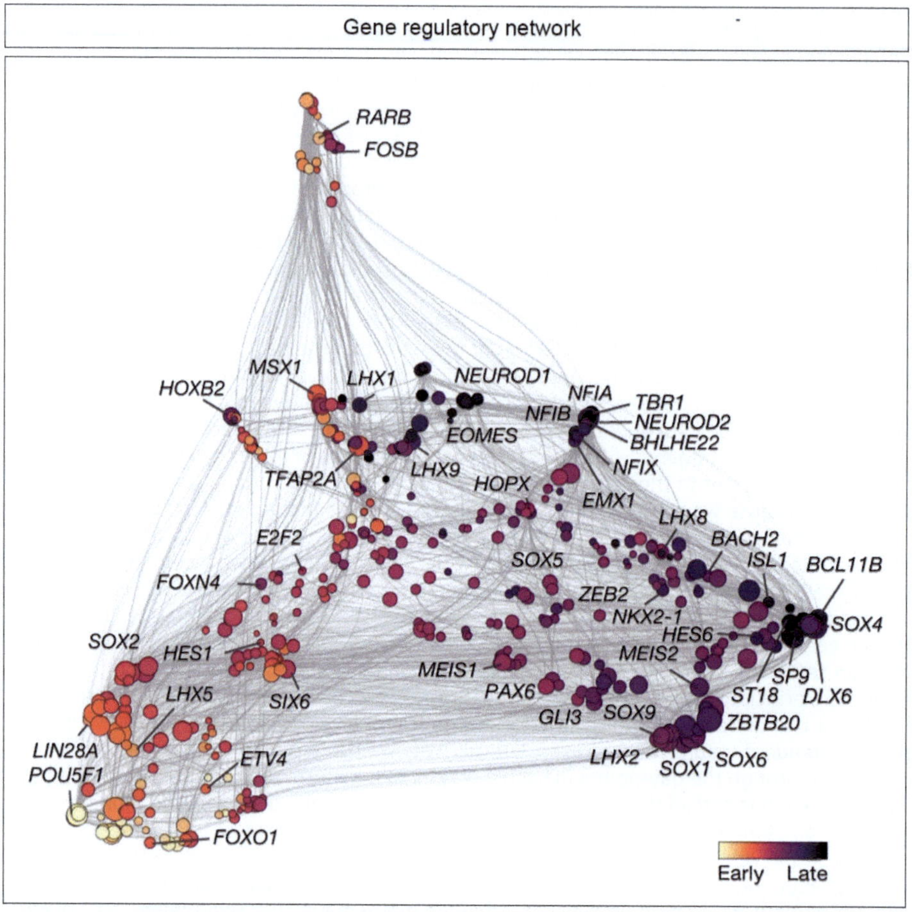

Figure 7.16 Gene regulatory networks. The complex interplay between transcription factors and chromatin is central to the mechanisms that control how different cells become specialized for their particular functions (or, in other words, become specific cell types). As researchers began to work out the details of some of these interactions and saw their complexity, they started referring to them as gene regulatory networks. By studying the networks of interactions controlled by transcription factors, we hope to understand not only how different cell types are created but also how disruption of gene regulatory networks can lead to developmental diversity and disorder. Gene regulatory networks are generally inferred using computational approaches combined with experimental data, such as those discussed in Figure 7.15. As I mentioned in the legend for Figure 7.15, single-cell technologies can be used not only to measure mRNA production but also to look at other parameters, such as how open different parts of the genome are (in other words, how available they are for transcription). We often refer to the simultaneous measurement of several aspects of each cell's activity as single-cell multi-omics (it can combine transcript*omics* with gen*omics*, or with analysis of other features of the cell such as what proteins or metabolites it contains, which we call prote*omics* or metabol*omics* respectively). The approach can provide powerful insights into how genetic regulation is occurring by telling us which genes are open for expression, which of them are actually being expressed and how active they are in each cell. It allows us to infer a cell's gene regulatory networks by pointing to where the transcription factors it is expressing are likely to be binding to its DNA to control the expression of its genes. I should point out that the methods involve various assumptions that I will not go into here, and you should always think of the networks they produce as being works in progress. The picture shows one

Figure 7.16 (Continued) example of a putative gene regulatory network, in this case for human neurons as they undergo the early stages of their development in culture. The researchers used knowledge of the genomic sequences to which specific transcription factors bind together with experimental data on gene expression and chromatin accessibility to draw this network: interacting genes are linked and the stage of development at which they act is shown by the colour (see key). The diagram traces transitions from an early state of pluripotency (marked by expression of genes such as *POU5F1* and *LIN28A*) through neuroepithelial induction (marked by, for example, *SOX2* and *HES1* expression) to brain regional neural progenitor cell specification (for example, marked by dorsal telencephalic expressed genes *EMX1*, *NR1D1* and *NEUROD6*). *Reproduced from Fleck et al.*[205] */ Springer Nature / CC BY 4.0.*

How Signalling Molecule Disruption Can Derail Cell Specification

From neural induction on, signalling molecules are intimately involved in the specification of cell types (Figure 1.5, Chapter 1). Variants affecting signalling might be of genes encoding signalling molecules themselves, their receptors or proteins involved in any other step in the intercellular signalling processes, including variants that disrupt primary cilia, second messenger cascades or the transcription factors that they target (Figures 7.3–7.5). Disorders caused by variants of genes involved in these processes include acrocallosal syndrome Schinzel type (*KIF7* variants), Greig cephalopolysyndactyly syndrome (*GLI3* variants) and Myhre syndrome (*SMAD4* variants).

KIF7 is a protein required for the proper function of primary cilia, which, as I explained earlier, are essential for several signal transduction pathways (Figure 7.5). KIF7 disruption causes a patterning defect that is particularly prominent in the parts of the brain between the two hemispheres, a region through which callosal axons must grow to connect the two sides. The upshot is partial or complete corpus callosum agenesis because the growing callosal axons are not guided properly through the incorrectly patterned tissue. Most likely, faulty processing of the transcription factor GLI3, a key target of the SHH signalling pathway, is a major cause of the patterning defect (Figure 7.5). Abnormalities are not restricted to the brain: they include other structures that SHH is also involved in patterning, notably the hands and feet, which often have additional fingers and toes (we call this polydactyly).

Variants directly affecting GLI3 typically cause developmental abnormalities that partially overlap those caused by *KIF7* variants, strengthening the conclusion that *KIF7* variants have their major effects on patterning by changing GLI3's activity. *KIF7* and *GLI3* variants do not cause fully overlapping defects for many reasons, such as the fact that *KIF7* variants are likely to impact more than just the SHH pathway and disrupting the GLI3 protein directly versus modulating GLI3 processing is likely to have somewhat different outcomes. Exactly how such differences arise remains unclear.

Myhre syndrome, whose symptoms include neurodevelopmental delay, mild to moderate intellectual disability and various behavioural difficulties as well as many outside the brain, is caused by variants of the *SMAD4* gene that render its protein hyperactive (we call this a gain-of-function). SMAD4 is a component of the transforming growth factor beta (TGFβ) signalling pathways activated by bone morphogenetic proteins (BMPs), which are members of the TGFβ family (Figure 7.3) and are important players in the early specification of brain cell types. Overactivity of these pathways is likely to cause patterning defects in ways that are not yet fully understood.

How the Development of Neuronal Connections Can Be Disrupted

Disruption of Axon Growth and Guidance

As I hope I made clear in the previous sections, many individual genetic variants are likely to affect more than one of the three principle cellular processes that create the brain: cell proliferation, cell migration and cell type specification. Sometimes this is because the affected gene is directly involved in more than one of these processes; sometimes it is because direct disruption of one or two of these processes has an indirect, knock-on effect on the other(s); most often it is probably a bit of both. The further into development one goes, the greater the scope for complexity in the ways in which variants can have their effects. The fact that axon growth and guidance start a little bit later in development enhances the potential for them to be affected indirectly via a direct impact on earlier events. Cell proliferation, migration and specification create the environment through which axons must navigate, and if that environment is disrupted, so is axon navigation likely to be. I gave an example of this in the previous section: *KIF7* variants incorrectly pattern tissue through which callosal axons must grow, causing partial or complete corpus callosum agenesis.

Variants might directly affect the ability of axons to grow and navigate by interfering with their **growth cones** (Figure 1.11, Chapter 1). They might affect their ability to generate mechanical forces or respond to signals from their environment. Examples of variants affecting growth cones' ability to grow are those causing Chudley–Mccullough syndrome (variants of *GPSM2*)[206]: GPSM2 protein accumulates in the tips of the filopodia of growth cones, where it appears to help drive their leading edge forward by aiding actin polymerization (which is involved in many cellular processes, as we saw in Figure 7.7). Examples of variants affecting the ability to navigate are those causing L1 syndrome (variants of *L1CAM*). As I discussed earlier in this chapter, L1 is a transmembrane neural adhesion molecule, and it is present in growth cones. One of its functions is to help guide the growth cone by binding to guidance molecules distributed in specific patterns in its environment. L1 acts as a link between these guidance molecules and the actin cytoskeleton, allowing extracellular molecular guidance information to be translated into physical forces and directional growth. Variants of *GPSM2* and *L1CAM* also affect many other processes including some (such as cell proliferation and migration, see above) that can indirectly influence axon navigation. Together, these effects likely explain why variants of these genes typically result in corpus callosum agenesis.[207]

Another way in which variants might affect the guidance of growing axons is to interfere with two of their important characteristics: their ability to branch along their length so as to connect with additional targets and their ability to navigate relative to each other. Regarding the latter, axons often navigate in large bundles (which we call fascicles), within which individual axons might grow in a parallel fashion without crossing over each other or might grow across each other, depending on what axonal pathway is being created. There is evidence that these behaviours can be controlled directly by signalling through primary cilia, as can other processes discussed above that might have indirect effects on axonal growth and guidance. This might explain the axon guidance defects leading to a cerebellar axonal malformation (called the molar tooth sign) characteristic of Joubert syndrome, which is caused by variants of many genes involved in primary cilia.[208]

Disruption of Dendrite and Synapse Formation

Most synapses form on the often highly extensive dendritic trees of neurons. Any variants that interfere with the development of these dendritic trees are likely to have consequences for the formation of neuronal circuits. The complexity of dendritic trees and the vast numbers of synapses they host makes it difficult to understand how effects of variants on dendrite formation and synapse formation translate into observable phenotypes. These phenotypes are most likely to be perceptual, cognitive or behavioural rather than detectable structural ones, and to reiterate a recurring theme, the variants are likely to affect many other processes as well.

For example, Down syndrome cell adhesion molecule (DSCAM) is encoded by a gene on chromosome 21. In Down syndrome (remember: caused by a person developing with an extra chromosome 21, or part thereof), DSCAM is overproduced. This is thought to be one possible cause of the reduced dendritic complexity observed in this syndrome, although it is likely to be only one of many contributing factors.[209] Moreover, the overproduction of DSCAM is likely to contribute to the cognitive symptoms of Down syndrome in other ways; for instance, DSCAM has been implicated in axon growth and guidance, among other processes.

Once growth cones have reached their targets, synapses can form. I will say more about the ways in which the functions of synapses can be affected in Chapter 8. Here I am focusing on how they develop in the first place. The major steps of synapse assembly are as follows. Initial contact between incoming axons and their targets is mediated by various cell adhesion molecules, including NLGN3, NLGN4X, NLGN1, NRXN1 and CADM1. This precipitates a cascade of molecular events that bring the proteins needed to build synapses to those contact points. These components are brought along microtubules by transport proteins, such as MAPT, KIF17 and GSK3B. Synaptic construction requires that synaptic components be held in places at the sites of nascent synapses, a task achieved by scaffolding proteins, such as SHANKs and CASK (I will say more about these molecules in Chapter 8).

Variants affecting the functions of proteins involved in these processes have been implicated as risk factors in various neurodevelopmental disorders and divergence, such as schizophrenia and autism. Specific examples include the following. Deletions of part of chromosome 22 that include *SHANK3* can cause developmental delay and intellectual disability. *CASK* variants cause X-linked intellectual disability Najm type and variants of many others of these genes are associated with polygenic conditions such as autism, schizophrenia and Tourette syndrome (I will discuss these in later chapters).[210] It is thought that deletions of part of chromosome 5 that cause cri du chat syndrome do so in large part because they include the *CTNND2* gene, which encodes a protein involved in cell adhesion at synapses (a note of caution: these large deletions involve other genes too, and this probably also contributes to the phenotype). Since many of the proteins involved in synapse formation continue to have important functions in mature synapses, it is often difficult to say when exactly their loss or dysfunction has its effect.[211]

Genetic Variants Can Affect Brain Development by Altering General Support Mechanisms

So far in this chapter, I have discussed how genetic variants and environmental factors can impact early prenatal developmental events by affecting the mechanisms that enact or control the major processes of brain development – cell proliferation, migration, specification and interconnection – in relatively direct and specific ways. We need to remember that these mechanisms are played out with the support of many other processes that maintain cell viability, provide energy to the cell and so on. These permissive activities, which are essential to all cells, provide the cellular infrastructure for the development of complex structures.

Another way of looking this might be the following. The major processes that create our brains are subject to expanding circles of control and influence. The closer a process is to the centre of each set of circles, the more direct is the influence. We can focus our expanding circles on whatever process we like – for example, cell proliferation, cell migration, cell type specification or synapse formation – or we could choose a subset of such events, such as cerebral cortical radial glial progenitor proliferation and so on. Multiple sets of circles focused on several different processes will overlap to some extent, since the same influences affect multiple processes. Up to now, I have been dealing mainly with variants acting relatively close to the centres of the circles, either directly influencing the process in question or only a few steps removed. Now I am moving to the more distant circles.

Genetic variants and environmental factors impacting generic processes operating in these distant circles can cause disorders of prenatal brain development. These generic mechanisms include those maintaining DNA integrity, ensuring a balance in rates of production and breakdown of proteins and other molecules, providing energy and so on.

Effects on DNA Integrity

Table 6.2 in Chapter 6 lists monogenic disorders that result from variants that affect DNA's ability to reproduce itself faithfully as cells divide. Interference with the functions of some genes, including *APTX*, *ATM*, *ERCC6*, *ERCC8*, *NBN*, *PNKP* and *SETX*, can lead to a build-up of defects affecting all types of cellular processes since these genes encode proteins that normally help repair damaged DNA, thereby minimizing the accumulation of genetic errors that might, at some point, affect development in some way. (I explained the importance of DNA repair and the ways that cells do it in Chapter 4.) For example, ATM and ERCC8 proteins help identify areas of damaged DNA, ERCC6 protein probably helps remove RNA polymerase from a damaged site to allow DNA repair and APTX and PNKP proteins modify the ends of broken DNA so they can rejoin. Interference with the functions of other genes, including *ATRX*, *ASXL1*, *ASXL3*, *CHD2*, *DNMT1*, *GATAD2B*, *KANSL1*, *KAT6B*, *KMT2A*, *NSD1*, *SETBP1*, *SMARCA2* and *SRCAP*, is likely to cause defects because these genes are involved in determining that the state of chromatin is suitable for the appropriate expression of many other genes. In some cases, the proteins that genes such as these encode have their effects on chromatin by regulating the methylation or acetylation of histones (Figure 4.11, Chapter 4). In other cases, they remodel chromatin by contributing to multi-protein complexes that bind to it, causing **nucleosome** incorporation and/or displacement and thereby altering DNA's openness for transcription. For example, SMARCA2 and SRCAP are components of two multi-protein chromatin remodelling complexes called SWI/SNF and SRCAP respectively.[212]

Effects on RNA Synthesis or Processing

Table 6.3 in Chapter 6 lists monogenic disorders that result from disruption of RNA synthesis or processing. For most of them, we know what process is disrupted, but it is unclear how this causes the observed phenotypes. For example, some disorders are caused by variants of genes such as *EFTUD2* and *RNU4ATAC* whose products, EFTUD2 protein in the former and a noncoding RNA called U4atac in the latter, contribute to the spliceosomes that remove introns from pre-mRNAs to generate mature RNAs (Figure 4.8, Chapter 4).[213–215] Others are caused by variants of genes such as *TSEN54*, which codes for a subunit of an enzyme called transfer RNA (tRNA) splicing endonuclease complex that **splices** pre-tRNAs to make mature tRNAs (tRNAs are described in Figure 4.9, Chapter 4); *RNASET2*, which codes for a protein called ribonuclease T2 (RNAse T2) which breaks down unneeded RNA; *DDX3X*, which codes for one of a family of so-called DEAD box proteins that play roles in most aspects of RNA metabolism (DEAD is not used here as a gloomy adjective, it refers to the fact that these proteins all contain a short sequence of amino acids, called a motif or box: aspartic acid – glutamic acid – alanine – aspartic acid, whose standard one-letter abbreviations are D-E-A-D).[216]

Effects on Proteostasis

One essential contribution to a cell's biochemical infrastructure is made by mechanisms that maintain the proteins of a cell at appropriate levels and in conformations that ensure they function properly.[217] We often say that these mechanisms are responsible for protein homeostasis, sometimes shortened to **proteostasis**. Homeostasis (from *homeo*, Greek for similar) describes a state of balance among all the elements of a system needed for it to survive and function correctly. Protein homeostasis is achieved through the regulation of protein synthesis, processing and degradation (Figure 7.17).

Brain development is likely to be affected by interference with the mechanisms of protein synthesis during translation or subsequent processing, which folds each protein in specific ways required for its proper function and adds functionally important molecular units such as acetyl, phosphoryl, glycosyl and methyl groups. Although we know of many variants that disrupt these processes, we have insufficient knowledge to suggest with any confidence how exactly they lead to the abnormal brain phenotypes with which they are associated (disorders are listed in Table 6.3 in Chapter 6). Examples of such variants involve genes such as *EIF2S3*, which encodes a part of a protein involved in constructing ribosomes (Figure 4.9, Chapter 4); *SIL1*, whose product helps fold newly made proteins; a long list of genes encoding enzymes that modify proteins, many by glycosylation, which involves the addition of sugar-based polymers (called glycans) to make glycoproteins.[218]

I will make a 'watch this space' aside here. It is likely that, in years to come, the subject of glycosylation and its disruption will become included into discussions of how genetic variants

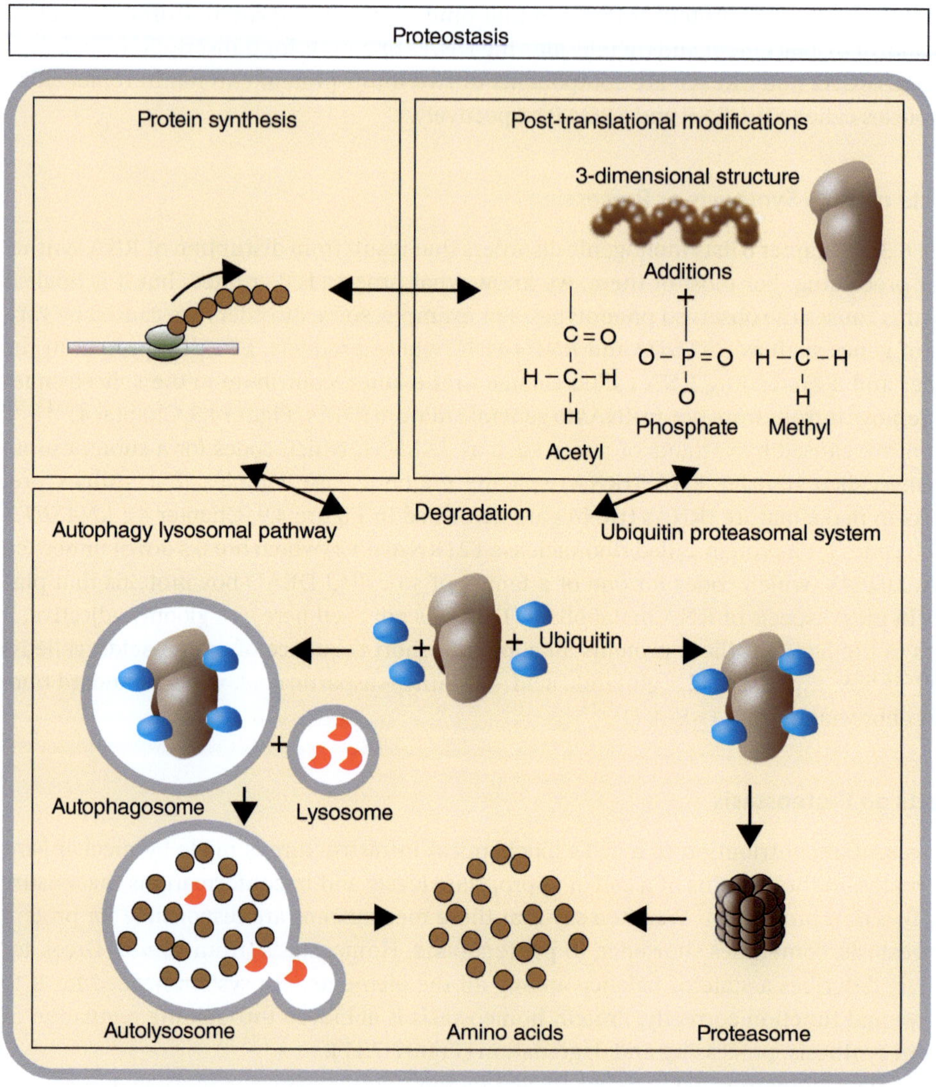

Figure 7.17 Proteostasis. The network of interactions that achieves proteostasis includes three major modules that work together intimately: protein synthesis, post-translational modification and degradation. Protein synthesis is discussed in Chapter 4; post-translational modifications include the creation of 3-dimensional structure (see Figure 7.12) and the addition of functional groups or other molecules to the side chains of amino acids or at the ends of proteins, thereby altering their structure and function (examples shown here are just common examples of some of the many possible additions). Proteins for degradation are tagged with ubiquitin proteins allowing them to be recognized and degraded through one of two pathways: the autophagy lysosomal pathway and the ubiquitin proteasomal system. In the autophagy lysosomal pathway, the ubiquitinated proteins become enclosed in vesicles called autophagosomes which then fuse with hydrolase-containing lysosomes to form autolysosomes, in which the target proteins are degraded to their constituent amino acids for recycling. In the ubiquitin proteasomal system, ubiquitinated proteins are degraded by a protein complex called the proteasome.

or environmental factors can interfere with the *control* of developmental processes, rather than just the infrastructure underpinning them. This is because evidence is accumulating to suggest that the way in which cell surface proteins are glycosylated can affect how cells respond to the signals they receive. The really interesting thing is that glycosylation can modify cell surface proteins in a huge variety of ways, since the glycan polymers that are added are made up of chains of various simple sugars whose arrangement has the potential for immense variability. Many researchers think that cells create specific glyco-codes on their surfaces that modulate their responses to intercellular signalling in specific ways, adding another layer of regulation to the mechanisms that control cellular development.[219]

Moving on to degradation, proteins that are no longer needed or are at inappropriately high concentration or that are faulty – for example, misfolded – are sensed by molecular mechanisms within the cell and marked for breakdown by ubiquitination, an enzymatic process that attaches the protein ubiquitin to the target protein (the process is reversed by de-ubiquitinating enzymes). Ubiquitinated proteins are broken down by two major biochemical pathways: the ubiquitin–proteasome pathway and the **autophagy**–lysosome pathway (Figure 7.17). In the former, ubiquitinated protein is moved to the cell's proteasome for degradation. In the latter, ubiquitinated protein is placed into spherical membrane-bound vesicles called autophagosomes that are destined for the lysosomes, where degradation takes place. Variants of *UBE3A* or *UBE3B*, which encode enzymes that catalyse ubiquitination, can cause prenatal micrencephaly as a part of Angelman syndrome and Kaufman oculocerebrofacial syndrome respectively. Vici syndrome, whose phenotype involves corpus callosum agenesis, is an example of a disorder caused by an effect on the autophagy–lysosome protein degradation pathway. It is caused by variants of the *EPG5* gene, whose protein is important for the transfer of materials to the lysosomes. There are many other examples of variants affecting these pathways that impact prenatal brain development, such as variants of *MCOLN1*, which codes for a transport protein in lysosomes, and of *TECPR2* and *ZFYVE26*, which code for proteins that help form autophagosomes. It is unclear exactly how the phenotypes caused by these variants are generated; micrencephaly is typical of some disorders, whereas corpus callosum dysgenesis is a hallmark of others.

Effects on Structural Proteins

Many of the proteins that cells make are exported to build their extracellular matrix (Figure 7.2). Variants affecting the construction of the extracellular matrix can disrupt the ability of neurons to function and survive. An example, which shows how both genetic and environmental factors can combine to produce pathology, is the effects of variants of the *COL4A1* gene that underlie cases of familial porencephaly. *COL4A1* encodes a type of collagen that makes a major contribution to basement membranes, which are specialized parts of the extracellular matrix that line tissues and provide physical strength and biochemical support to adjacent cells (Figure 1.11, Chapter 1). Harmful *COL4A1* variants have direct effects mainly on blood vessels, which are weaker than normal leading to a likelihood of haemorrhage, which can cause damage and cell death among the brain's neurons and the formation of fluid-filled cysts that are characteristic of porencephaly (Chapter 3, Figure 3.1). It is likely that this is exacerbated by pressure on the head before or during birth.

Effects on the Metabolic Processing of Molecules Other Than Proteins

There are many ways in which metabolic disruption might impact the prenatal development of the brain. In most cases, it remains unclear whether and, if so, how the metabolic defect results in prenatal abnormalities. I will give a couple of examples where there is at least some tangible evidence.

Variants that disrupt the processing of nucleotides are known to cause prenatal neurodevelopmental disorders. For example, adenylosuccinase deficiency (caused by variants of *ADSL*) results from a deficiency in an enzyme called adenylosuccinate lyase, which is involved in the production of purine nucleotides, which contribute to DNA and RNA, and molecules such as **adenosine triphosphate** (**ATP**), which serve as energy sources in the cell (see Figures 8.6 and 8.7 in Chapter 8). If adenylosuccinate lyase deficiency prevents these biochemical pathways from working properly, substrates are processed though alternative ones that produce harmful levels of toxic molecules. These result in the micrencephaly characteristic of this syndrome by interfering with cell proliferation and migration.

Many variants can interfere with the predominantly energy-producing functions of mitochondria. In many cases, symptoms emerge postnatally and I will, therefore, discuss them further in the next chapter. One example of a disorder affecting prenatal development that is caused by variants of a mitochondrial enzyme is ethylmalonic encephalopathy (caused by *ETHE1* variants). The enzyme helps breaks down sulphide (H_2S), which, at high levels, is toxic to many cellular activities including mitochondrial energy production. Symptoms include spasms and seizures present from birth, indicating disruption of the developmental processes I have discussed throughout this chapter.

Summary

This chapter highlights the difficulties of navigating from ultimate causes through the many possible proximate causes to phenotypes. Even in the case of prenatal conditions with a single major identifiable cause, all sorts of cellular processes are affected either directly or indirectly. Genetic or environmental perturbations have consequences that cascade down through vast numbers of molecular and cellular events driving and supporting the brain's development from within or around it. The symptoms we observe are the end result of all this complexity. We shall see in the following chapters that things become progressively even more challenging to untangle as birth ushers in the opening of many new avenues through which ultimate causes can affect phenotypes. Key bullet points from this chapter are as follows.

- A large proportion of protein-coding genes are expressed prenatally, making the prenatal brain highly susceptible to the effects of genetic variation.
- High cell proliferation rates in the developing brain make it vulnerable to the effects of harmful environmental factors.

- Genetic and environmental factors can cause neurodevelopmental conditions by affecting some or all of the key cellular processes involved in brain development, including neural induction, neurulation, cell proliferation, cell migration, cell type specification and the formation of neural connections.
- Genetic and environmental factors can have direct and indirect effects on molecules operating between cells, on their surfaces or inside them to mediate cell adhesion, intercellular signalling, cell shape changes and synapse formation.
- The proximate causes of neurodevelopmental conditions can include effects on mechanisms that support the developmental processes listed above by maintaining cell viability, providing physical support to cells, generating energy for cells, removing waste products from cells and so on.

8

Proximate Causes of Childhood Conditions

Most babies are born with no evidence of brain disorders. Unfortunately, a few will have abnormalities that lurk undetected until they reveal themselves sometime after birth, usually by causing functional problems, such as seizures, difficulty with movements and so on. When symptoms appear, it can be very difficult or impossible to determine their ultimate causes. If the cause is genetic, for example, its discovery usually relies on genetic testing that might not be available, depending where in the world the person lives, and even if it can be done it is very often inconclusive (remember, as I discussed in Chapter 6, that working out whether a particular variant is causative is not straightforward). It can also be very difficult to identify proximate causes. Many disorders of brain function occur without there being clearly detectable underlying pathology, although it must exist at some level, however microscopic. Rarely, a child might have a condition that is degenerative, with major structural and functional abnormalities appearing and worsening as the child grows.

Neurodivergence is usually detected as a child's perception, cognition and behaviour develop and they acquire language and social skills. These skills and attributes might develop in ways that are either clearly atypical, when compared with those of most other children of similar age, or cause difficulties and concerns of some sort for the child, their family or others involved with them.

An Overview of This Chapter's Content and Layout

In this chapter, I will discuss the known proximate causes of disorders and divergence that are first diagnosed in the early years of postnatal life. Please bear in mind that the origins of many of them are very likely to be prenatal, at least to some extent, given that most of our protein-coding genes are already expressed in our fetal brains, as I emphasized at the start of Chapter 7. Nevertheless, I have included them in this chapter if there is a lack of evidence as to whether and to what extent this is the case. As with Chapter 7, aspects of this chapter build on the lists of disorders and, where known, their ultimate causes in the tables in Chapter 6. It is important to stress again that, even in the best-studied cases, our current outstanding of the link from ultimate causation to specific phenotypes is far from complete.

This chapter will focus primarily on brain *function*, particularly the activities of its electrical components, since many detectable disorders of brain structure have already appeared by birth. Our brains are electrical devices. If their microscopic components develop with an impaired ability to transmit signals, either as electrical impulses traveling along nerves or as bursts of neurochemicals released at synapses, symptoms of their disordered function are likely to emerge as a child grows. As I discussed in Chapter 2, synapses are dynamic structures. The magnitude of the effects of **presynaptic** electrical activity on **postsynaptic** membranes – which we refer to as synaptic strength – varies not only from synapse to synapse but can also change at the same synapse from time to time in response to electrical activity. This plasticity is a critical property of the brain, giving it flexibility and adaptability. Because the brain's neurons can change their interactions in response to the electrical activity they experience, they learn. Factors that affect this dynamic property of synapses are likely to impact the brain's cognitive functions. The term synaptopathy is often used to describe disorders thought to arise primarily either from dysfunction of the basic processes that underpin synaptic transmission or from interference with the mechanisms of **synaptic plasticity**. In practice, you are most likely to encounter the term in discussions of the possible causes of conditions affecting higher cognitive functions.

Genetic variants and environmental factors can cause disorder and divergence because they affect the molecules that themselves directly participate in the development and plasticity of the brain's electrical components or because they act indirectly through their effects on other molecules that then, in turn, influence the direct participants. As we saw in the previous chapter, indirect effects can be initiated by factors acting at a range of steps (or circles of influence, if you like) removed from the direct participants. They range from factors impacting the mechanisms that produce the direct participants to those affecting the biochemical infrastructure of cells, such as the processing of proteins and other molecules, including those that produce energy.

In this chapter, I will first discuss how specific genetic variants can act directly or indirectly to affect the development, function and plasticity of the brain's electrical components: some of these genetic variants might cause monogenic conditions, some might contribute to polygenic and multifactorial conditions. I will illustrate the types of mechanisms involved by selecting some of the better understood examples of the proximate causes of some of the many conditions listed in Tables 6.1–6.7 in Chapter 6. I will then discuss how combinations of genetic variants and environmental factors might cause common complex multifactorial conditions (including both disorders and types of neurodivergence) that typically emerge early in postnatal life.

How Genetic Variants Can Directly Affect the Brain's Electrical Components

Effects on Electrical Transmission Along Nerves

The ability of nerves to transmit electrical impulses depends on the nature of the proteins in their cell membranes, as explained in Chapter 2. Disorders are known to be caused by genetic variants that disrupt the functions of some of these proteins, including **ion**

channels and transporters (or **ion pumps**) that move ions across membranes (Table 6.4, Chapter 6). These variants cause the kinds of symptom that one might expect to arise from disruption of a biological machine whose modus operandi relies on controlled and coordinated electrical activity. Variants might cause decreased neuronal activity in the brain, resulting in symptoms such as paralysis or a loss of control of movement and balance (which might include difficulty with activities such as speaking, eating and drinking) or they might cause increased neuronal activity in an abnormal and uncontrolled way, leading to seizures. As we shall see in the following examples, selected to illustrate a range of possibilities, the symptoms they cause depend on where and how the genes they affect normally operate. In all cases, we are left with much still to learn about why particular gene variants tend to cause the particular sets of symptoms that are associated with specific clinical disorders.

Variants of *ATP1A3*, which encodes a subunit of a membrane protein, the sodium pump (which we also call the Na^+/K^+ ATPase protein), can disrupt its ability to use energy (in the form of **adenosine triphosphate**, or **ATP**) to pump sodium ions (Na^+) out of cells and potassium ions (K^+) into them. This ionic pumping is essential for creating the resting **membrane potential** of neurons (Chapter 2). If resting membrane potentials are not maintained, then neurons will lose their excitability (non-excitable cells have much lower resting membrane potentials than excitable ones). If the ability of neurons to generate normal levels of activity is compromised, this can cause syndromes such as alternating hemiplegia of childhood, which is characterized by periods of temporary paralysis – often on just one side of the body (which is what hemiplegia means) for reasons that are unclear – and cognitive problems.

Depolarization of a neuron causes the sodium channels in its cell membrane to open allowing Na^+ back into the cell to generate an **action potential** (see Chapter 2). Variants of the *SCN1A* gene, which encodes a sodium channel protein, disrupt this process and will, therefore, lower the ability of neurons to generate action potentials and hence normal levels of activity. Paradoxically, however, this causes seizures, which we usually think of as products of abnormally *increased* excitability in brain neurons. The best current explanation for this illustrates the fact that where and how these gene products normally operate is crucially important. Researchers think that that *SCN1A* variants have their greatest effect in lowering activity among **inhibitory** neurons, since the gene is particularly highly expressed in them. If their activity is reduced, this will remove the normal inhibitory constraints on the excitability of **excitatory** neurons, predisposing the individual to seizures.[220]

Seizures can also be caused by variants affecting potassium channels. Those encoded by the *KCNQ2* and *KCNQ3* genes allow positively charged potassium ions to make a controlled exit from the cell. When a neuron in the brain is not making an action potential, its inside maintains a negative charge compared to its outside, and a small controlled exit of positively charged ions (which we refer to as the M-current) tends to keep it that way (note that I did not mention this detail in Chapter 2, but if you need to you can refer back to Figure 2.3 to see how this would help). The effect of the M-current is to hold the resting potential stable and make neurons a bit less likely to activate action potentials accidentally. If the M-current is disrupted by variants of the *KCNQ2* or *KCNQ3* genes, neurons become excessively excitable, creating a risk of seizures.

The resting potential of neurons is also dictated by a process that we call potassium spatial buffering (Figure 8.1). This is carried out by potassium channels, encoded by the

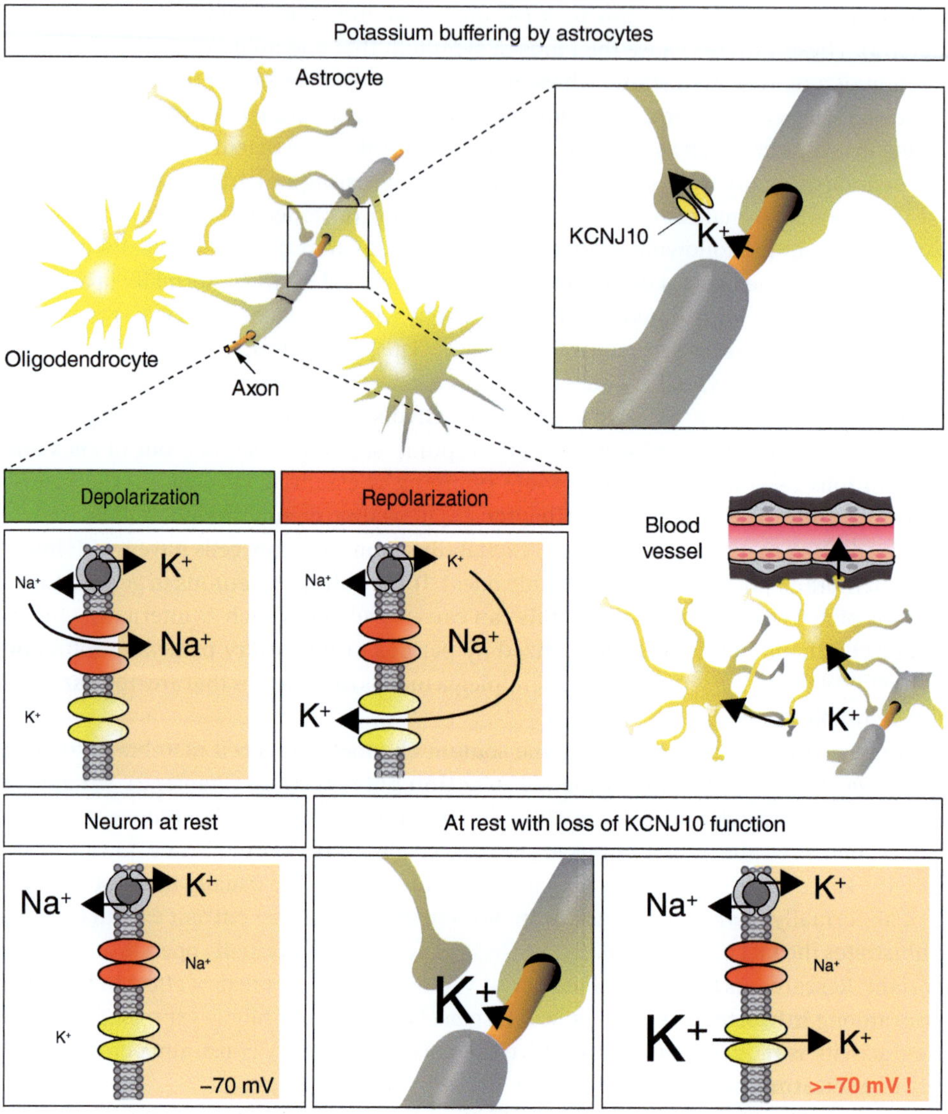

Figure 8.1 **Buffering of extracellular potassium by astrocytes and the effect of KCNJ10 loss of function.** Astrocytes around the exposed parts (or nodes) of axons between segments wrapped with myelin can take up potassium ions through their KCNJ10 channels. When an action potential travels along the axon, jumping from one exposed node to the next (remember, this is called saltatory conduction), potassium ions are released into the extracellular space around the nodes (diagrams of depolarization and repolarization are reproduced from Chapter 2, Figure 2.3). The astrocytes help to keep levels of extracellular potassium ions around the nodes from rising too much by distributing them to other astrocytes and to the blood. If the astrocytes' potassium channels are not working properly, then potassium ions will build up around the nodes and, because axon membranes let potassium ions leak in (or, in the jargon, they are relatively permeable to potassium ions), this will tend to depolarize the axons by adding more positive ions internally (that is, making their resting membrane potential greater than −70 mV). This makes the axonal circuits more likely to fire action potentials – or more excitable – and therefore more likely to produce seizures.

KCNJ10 gene, that are located in the membranes of glial cells. These channels allow extracellular potassium ions to move into glial cells, which counteracts their potential build up in the extracellular space around neurons. This build up is liable to occur as a result of repeated action potentials, during which potassium ions exit the cell; glial cells can effectively mop up the excess potassium ions. Because neuronal membranes are relatively more permeable to potassium ions than to other ions, an excess of potassium around a neuron would result in a disproportionately large number coming into the cell, and this net entry of positively charged ions would depolarize the neuron's membrane and make it more excitable. This provides an explanation for why variants of *KCNJ10* causing defective spatial buffering result in seizures.[221]

Effects on Myelination

One of the most important features of most axons in the brain – exceptions being those with the smallest diameters – is that they are **myelinated**. This greatly increases the speed at which they transmit their action potentials and, therefore, the brain's ability to function as effectively and efficiently as possible. Variants that interfere with myelination will have a major impact on the ability of the brain to function. One well-known example of a condition in which this happens is Pelizaeus–Merzbacher leukodystrophy, which is characterized by developmental delay, spasticity, **ataxia** and other abnormalities of movement, including speech. This condition is caused by variants of the *PLP1* gene, which encodes a proteolipid that is a major component of myelin. While this might seem a relatively straightforward example of proximate causation and almost certainly provides much of the explanation, it is unlikely to be the full story. There are probably many other paths linking variants with phenotypes since *PLP1* is also expressed in many other cell types not involved in myelination, which are also likely to be affected.

Effects on Synaptic Transmission

As I discussed in Chapter 2, synapses are specialized points of contact between neurons at which electrical impulses coming down an axon cause the release of **neurotransmitters** that cross a narrow gap to affect the activity of the neuron on the other side. They might affect the postsynaptic neuron by depolarizing its membrane (making it more likely to make action potentials, or exciting it, if you like) or by hyperpolarizing it (making it less likely to make action potentials, that is, inhibiting it). Some neurotransmitters are also neuromodulators, meaning that they can acquire a subtly different function in which they diffuse more slowly and have more sustained effects, essentially tuning overall neural activity at synapses. In their neuromodulatory capacity, they can influence a broad range of postsynaptic activity and synaptic plasticity. Examples include serotonin (also known as 5-hydroxytryptamine), noradrenaline (also known as norepinephrine) and dopamine (Figure 8.2), with dopamine being arguably the most extensively studied in synaptic plasticity, emotions and motor control, as we shall also see in Chapter 9.

Some genetic variants cause problems by disrupting the synthesis of neurotransmitters. For example, the *TH* gene encodes the enzyme tyrosine hydroxylase, which helps convert the amino acid tyrosine to the neurotransmitter dopamine (Figure 8.2). Variants of the *TH*

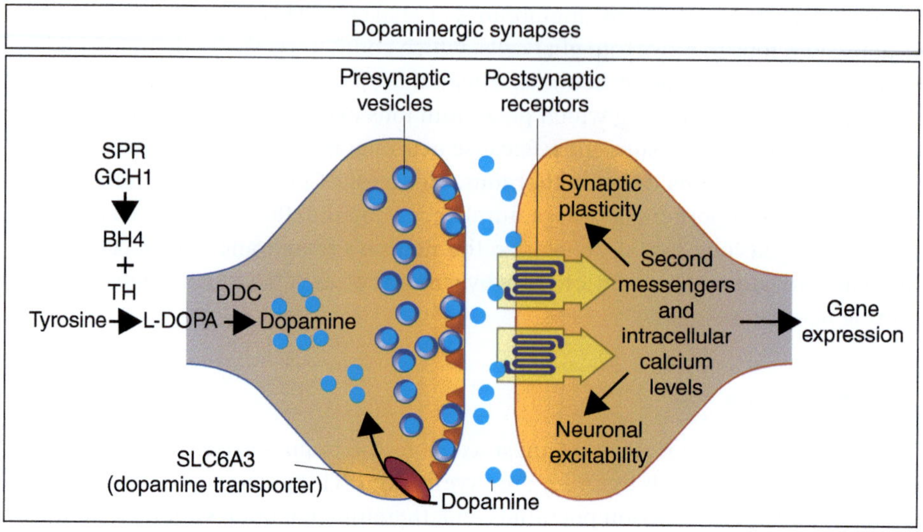

Figure 8.2 Dopaminergic synapses. Dopamine is a major neurotransmitter and neuromodulator in the brain. It is a monoamine synthesized from tyrosine via the intermediary, L-DOPA. Tyrosine is converted into L-DOPA by the enzyme tyrosine hydroxylase (TH), with co-factor tetrahydrobiopterin (BH4), which is produced by the enzymes SPR and GCH1. L-DOPA is converted into dopamine by the enzyme DOPA decarboxylase (DDC). Newly synthesized dopamine and recycled dopamine taken back into the cell by the dopamine transporter SLC6A3 are enclosed in vesicles for release into the synaptic cleft on depolarization. Dopamine binds its metabotropic G protein-coupled receptors (characterized by having seven domains that span the membrane); there are five subtypes (called D1–D5). The receptors can be divided into two families: D1-like (D1 and D5) and D2-like (D2, D3 and D4). D1-like activation (of D1 and D5) can cause excitation by triggering opening of sodium channels or inhibition by triggering opening of potassium channels; D2-like activation (of D2, D3 or D4) usually causes inhibition of the target neuron. Dopamine binding has these and many other effects via its activation of complex second messenger systems.

gene that reduce the activity of the tyrosine hydroxylase enzyme result in the production of less dopamine (as well as its derivative noradrenaline). Dopamine is a major neurotransmitter and neuromodulator in the **basal ganglia**, which you might remember from Chapter 2 are involved in the control of movement. The phenotypic consequences of reduced dopamine production include involuntary muscle contractions (dystonia), tremors and other uncontrolled movements. These symptoms usually respond to treatment with L-DOPA (l-3,4-*d*ihydr*o*xy*p*henyl*a*lanine), which is a precursor of dopamine (Figure 8.2). Variants that disrupt the function of the *SPR* or *GCH1* genes, which encode enzymes involved in producing tetrahydrobiopterin (also known as BH4), which is an essential co-factor for tyrosine hydroxylase activity, have similar effects.

Variants can have effects postsynaptically, for example on receptors. Variants of *GRIN2B* cause *GRIN2B*-related neurodevelopmental disorder, whose symptoms include intellectual disability, delayed development of speech and movement skills, such as sitting and walking, and seizures in some individuals. GRIN2B is subunit of **NMDA receptors**, which, as we saw in Chapter 2, are essential components not only of **glutamatergic** synaptic transmission but also of synaptic plasticity.

Both synthesis and recycling are important at synapses. The *SLC6A3* gene encodes a membrane protein that recycles dopamine in dopaminergic neurons. Its job is to collect dopamine that has been released into the synaptic cleft back into the presynaptic cell for reuse (Figure 8.2). Its impairment results in dopamine transporter deficiency syndrome, whose symptoms include difficulty speaking, eating, drinking and walking. The *DNM1* gene encodes a cytoskeletal binding protein involved in the recycling of synaptic vesicles after they have released their neurotransmitter. Variants of *DNM1* can cause DNM1 epileptic encephalopathy, whose symptoms include profound intellectual disability, **hypotonia** and epilepsy starting with infantile spasms. There is evidence that variants affecting synaptic recycling also contribute to the phenotypes of polygenic and multifactorial conditions such as Down syndrome.[222,223]

Neurotransmitter breakdown is also important for synaptic function. For example, the *MAOA* gene produces a protein called monoamine oxidase A that breaks down monoamines, including dopamine, serotonin and noradrenaline. Genetic variants that result in *MAOA* losing its function cause a build-up of these monoamines in the brain and, since they are involved in neural circuits that regulate a person's mood and emotions and control their responses to stress, this build-up can lead to unusually aggressive behaviours. These behaviours can be made worse if an affected individual takes food high in monoamines, since monoamine oxidase A also helps break down monoamines from the diet. In the long term, loss of MAOA function might alter how neuronal connections in some parts of the brain develop, and this could contribute in ways that we do not understand to the intellectual and behavioural symptoms shown by individuals.[224]

Effects on Synaptic Plasticity

I have discussed the principles of synaptic plasticity in Chapter 2. Variants that modify the critical ability of synapses to modify their strength as their activity changes can cause, or contribute to, symptoms associated with neurodevelopmental disorders or all types of neurodivergence. Because synaptic plasticity is so important for the brain's ability to learn from experience, these symptoms are likely to include developmental delay, intellectual disability, learning disability and/or more complex behavioural features (that I discussed in Chapter 3).

Much research has focused on variants affecting the postsynaptic density, which I introduced in Chapter 2.[225,226] This dense meshwork of well over 1,000 different types of proteins sitting underneath the postsynaptic membrane of glutamatergic synapses contains molecules that are critical for anchoring neurotransmitter receptors and mediating the long-term biochemical response of the synapse to activity (Figure 8.3). It comprises a core set of 30–60 different proteins around which its detailed composition varies within and between different regions of the brain (only a few of these proteins are shown in Figure 8.3). This creates considerable synaptic diversity, including differences of protein composition, arrangement and turnover, that is likely to contribute to the brain's cognitive capacity. Roughly a third of its core proteins provide structural support (we call these scaffold proteins), and the others include enzymes of various sorts involved in **second messenger** pathways, protein degradation and other functions. This dynamic system has evolved to provide an essential link between synaptic activity and synaptic structural changes

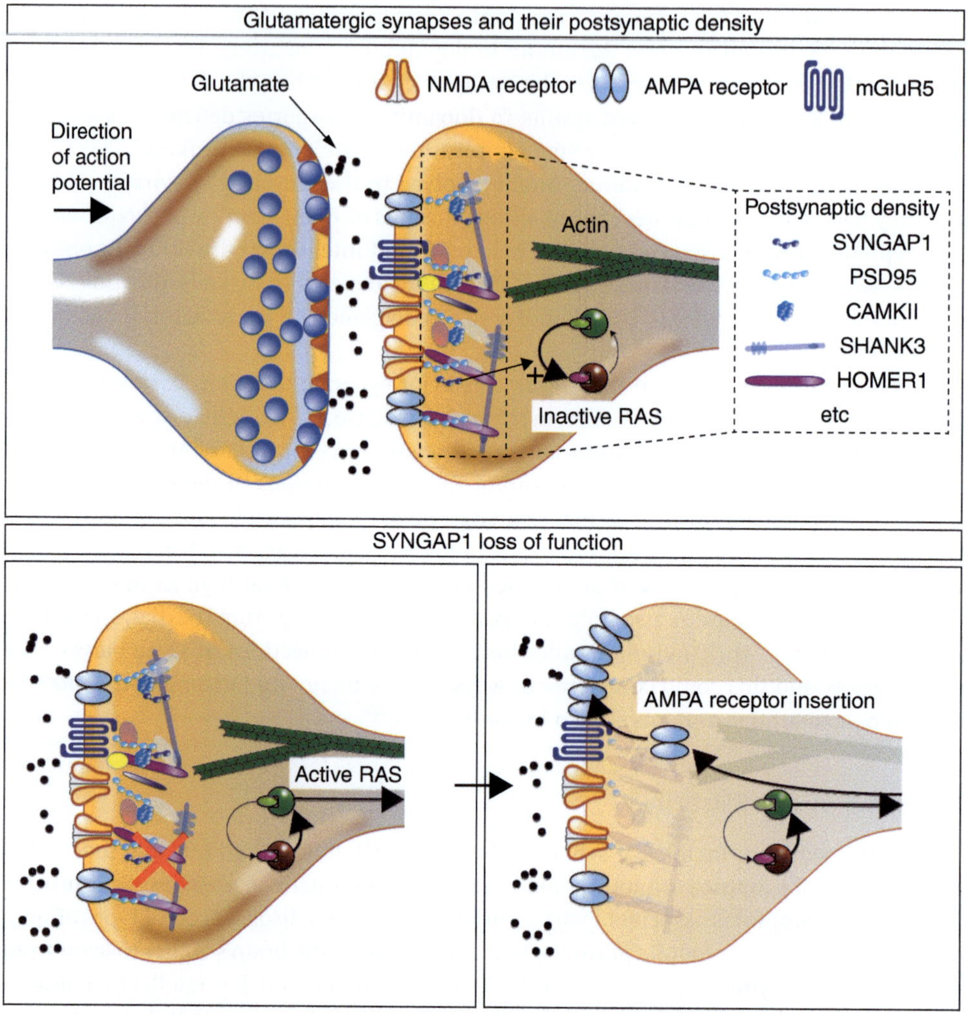

Figure 8.3 The postsynaptic density and possible effects of SYNGAP1 loss of function. The postsynaptic density of glutamatergic synapses has several critical functions. It helps to regulate the activity of neurotransmitter receptors; it is involved in synaptic plasticity; it anchors other proteins; its connection to the cytoskeleton provides structural support and enables dynamic changes in the synapse. It is composed of a complex network of proteins, including neurotransmitter receptors, scaffolding proteins, **signalling molecules** and cytoskeletal components. The identities of a few of these proteins are shown here. Scaffold proteins include HOMER1–3, SHANK1–3 and PSD95. CAMKII is crucial for synaptic function and plasticity due to its regulation of processes including receptor trafficking, localization and activity and actin cytoskeletal dynamics, affecting particularly long-term potentiation and long-term depression (see Chapter 2). SYNGAP1 also plays a crucial role in regulating synaptic plasticity. One of its functions is to inhibit the activation of small GTPases such as RAS by promoting the formation of its inactive form (Figure 7.4 in Chapter 7). It is thought that variants that result in SYNGAP1 loss of function cause increased active RAS formation, which stimulates processes including translation to produce more AMPA receptors than usual, the effect of which is to increase synaptic excitation in neural circuits.

affecting synaptic strength. Although we know the identities of most of the proteins that it contains, we still have much to learn about how these proteins work together to bring about functional synaptic changes.

Variants of many of the genes that encode these proteins have been implicated in neurodevelopmental disorders and neurodivergence. These variants are often risk factors for complex multifactorial conditions. I will say more about these conditions later in this chapter. In some cases, variants of individual genes have been identified as the causes of specific syndromes characterized by symptoms that might include developmental delay, intellectual disability, seizures, other movement disorders and behavioural difficulties of various types. A well-studied example is SYNGAP1-related intellectual disability, caused by variants of *SYNGAP1*.[227,228]

SYNGAP1 protein, which is encoded by the *SYNGAP1* gene, is very abundant in the postsynaptic densities of glutamatergic excitatory synapses. The protein is large and complex with many isoforms (protein isoforms are different versions of a protein made from the same gene, arising through alternative mRNA **splicing**, which I explained in Chapter 4). The function of one of its parts, or **domains**, is to activate other proteins called GTPases (hence, SYN, standing for *syn*aptic, and GAP, standing for *G*TPase-*a*ctivating *p*rotein) (Figure 8.3). This domain of SYNGAP1 is the one that usually receives most attention when it comes to explaining its major functions at the synapse. GTPases are enzymes that catalyse the hydrolysis of guanosine triphosphate (GTP) to guanosine diphosphate (GDP), thereby impacting the activity of second messenger pathways (see Figure 7.4 in Chapter 7). GTP is the active form and GDP the inactive one. A popular theory proposes that variants of *SYNGAP1* that interfere with the ability of its protein to convert GTP to inactive GDP are likely to elevate biochemical activity in the postsynaptic neuron in such a way as to strengthen the synapse, resulting in hyperactivity (Figure 8.3). This might explain symptoms such as seizures and possibly others that emerge because activity-dependent changes at synapses are not appropriately controlled. If this sounds rather hand-waving, it is because it is still very unclear how symptoms are brought about. In fact, some researchers have suggested that SYNGAP1's action on GTP is *not* the aspect of its function that is most important for synaptic plasticity.[229] Furthermore, there is evidence that SYNGAP1 might play a non-synaptic role much earlier in brain development than previously thought, in the early prenatal formation of the cerebral cortex.[230] You should keep an open mind.

Indirect Ways in Which Genetic Variants Can Affect the Brain's Electrical Components

Any variants that impact the perceptual, cognitive and behavioural functions of the brain must, in the end, do so by affecting its electrical components in some way. So far in this chapter, I have considered variants that directly affect the ability of these components to function individually and collectively. While axons and synapses may be the essential elements of brain function, their existence and ability to work properly relies on vast numbers of other biochemical processes, some of which are many steps removed from the biology of the components themselves. In this section, I will turn to variants that influence these more distant mechanisms.

I will start by discussing how problems can be caused by disruptions that affect the mechanisms regulating which genes are expressed by cells. I will go on to describe the impacts of genetic variants affecting molecules responsible for regulating the balance and functionality of the proteins that make up our cells, that is, **proteostasis**, which I introduced in Chapter 7. I will then consider the consequences of disrupting specific proteins, including those that provide structure and strength to our tissues and that act as enzymes controlling the basic metabolic processes of our cells, usually not only in the brain but also elsewhere throughout the body.

For conditions caused indirectly, the routes from ultimate causes to phenotypes are likely to be particularly numerous and circuitous. In most cases, we have only a patchy understanding of the many steps involved. This complexity is compounded by the fact that many variants disrupt processes both inside and outside the developing brain, with the latter often having knock-on consequences for the developing brain.

Variants Affecting Gene Expression

In Chapter 7, we saw how the correct maintenance and regulation of DNA is critical for creating the brain's structure and function before birth. These processes need to continue after birth. It is crucial for all cells throughout life that they maintain their DNA's integrity and allow their genome to replicate effectively (if and when they divide, which some neuronal stem cells and many non-neuronal brain cells continue, or start, to do after birth). Underpinning all of this, correct regulation of gene expression must also continue after birth. If these processes fail as a child grows, the results are likely to affect the brain's electrical components in some way (Table 6.2 in Chapter 6 lists some specific conditions that can result).

The development of the brain's complement of cell types relies critically on the regulation of gene **transcription** by **epigenetic** mechanisms, as I outlined in Chapter 4. In summary, these mechanisms are of two main types: some control gene expression by regulating **chromatin** configuration, determining which parts of the genome are open and which are closed for transcription in any given cell; others involve the actions of the cell **lineage**-specific **transcription factors** in any given cell on the available gene **regulatory elements** in that cell. These two factors are closely interrelated, since cell lineage-specific transcription factors can participate in modifying chromatin configuration and chromatin configuration needs to be appropriate to allow transcription factors to access the specific genomic sites from which they affect gene expression. There is evidence that, as developing cells transit to progressively more differentiated states, their chromatin undergoes an overall change from a relatively open configuration to a more compact one. Intuitively, this makes sense: increasing the amount of compacted chromatin is likely to limit a cell's options for further development, putting the cell into a more stable state. If you like, you can think of it this way: many of the earliest events of prenatal brain development that are initiated by the nimble actions of lineage-specific transcription factors on relatively open chromatin are progressively consolidated by long-term control through chromatin modification. No doubt, this is a major simplification, not least because variants of some chromatin modifiers do affect very early development by facilitating or impeding the DNA binding and thereby the actions of transcription factors, as I outlined in Chapter 7 (Figure 7.6). Nevertheless, this generalization might help us understand why some of the variants that affect chromatin modification cause detectable disorders mainly relatively *late* in development, during postnatal life, as I will now explain.

Some of the most extensively studied variants to cause postnatal disorders of brain development are of the *MECP2* gene.[231,232] It encodes the protein MECP2 (methyl-CpG-binding protein 2). Variants that disrupt its function cause Rett syndrome and a few other less prevalent related disorders (Table 6.2, Chapter 6). *MECP2* is on the X chromosome and the ultimate cause of most cases of Rett syndrome is inactivation of one copy of the gene. If you think back to the principle of X inactivation (Chapter 4), you will appreciate that only some (on average about half) of a female's cells will express the variant form, which mitigates the overall impact. If a male carries a deactivating variant, all cells will be affected and the overall impact is likely to be sufficiently severe to cause miscarriage in many cases. This probably explains why Rett syndrome appears more commonly in females than in males. In an associated condition, *MECP2* duplication syndrome, MECP2 is overproduced. The overproduction is usually due to the *MECP2* locus being copied within the X chromosome, and males are affected more than females because all their cells produce extra MECP2. In females, the overall impact is less severe because only some cells use the X chromosome carrying the duplicated locus; furthermore, it appears that X activation becomes skewed to make the abnormal X chromosome much more likely to become inactivated, which effectively silences the duplicated locus. These observations illustrate two important principles that apply to a number of genes: first, there is a sweet spot for the activity levels of some genes, with both too much or too little causing problems; second, biological systems can often compensate for potentially deleterious changes.

Rett syndrome and *MECP2* duplication syndrome have many symptoms in common, including developmental delay, intellectual disability and gross and fine movement difficulties including delayed or absent speech, epilepsy and behavioural problems. Their severity varies considerably for reasons that include the person's biological sex, as I explained above. The next question is: how do variants of *MECP2* cause these symptoms?

MECP2 binds specifically to methylated DNA (Chapter 4, Figure 4.11) using a domain that is approximately 90 amino acids long called the methyl-CpG-binding domain, which is affected by many variants causing Rett syndrome (Figure 8.4). These variants almost invariably interfere with binding to methylated DNA. MECP2 acts as a bridge linking methylated DNA with a complex of proteins that repress gene expression. The best known of these are called nuclear corepressor 1 and 2 (or NCOR1 and NCOR2 for short), which interact with MECP2 via its NCOR interaction domain (or NID), which is another domain often affected by variants that cause Rett syndrome. Other proteins in the complexes include transducin beta-like 1 Y-linked (TBL1Y) and TBL1X/Y-related 1 (TBL1XR1) and histone deacetylases (HDACs), which deacetylate **histones**, resulting in the tightening of chromatin and, thereby, the inhibition of gene expression (Chapter 4, Figure 4.11). Variants of many of these proteins are also linked to intellectual disability and other neurodevelopmental disorders and neurodivergence. (As you will have realized by now, the mechanisms of development are never simple, and so it should come as no surprise to learn that the NCOR complex can be recruited to the genome in other ways too, for example by DNA-binding proteins such as some transcription factors.)

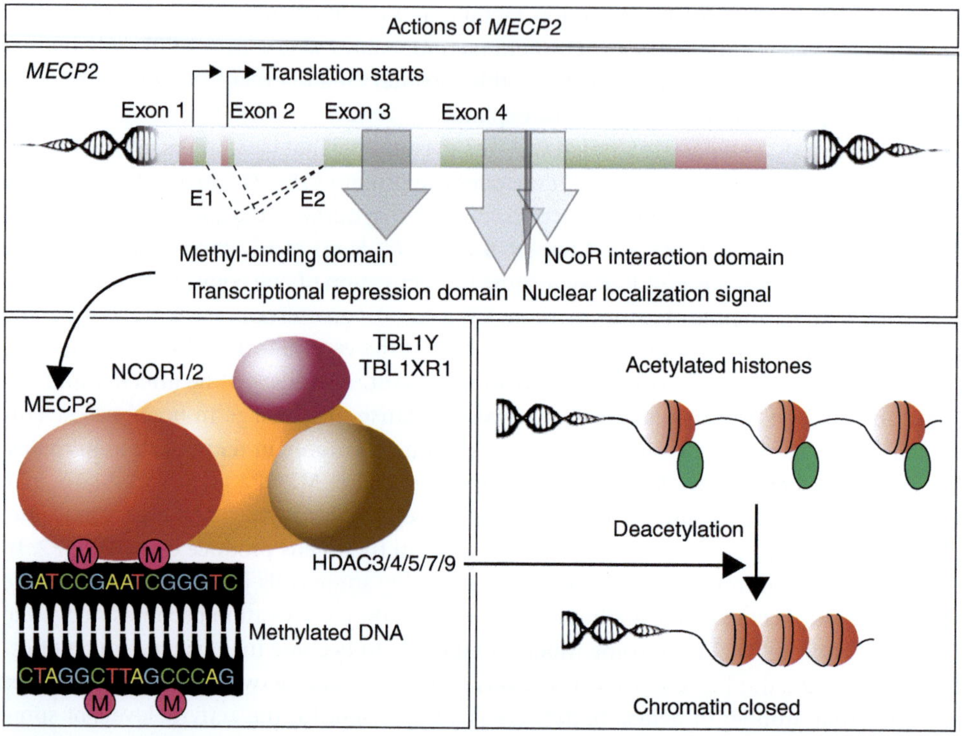

Figure 8.4 **Schematic of *MECP2* gene organization and function.** Two isoforms (E1 and E2) of *MECP2* produce proteins from two different start points. Both contain several regions coding for important functional domains in the protein, enabling binding to methyl groups, transcriptional repression, interaction with NCOR proteins and the nuclear localization of MECP2. MECP2 acts as a bridge from methylated DNA to a complex of proteins, including nuclear corepressor 1 and 2 (NCOR1 and NCOR2), transducin beta-like 1 Y-linked (TBL1Y) and TBL1X/Y-related 1 (TBL1XR1) and histone deacetylases (HDACs), which deacetylate histones resulting in tightening of chromatin inhibiting gene expression.

Interference with *MECP2* function is likely, therefore, to limit its repression of other genes. Faulty gene repression is as likely as faulty gene activation to disrupt development. If you look back to Waddington's epigenetic landscape model in Chapter 1 (Figure 1.5), you can imagine repression as being the closing of valleys, thereby blocking the availability of undesirable options. Stopping cells adopting fates they should not is a critical control mechanism. If repression is impaired by *MECP2* variants, cells might develop in inappropriate ways, whereas enhanced repression due to *MECP2* duplication might remove the possibility of desirable options. But, as ever, you need to keep an open mind to other possibilities. Although I have emphasized the repressive functions of MECP2, there is evidence suggesting that MECP2 is also involved in a range of other processes via interactions with many other possible binding partners. As well as transcriptional repression, these processes might include transcriptional activation, alternative splicing, chromatin remodelling and miRNA processing.[231,233]

Exactly how variants of *MECP2* lead to neurodevelopmental disorder is not yet known. One hypothesis is that they have a relatively direct impact on the brain's functional components, interfering with synaptic function and plasticity. Another is that they have a more indirect effect via disruption of metabolic processes that provide the infrastructure required to support specific developmental events. For example, metabolic defects in glucose and cholesterol metabolism are often found in people with Rett syndrome. There is evidence that variants can cause glial and **microglial** dysfunction, which might also contribute indirectly to symptoms. It is very likely that many or all of these possibilities, and others too, are involved, given the widespread nature of MECP2's actions on the genome. Whatever the critical defects are, it appears that they do not stem primarily from cell death and that they are potentially reversible if MECP2 were to be restored to normal levels sometime after birth.[234] Scientists are currently trying to use this insight to develop new approaches to treat people with Rett syndrome.

These findings give good reason to think that variants of *MECP2* cause disorders that arise primarily during postnatal development. Although *MECP2* is expressed before birth, with levels increasing from around the middle of gestation, its expression increases substantially throughout early childhood when the brain's neurons are maturing, growing their **dendrites** and forming and refining their synaptic connections. This pattern is likely to help explain why the early development of people with Rett syndrome is apparently unaffected. Some caution is needed, however: some human studies have shown neurological alterations within the first months of life and research in other species has suggested that variants of the gene might interfere with neurogenesis and neuronal differentiation before birth.[235] Also, as I mentioned earlier, males with loss of *MECP2* function are liable to die prenatally, suggesting that *MECP2* does have important roles to play before birth.

As well as causing prenatal developmental disorders (discussed in Chapter 7), variants affecting transcription factors can cause conditions that emerge in postnatal life. Unlike chromatin modifiers such as MECP2, transcription factors regulate gene expression by binding to specific **promoter** and **enhancer** sequences in genomic DNA (Figure 4.10 in Chapter 4). One example of a condition caused this way is HIVEP2-related intellectual disability. Early symptoms typically include hypotonia and delayed development of skills such as sitting, standing, walking and speaking, followed by intellectual disability and impaired communication and social interactions. There is evidence that one of the many genes HIVEP2 regulates is *SLC6A3*, which is important for the function of dopaminergic neurons (Figure 8.2). Dopaminergic neurons are involved in many of the functions that are disrupted in this disorder. Overall, however, it would be a mistake to assume that this is the sole proximate cause of the symptoms, since HIVEP2 is likely to affect the expression of hundreds of other genes. Moreover, its effects might start prenatally; we just do not know enough to say with any confidence.

Variants of other genes can affect gene expression by altering the activity of proteins that do not themselves bind to genomic DNA but, instead, bind to transcription factors to modulate what happens when they bind. In other words, they are co-factors (Figure 4.10 in Chapter 4). For example, the protein encoded by the *ATN1* gene appears to interact with transcription factors to suppress the expression of the transcription factors' target genes. Variants that disrupt ATN1 function often have a **trinucleotide repeat** that is two to three times longer than usual (Chapter 4). This causes a condition called dentatorubral-pallidoluysian atrophy. The symptoms include epilepsy, ataxia and other movement disorders and early onset dementia. It is a progressive disorder in which neurons gradually die.

One possibility is that the repeat expansion interferes with ATN1's ability to modulate transcription factor activity. Another is that the extended repeat slows the degradation of the protein, which builds up in the cytoplasm and nucleus of cells causing their dysfunction and death. More work is needed to establish the validity of these hypotheses.

Variants Affecting Proteostasis

The functioning of the brain's electrical components can be affected by variants of genes whose products help regulate mRNA **translation** or other processes involved in maintaining the production, levels, structures and functions of proteins. A well-studied example of a condition caused in this way is fragile X syndrome (FXS), one of the commonest forms of inherited intellectual disability. It is caused by variants of the *FMR1* gene.

FMR1 is located on the X chromosome and encodes the RNA-binding protein FMRP, an acronym that was formerly taken to stand for fragile X mental retardation protein and is now taken to stand for fragile X messenger ribonucleoprotein. This change in nomenclature is welcome for several reasons: it removes the widely disliked phrase mental retardation, which has a history of being used in a derogatory way, and serves as a constant reminder that the main biological function of the protein is to interact with mRNA and regulate its function. When FMRP binds to mRNA, it limits its translation, effectively acting as a brake on protein production by slowing the movement of the ribosome along the mRNA.[236] I should emphasize here that we do not yet have a comprehensive knowledge of all the more than 1,000 mRNAs that FMRP interacts with and, therefore, all the processes in which it is involved. What we do know indicates that it has synaptic and non-synaptic actions on neurons, including regulation of the ion channel composition of dendritic and axonal membranes, and on non-neuronal cell types, including **oligodendrocytes.**[237]

Almost all cases of fragile X syndrome are caused by the expansion of a trinucleotide repeat within the gene. The three nucleotides that are repeated are CGG; normally, the *FMR1* gene has a limited number of CGG repeats (less than 55 in the 5′UTR; Figure 8.5). If the trinucleotide becomes repeated over about 200 times in the 5′UTR, then the *FMR1* gene becomes hypermethylated to the point where it cannot produce FMRP (remember, from Chapter 4, that DNA methylation occurs on C bases that are followed by G bases). Individuals with an intermediate expansion of 55–200 repeats can have a mild version of the syndrome which might include emotional problems such as anxiety or depression, learning disabilities or atypical behaviour. As discussed for other **X-linked** syndromes, such as Rett syndrome in the previous section, males having only one X chromosome will be affected more severely by a harmful variant of the *FMR1* gene than a female with a similar variant in one of her X chromosomes.

Since FMRP limits the rate of translation, its loss of function is liable to result in excessive protein synthesis. Much research on FMRP has focused on the possible consequences of this for synapses, where it is highly enriched. Variants that reduce FMRP function appear to interfere with synaptic plasticity by enhancing the magnitude of long-term depression (LTD) and reducing that of long-term potentiation (LTP) (look at Figure 2.6, in Chapter 2 if you need reminded what LTD and LTP are). In Chapter 2, I mentioned that **metabotropic** glutamate receptors might be involved in LTD. One theory proposes that this is because these receptors have the potential to activate the translation of mRNAs

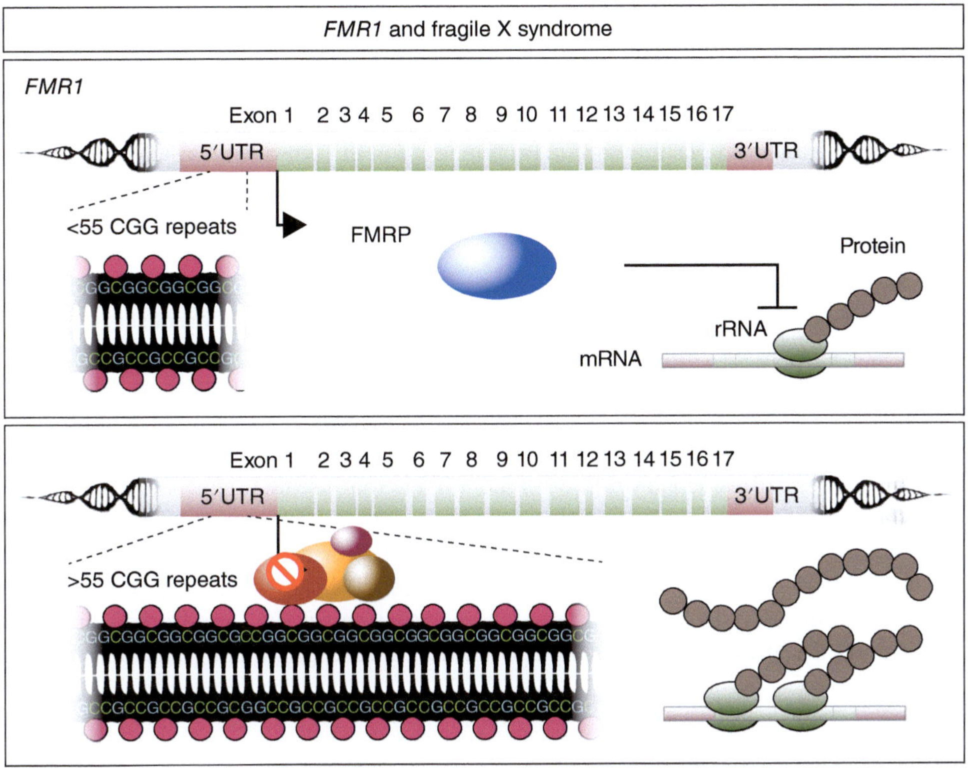

Figure 8.5 Hypermethylation of *FMR1* in fragile X syndrome. The fully functional *FMR1* gene has less than 55 CGG repeats in its 5′-untranslated region (UTR) and produces FMRP protein, which limits the translation of the mRNA of many other genes. In fragile X syndrome, the CGG repeat in the 5′UTR expands leading to the hypermethylation (purple circles) of the upstream promoter region of the *FMR1* gene. This reduces the activity of, or silences, the *FMR1* gene. The consequences of lowered or absent FMRP include increased translation of mRNAs.

whose products enhance LTD, a potential that is normally kept in check by FMRP.[238] Based on this idea, one might predict that loss of FMRP function would allow levels of synaptic proteins responsible for LTD to increase. However, one line of research has suggested that this does *not* occur because protein degradation increases to compensate. Indeed, this compensatory increase in degradation might itself contribute to pathological changes.[239] We have much more to discover about the proximate causes of fragile X syndrome, but what we have learned so far suggests that things are not as straightforward as first suggested. Personally, I am not surprised; this fascinating story is a great example of how science progresses through constantly evolving hypotheses rather than the discovery of incontrovertible facts.

In Chapters 3 and 7, I introduced a condition known as tuberous sclerosis caused by variants of *TSC1* or *TSC2*, in which benign tumours appear in the developing brain, starting prenatally (Figure 3.2 in Chapter 3). It has been suggested that these variants might cause the intellectual disability and other functional symptoms that are associated with tuberous sclerosis through effects on protein synthesis at the synapse. In the case of tuberous sclerosis, the evidence points to a decrease rather than an increase in LTD, suggesting that rather different underlying synaptic mechanisms might converge to produce somewhat similar cognitive symptoms in fragile X syndrome, tuberous sclerosis and possibly other such conditions. How LTD is decreased in tuberous sclerosis remains unclear, although it might be due to an effect of some sort on protein levels at the synapse since TSC1 and TSC2 are involved in a signalling pathway that links cell surface receptors to translation.[240]

Although tuberous sclerosis involves pathology that develops prenatally, it is possible that the cognitive difficulties that people experience develop relatively independently after birth, due to the problems with synaptic plasticity I just outlined. Indeed, we need to be alert to this possibility as we learn more about other syndromes too: different sets of symptoms might emerge for different, relatively unconnected reasons at different stages of development. In the case of fragile X syndrome, newborns do not usually show symptoms, and neurodevelopmental delay is followed by intellectual disability, symptoms of anxiety, hyperactivity and attention deficit, difficulty with communication and social interactions and sometimes seizures. However, it does seem likely that prenatal effects of *FMR1* variants might lie undetected at birth. FMRP is expressed in cerebral cortical **progenitor** cells and in neurons during prenatal development, and there is evidence that variants of *FMR1* can impair mitochondrial functions and cause neuronal hyperexcitability prenatally.[241] I suspect that the results of ongoing research will result in fragile X syndrome being best discussed in the previous chapter as a condition whose proximate causation begins prenatally.

Variants Affecting Structural Proteins

As we have seen in previous chapters, many of a cell's proteins are the building blocks of the **cytoskeleton**, the internal structure that generates a cell's shape and gives it strength and support. There are many examples; here I will mention one, glial fibrillary acidic protein (GFAP), which is produced mainly by **astrocytes**. Genetic variants that disrupt the functions of this protein have intriguing effects. They cause disorders of brain development, but it is not yet clear how. Particularly puzzling is the fact that disruption of GFAP causes a **leukodystrophy**, which means that its effects include the underproduction, loss or both of myelin, which is made by oligodendrocytes and not astrocytes. In some way, astrocyte disorder resulting from GFAP disruption has a secondary, knock-on effect on oligodendrocytes; there is some evidence that it might be a result of changes in the molecules secreted by astrocytes.[242]

Variants Affecting General Metabolic Processes

Some proteins are enzymes controlling cells' generic metabolic activities, which are numerous and include making specific types of molecules, moving them within and between cells and breaking them down when they are no longer needed.

Some variants affect the synthesis of molecules. These include enzymes, such as HPRT1 (hypoxanthine-guanine phosphoribosyltransferase 1) and PRPS1 (phosphoribosylpyrophosphate synthetase 1), that are involved in the biochemical pathways that make and recycle purines and pyrimidines, which are the building blocks of DNA and RNA and major

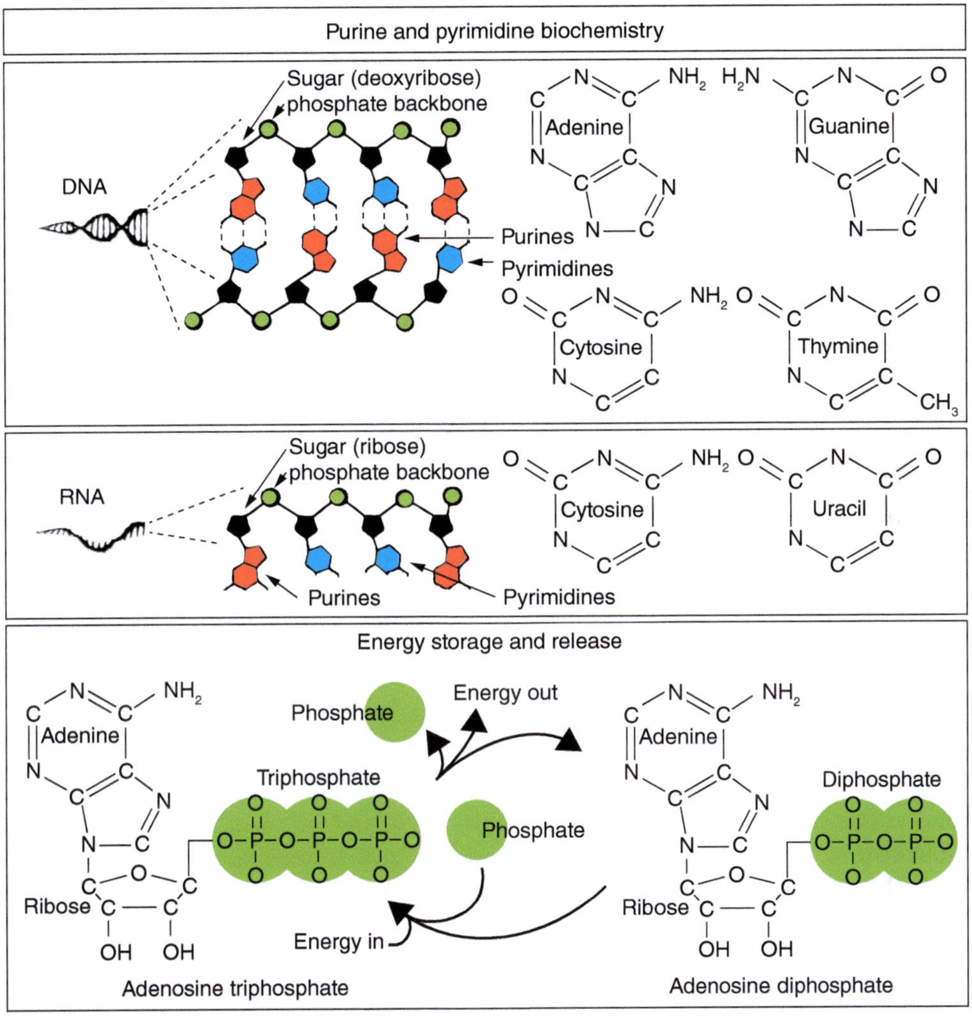

Figure 8.6 Purine and pyrimidine biochemistry. Purine and pyrimidine nucleotides are not only the building blocks of nucleic acids but also store and release energy, act as signalling molecules and are components of essential coenzymes (we have already seen examples of guanosine triphosphate (GTP) acting in signalling pathways: Chapter 7, Figure 7.4). Adenosine triphosphate (ATP) plays a particularly crucial role in energy transfer within cells. The diagrams at the top show the molecular structures of DNA and RNA: both have a sugar (deoxyribose in DNA, ribose in RNA) and phosphate backbone; adenine and guanine are purines found in DNA and RNA; cytosine and thymine are the pyrimidines in DNA; cytosine and uracil are the pyrimidines in RNA. The diagram at the bottom shows the adenosine triphosphate (ATP) – adenosine diphosphate (ADP) cycle, the predominant mechanism for storage and release of cellular energy.

energy-carrying molecules (Figure 8.6). Such variants can cause movement and cognitive disabilities as well as disorders of other organs. Other variants can cause neurodevelopmental disorders by affecting enzymes such as ASPA (aspartoacylase) or PAH (phenylalanine hydroxylase) that make or modify amino acids. For example, Canavan disease is caused by disruption of ASPA function in oligodendrocytes, interfering with their production of the amino acid aspartic acid and, ultimately, their ability to generate myelin sheaths. Canavan disease is an example of a leukodystrophy. Another example is fatty acid hydroxylase-associated neurodegeneration, in which the function of an enzyme that modifies fatty acids (which are used to make lipids) is disrupted. In these conditions, infants often appear normal in the months after birth, but fail to develop motor skills such as control of head movement and sitting unsupported, can go on to develop problems such as seizures and can lose skills that they had mastered. This occurs because the lack of myelination prevents the rapid transmission of action potentials required for the brain's functions to develop.

Some variants affect cellular transportation.[108] The solute carrier (SLC) family is the most abundant class of transmembrane transporters. There are several hundred of them. We met one of them, the dopamine transporter SLC6A3, in Figure 8.2. They are responsible for translocating various substances, including hormones, nutrients and so on, across cell membranes. They operate throughout the body including in the developing brain. They include SLC16A2, which transports the hormone triiodothyronine into cells to regulate the rates of metabolic processes; SLC2A1, which transports glucose – a vital source of energy – into cells; SLC22A5, which transports carnitine obtained from food into cells, where it moves fats into mitochondria to provide energy; SLC52A2 and SLC52A3, which transport riboflavin into cells, and SLC6A8, which transports creatine into cells (both riboflavin and creatine are required for energy production and storage); SLC30A10, which transports manganese out of cells to prevent its toxic accumulation; SLC19A3, which transports a vitamin called thiamine (or vitamin B1) into cells; SLC17A5, which transports sialic acid, a product of fat and protein breakdown, across the membranes of lysosomes. Variants of the genes that encode SLC and other transporters have been identified as the causes of monogenic syndromes with a range of neurological and non-neurological symptoms (Table 6.4 in Chapter 6). They are often associated with disorders including epilepsy.[243] Although these disorders often reveal themselves postnatally, it is likely that at least some have their origins in prenatal development.

Some variants affect cellular degradation processes. As I outlined in Chapter 7, degradation of proteins occurs via the ubiquitin-proteasome or the **autophagy**-lysosome pathway (Figure 7.17). Although some disorders caused by variants affecting these processes clearly have their origins prenatally, others do not rear their heads until after birth, making it uncertain when they start; my hunch is it will be prenatally in at least many cases. For example, variants disrupting the function of the gene *GAN*, whose protein product acts in the ubiquitin-proteasome pathway to break down neurofilaments, cause giant axonal neuropathy. Neurofilaments are long flexible protein polymers that form key components of a neuron's cytoskeleton; they are composed of three major protein subunits – called neurofilament light, neurofilament medium and neurofilament heavy – which assemble to give structural stability and organization to axons. In giant axonal neuropathy, the accumulation of neurofilaments in axons makes them bigger than they should be and prevents them from transmitting action potentials properly. Many neurons eventually die. Symptoms such as problems with movement start early in childhood and progress with age to include intellectual disability.

Lysosomes break down not only proteins but also carbohydrates and lipids. Variants affecting the degradation of any of these types of molecules can cause neurodevelopmental disorders. For example, alpha-mannosidosis, aspartylglycosaminuria, fucosidosis, Gaucher disease, GM1 gangliosidosis, mucopolysaccharidosis types I (also known as Hurler syndrome), II (also known as Hunter syndrome) and III, neuronal ceroid lipofuscinosis, Schindler disease type 1 and Tay–Sachs disease (Table 6.6 in Chapter 6) are syndromes caused by variants that disrupt proteins involved in the breakdown of carbohydrates and/or lipids. You will see these conditions referred to as lysosomal storage disorders (there are about 50 of them). The accumulation of undegraded or partially degraded carbohydrates enlarges the lysosomes – which might explain the enlargement of organs including the brain often seen in some of these disorders – and interferes with their function. The upshot is usually the onset of seizures and a progressive deterioration in the function of the postnatal developing brain that is sometimes summarized as childhood dementia.

Some variants affect mitochondria (Table 6.7 in Chapter 6).[244] Mitochondria provide the bulk of the energy needed by eukaryotic cells in the form of adenosine triphosphate (ATP) by a biochemical process known as oxidative phosphorylation (Figure 8.7). Oxidative phosphorylation involves the electron transport chain and ATP synthase. The electron transport chain comprises four complexes (I–IV) which drive a fifth complex (V), the ATP synthase. There are many ways in which this supply can be disrupted by variants of mitochondrial component-encoding genes. Some of these genes are in the cell's nuclear genome and others are in the genome of mitochondria themselves. Of the approximately 90 proteins that make up the oxidative phosphorylation system, 13 are derived from the mitochondrial genome and the rest are from the nuclear genome. Variants of genes in nuclear DNA that make mitochondrial proteins will affect all mitochondria in the cell. Variants in

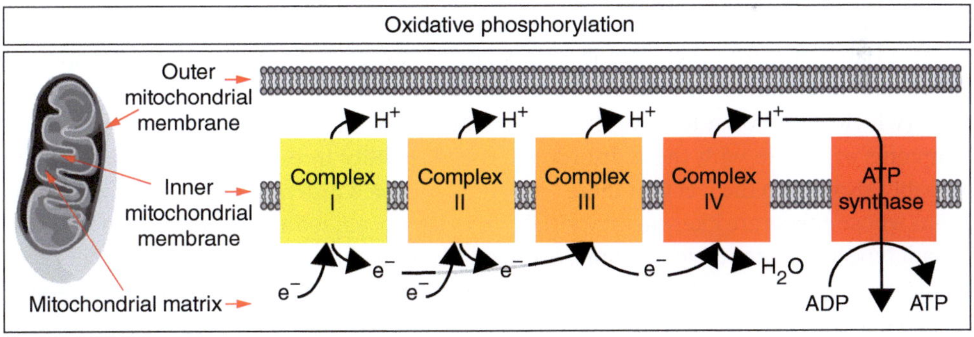

Figure 8.7 Oxidative phosphorylation in mitochondria generates adenosine triphosphate (ATP). The electron transport chain in the mitochondria involves five main protein complexes located in the mitochondrial inner membrane. These are labelled Complexes I, II, III, IV and ATP synthase. Electrons donated to Complexes I and II by two electron carriers (NADH and $FADH_2$, which are produced by the breakdown of glucose, which we call glycolysis) are passed along to Complex III and then on to Complex IV. This process generates energy which is used to pump hydrogen ions (protons) into the space between the outer and inner mitochondrial membranes, creating an electrical and chemical gradient of hydrogen ions between the intermembrane space and the mitochondrial matrix. The main route for proton re-entry to the matrix is through ATP synthase (also often called Complex V). Energy generated by this re-entry creates ATP from adenosine diphosphate (ADP). The electrons that have been passed along the chain combine with hydrogen ions and oxygen to form water in Complex IV. As you can see, the process involves oxygen (hence, it is oxidative) and results in phosphorylation of ADP to make ATP.

mitochondrial DNA are less likely to affect all mitochondria because, unlike nuclear DNA, cells contain many mitochondrial DNA molecules and they can carry different variants. The impact of variants in mitochondrial DNA on a cell depends in large part on the proportion of mitochondrial DNA molecules carrying the variant (as I discussed in Chapter 4).

These points are well exemplified if we consider the effects of variants that affect the POLG complex, which are the commonest cause of inherited mitochondrial disorders. The *POLG* gene is located in the cell's nuclear genome and encodes a catalytic subunit of DNA polymerase gamma, which is the core of a complex of proteins that replicates mitochondrial DNA. Other proteins in this complex, by providing additional functions such as proofreading and repair of replicated mitochondrial DNA, ensure that its reproduction occurs with high fidelity. Variants can have two sorts of impact. Some cause disordered phenotypes simply by limiting the numbers of mitochondrial DNA molecules that cells can produce – these are referred to as mitochondrial depletion syndromes. They generally appear early in infancy. Others result in replication becoming error-strewn, which leads to the accumulation of mitochondria with deletion errors – these are referred to as mitochondrial deletion syndromes. Unlike depletion syndromes, deletion syndromes can appear as late as adulthood since it takes time for the proportion of mitochondrial DNA molecules with particular deletions to build up to a level at which they start to have a detectable effect (so I could have put them into the next chapter, but I have kept all these *POLG*-related conditions together here).

POLG variants cause a broad continuous range of phenotypes (Table 6.7 in Chapter 6).[245] In part, this is because the way in which variants affect phenotypes depends on what proportion of molecules carry a particular variant in each of the cells in a population. This proportion is likely to depend both on chance and on any number of variables, both internal and external to the cell, that can influence the replication of its mitochondrial DNA. These variables include other genetic factors, such as variants of modifier genes (Chapter 6) in the nuclear or mitochondrial DNA, and environmental effects such as toxins or infections (which are known to have an adverse effect on mitochondria and exacerbate the symptoms of people with these disorders). The variability in the phenotypes observed is probably also due to the fact that there are several hundred different pathogenic variants of *POLG* whose effects are likely to vary, although we do not yet see a clear correlation between variant type and phenotypic outcome. Six main disorders have been described. Alpers syndrome is one of the most severe and is characterized by childhood-onset progressive encephalopathy and epilepsy; childhood myocerebrohepatopathy spectrum presents with developmental delay or even childhood dementia; myoclonic epilepsy myopathy sensory ataxia and ataxia neuropathy spectrum, whose symptoms include those included in their cumbersome names; **autosomal recessive** or **dominant** progressive external ophthalmoplegia, which include a variable degree of psychological depression, ataxia and other problems with movement including speech.

The proximate causes of these *POLG*-related conditions stem from the loss of energy supplies in brain cells. This is also the case for other disorders caused by variants of other mitochondrial genes located in the nuclear chromosomes, such as *MTPAP* and *HSD17B10*. These genes, which encode, respectively, a DNA polymerase that makes polyA tails (Figure 4.8, Chapter 4) on the ends of mitochondrial transcripts and an enzyme involved in the production of mitochondrial transfer RNA (tRNA). The loss of segments of the *mitochondrial* DNA (mitochondrial deletion disorders) can also affect energy generation by disrupting mitochondrial-encoded enzymes involved in oxidative phosphorylation: this will lead to variable symptoms such as ataxia, movement disorders and so on, once the deleted mitochondria are present in high enough proportions to have an impact.

Brain cells have particularly high energy demands, but the details of how the complex and variable effects of an inadequate energy supply play out to produce the symptoms we observe in particular syndromes is far from clear. In general, we can predict that at least most developmental processes are likely to be affected to some extent, resulting in an accumulation of negative consequences for the brain's electrical components and neuronal networks that eventually disrupt their functions in detectable ways.

Mitochondria play other crucial roles in cells, aside from their production of energy. For example, they maintain calcium homeostasis within cells, acting as calcium buffers by storing and releasing Ca^{2+} to regulate calcium concentrations in the cytoplasm. This buffering is essential for a range of cellular functions, including intercellular signalling and neuronal excitability. Many researchers think that, if mitochondrial functions are compromised, a phenomenon known as excitotoxicity, in which excessive excitation causes cells to die, is likely to occur (Figure 8.8).[246] This is because interference with mitochondrial function is likely to affect intracellular Ca^{2+} regulation.[247] For example, in Gaucher disorder, which results from a deficiency in the enzyme glucocerebrosidase (found in mitochondria as well as other organelles, such as the lysosome), the ability of mitochondria to take up calcium from around themselves in the cell is impaired. Increasing cytoplasmic Ca^{2+} levels is likely to predispose to excitotoxicity because increasing positive ions inside the cell will tend to depolarize it (see Chapter 2), allowing yet more Ca^{2+} into the cell. Cell death due to the Ca^{2+}-induced activation of degradative enzymes is a likely explanation for the variable neurological symptoms including tremor, seizures, muscle stiffness, coordination issues and neurodevelopmental delays (note there are also many non-neurological symptoms in Gaucher disorder).

The Proximate Causes of Polygenic and Multifactorial Conditions in Childhood

So far in this chapter, I have talked a lot about the proximate causes of phenotypes that result predominantly from a *single* genetic variant (that is, they are monogenic; I say *predominantly* here to allow for the possibility that other factors probably modulate the phenotypes to some extent, as I discussed before). Now, I will consider the proximate causes of neurodevelopmental disorders and neurodivergent conditions whose ultimate causes are often a more balanced combination of multiple genetic and environmental risk factors. Remember from Chapter 4 that we refer to these conditions as polygenic (involving variants of multiple genes) and multifactorial (involving both genetic and non-genetic factors).

Sometimes the same or similar factors might cause monogenic disorders and contribute to multifactorial disorders. This is likely to occur because different variants of the same gene can have different strengths of effects or even different effects altogether, and environmental factors can vary in strength. It follows, therefore, that many of the developmental

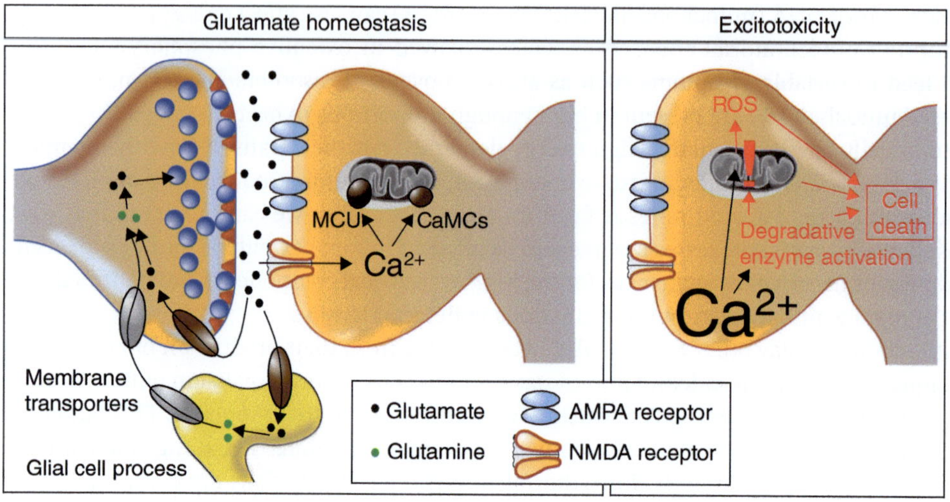

Figure 8.8 Glutamate homeostasis at synapses and excitotoxicity. At glutamatergic synapses, glutamine is converted into glutamate. Much of this glutamine arrives from glial cells, which participate in synaptic functions. Its delivery is mediated by various membrane transporters. Glutamate is stored in presynaptic vesicles until its release when electrical activation causes vesicles to fuse with the presynaptic membrane. Once in the synaptic cleft, glutamate can bind to postsynaptic **ionotropic** receptors (such as AMPA and NMDA receptors; Figure 2.7, Chapter 2). After stimulation, membrane transporters, many of which are on glial cells although some are on neurons, remove released glutamate from the synaptic cleft in preparation for subsequent stimulation. The recovered glutamate is converted into glutamine, completing the cycle. Glutamate binding to ionotropic receptors depolarizes the postsynaptic membrane and causes an influx of calcium through the NMDA receptors. Ca^{2+}-activated mitochondrial transporters (CaMCs) and the mitochondrial calcium uniporter (MCU) regulate Ca^{2+} movement into mitochondria, allowing them to act as buffers against the excessive build-up of Ca^{2+} in the cytoplasm. Dysfunction of these mechanisms for regulating synaptic glutamate and postsynaptic Ca^{2+} levels can lead to excitotoxicity. High levels of Ca^{2+} in the cytoplasm activate a variety of enzymes – such as phospholipases, endonucleases and proteases – that break down a variety of molecules including phospholipids, DNA and proteins, leading to cell damage and death. Ca^{2+} excess in the cytoplasm and mitochondria can also disrupt mitochondrial function. As well as compromising energy production, this can lead to the production of harmful **reactive oxygen species** (Chapter 4).

processes I have been discussing so far throughout this and the previous chapter are likely to be affected to some extent in polygenic and multifactorial conditions. For example, previously in this chapter I discussed how variants of histone deacetylases (HDACs) can cause monogenic disorders. Alterations in HDAC activity or expression have also been implicated in a range of multifactorial neurodevelopmental conditions, including autism.[248] In the polygenic 2q37 deletion syndrome, in which a piece of chromosome 2 containing about 200 genes including *HDAC4* is missing (from the long arm, as signified by the letter q; Figure 4.7, Chapter 4), evidence suggests that loss of the *HDAC4* gene might account for many of the syndrome's characteristic symptoms, such as intellectual disability. However, many other genes are also affected, and the main challenge is to work out how the proximate effects of genetic and non-genetic factors combine to produce the phenotypes in these sorts of conditions.

I will discuss what we know about the proximate causes of several prominent conditions: epilepsy, autism, attention deficit hyperactivity disorder (ADHD), Tourette syndrome and obsessive-compulsive divergence (OCD). In this and previous chapters, I have discussed how these clinically defined conditions can sometimes be caused by a single identifiable genetic variant or environmental factor, but more commonly they are caused by a significant contribution of both genetic and non-genetic factors. Since many of their ultimate causes are unknown, it is difficult to work out the proximate causes in any comprehensive way or with any degree of certainty.

Epilepsy

Epilepsy can result from disturbances of the large number of processes determining how the brain's electrical components work that I discussed above. For example, these disturbances might affect the distribution of charged ions around and within neurons and, therefore, the excitability of neuronal membranes, or they might affect excitability by changing receptor or neurotransmitter functions, or they might affect metabolic processes such as energy production. They might be of genetic or environmental origin. Whereas some cases of epilepsy are caused by single genetic variants (we have seen plenty of examples in this and Chapter 6 and 7), a larger proportion is almost certainly due to the cumulative effects of multiple common genetic variants combined with environmental factors. In about half of all cases of epilepsy, the ultimate causes remain unknown.

Despite major differences in aetiology, the phenotype of synchronized activation of cortical neurons that is the hallmark of epilepsy appears similar and leads to overlapping consequences in most cases. Even if we remain uncertain of how epilepsy was initially triggered in a given individual, which usually occurs early in postnatal life, we know a bit more about cellular events that might occur in the brain once it has started. These events can progress to cognitive dysfunction and an increased risk of dementia, due to the degeneration of synaptic structure and function, compromised neuronal structure and brain cell death. A current hypothesis is that an altered balance between the activities of excitatory and inhibitory circuits (due to disturbances such as those outlined in the previous paragraph) predisposes to a first seizure, which then elicits cellular and molecular changes that lower thresholds for triggering subsequent episodes, creating a vicious circle of decline. These changes might include alterations in the subunit composition and distribution of **AMPA receptors** and NMDA receptors. Furthermore, by elevating glutamate levels at excitatory synapses, seizures are likely to contribute to excitotoxic damage to brain cells and, eventually, to their death (Figure 8.8).

Autism

As I discussed in Chapter 6, autism is much harder to define in a clear way than conditions such as epilepsy. There is a large grey area where the spectrum of phenotypes manifested by individuals diagnosed with mild autism merges with the spectrum shown by neurotypical people. Research on the causes of clinical conditions often requires the categorization of individuals as either having or not having the condition, allowing potential causative factors to be compared in the two groups. The clinical diagnosis of autism can be driven as much by practical imperatives – such as gaining access to opportunities for help with particular social needs – as by any rigorous diagnostic methodology (Chapter 6). In most cases, this makes it difficult to work out its ultimate causes, let alone its proximate ones.

All evidence indicates that autism has a strong genetic component, in common with related conditions such as ADHD, epilepsy, depression and anxiety, combinations of which often co-occur. Environmental factors also make important contributions to causing autism: for example, current evidence identifies strong effects of advanced parental age or birth trauma, ischemia or hypoxia, but there are likely to be many more subtle ones. There is evidence that most cases of autism are part of a single continuum sharing common causative factors with related conditions (see Chapter 6).

For about 95% of people diagnosed with autism, the ultimate causes are unknown. In the approximately 5% of people diagnosed with autism who carry identified causative genetic variants, sometimes of a single gene, the effects of these variants are likely to be on early prenatal development. They are also likely to cause more than autism. For example, variants of the transcription factor TCF4 can cause Pitt–Hopkins syndrome, which is often described as an autism spectrum disorder, include developmental delay, intellectual disability and epilepsy. Its proximate causes are likely to include reduced proliferation in the prenatal brain (which probably explains why **micrencephaly** is often seen) and disruptions to some or all of cell migration, specification, differentiation, maturation, morphology, excitability and plasticity in neuronal, oligodendrocytic and astrocytic lineages. Similarly, 16p11.2 deletion syndrome, caused by the removal of about 30 genes on chromosome 16, is characterized by developmental delay and intellectual disability and is described as an autism spectrum disorder. Its proximate causes are likely to be very complex, since the deleted region contains many genes, but current evidence suggests that, like Pitt–Hopkins syndrome, its origins lie in the processes that create the brain prenatally.

Just because some cases of autism might be caused by events occurring before birth does not mean that the origins of the vast majority of cases of autism lie in the prenatal period. The ways in which the phenotypes of most autistic people arise are likely to be diverse. One broad unifying possibility is that neurodivergent conditions such as autism are the result of genetic and environmental variations that somehow conspire to reduce the constraints that limit the potential for cellular variability in the developing brain. Bear in mind that the mechanisms of brain development have not evolved to ensure precisely the same outcome every time a new brain is made. Instead, these mechanisms have evolved ways of constraining development so as to produce outcomes that lie within ranges (that are compatible with the ability to thrive and survive). It seems likely that some variants that are risk factors for autism might reduce these constraints, thereby increasing the *potential* for variability in how the brain develops. Essentially, the variants that some brains possess might increase their freedom to develop in a wide variety of ways, some of which might lead to autism. How this change of potential plays out in any given individual, and whether that person does or does not become autistic, is likely to depend on a large array of other genetic and environmental variables.

It is likely that some genetic and environmental causes of – or risk factors for – autism have either direct effects or indirect effects that converge on postnatal mechanisms of synaptic function and plasticity.[249] You will sometimes see autism described as a synaptopathy, which might be appropriate in a few cases of autism with a defined cause that results in synaptic pathology, but is unlikely to be appropriate in many cases. In most people with autism, there is no evidence of synaptic pathology as an underlying cause of their ways of thinking and behaving. It is often claimed that an imbalance between the excitatory and inhibitory inputs onto neurons in the brain – which, as I discussed in Chapter 2, would affect how postsynaptic neurons respond to their inputs – is a key proximate cause of autism. This sweeping generalization might be true in some cases, for example those caused by variants affecting glutamatergic or GABAergic neurotransmission, but extrapolation to most other cases of autism seems to go well beyond current evidence.[250,251]

There is evidence suggesting that effects on the brain's non-neuronal components can contribute proximate causes. These components include those involved in myelination and immune responses involving microglia (Figure 1.9, Chapter 1).[252,253] Alterations in the functions of microglia – which are the primary immune cells of the brain – might include either their excessive activation or their impaired function. Effects on microglia might alter the ways in which the brain reacts to harmful environmental factors such as infection, injury or illness – which we call its inflammatory response. These inflammatory responses involve the recruitment of immune cells to the affected areas of the brain, and if the microglia are compromised in some way, then the course of subsequent brain development might be more variable than it otherwise would be.

In summary, understanding the proximate causes of the differences in brains of autistic people will require much deeper knowledge of the complex interplay between a plethora of genetic and environmental factors, many of which have not yet been identified or are only poorly defined. The identification of these factors is likely to require a much clearer definition of the condition itself.

ADHD

If autism is hard to define, ADHD is arguably even harder, for pretty much all the same reasons. **Genome wide association studies** (GWAS) have shown the development of ADHD involves potentially thousands of variants, most of which also contribute to other psychiatric disorders. The proximate causes of ADHD are, therefore, very likely to involve many of the same types of processes outlined in the previous section on autism. I will cherry-pick just one example of how the symptoms of ADHD might be explained. At this stage, this should be viewed as strongly hypothetical.

Some genetic variants implicated in ADHD probably impact synaptic function in frontal cortex. Multiple studies have reported atypical findings in the frontal lobes of the cerebral cortex as well as a region of the basal ganglia known as the striatum (Figure 9.2, Chapter 9) and the neural circuits that interconnect them, which are called the frontostriatal circuits. Moreover, there is evidence linking ADHD with dysregulation of dopaminergic systems, which play a crucial role in attention, motivation and reward processing in these parts of the brain. It seems possible that variation in these frontostriatal circuits would lead to variation in the cognitive and behavioural functions that they mediate, which include executive decision-making. If variation is at the extreme of the range in an individual, then the behaviours that person manifests, including the decisions they make, might lead to them being diagnosed with ADHD.

It will probably come as no surprise to learn that these findings come with the caveat that, as for autism, neither structural nor functional brain imaging of people diagnosed with ADHD have resulted in complete consensus on possible causative alterations, most likely because ADHD is so heterogeneous and hard to define precisely.

Tourette Syndrome

Researchers think that the ultimate causes of Tourette syndrome are a combination of multiple gene variants, each of which probably makes a small contribution, and other factors such as infections, the body's immune responses and stress.[173] In a few cases, the cause might be a variant of a single gene involved in synapse formation and function, such as *CASK*, which encodes a scaffold protein; *NLGN4X*, which encodes a cell adhesion molecule; *SLITRK1*, which encodes a cell membrane protein.

While this suggests that the proximate causes of Tourette syndrome can be a direct effect on synapses, GWAS have also implicated variants of genes involved in a much broader range of processes. As with other perceptual, cognitive and behavioural conditions that are hard to define unambiguously, research is complicated by the probability that Tourette syndrome is at the extreme of a continuum of phenotypes that include mild self-limiting childhood tics and other types of obsessive or compulsive behaviour. This is supported by evidence that Tourette syndrome is related genetically to some forms of ADHD and OCD (with which it often co-occurs) more closely than to other perceptual, cognitive and behavioural conditions. Neuroimaging studies have indicated that, like ADHD, some parts of the brain are more involved than others in the proximate causes of Tourette syndrome, with suspicion falling particularly on disruption of the frontostriatal circuits. Much more research will be needed to assess the validity of such ideas and their functional significance.

OCD

OCD has a genetic component but finding specific genetic variants that are risk factors for OCD has proven difficult. Like many such behavioural conditions, OCD probably represents the extreme of a continuum of behaviours that we all experience to a greater or lesser degree but do not usually reach a threshold for a diagnosis of OCD.[254] This is supported by GWAS, which have indicated that subthreshold (or subclinical) obsessive or compulsive symptoms and diagnosed OCD are associated with a shared set of risk factors. As with ADHD and Tourette syndrome, brain imaging has suggested that variability in the functioning of the frontostriatal circuits is likely to make an important contribution to variability in the obsessive and compulsive behaviours manifested by individuals.[255] The proximate causes of OCD are very likely to involve many of the same types of processes outlined in the previous sections on autism, ADHD and Tourette syndrome.[256]

Summary

- Neurodevelopmental conditions first detected in childhood are often diagnosed because they affect the functions of the brain rather than because they have detectable effects on its anatomical structure.

- The proximate causes of many of these childhood conditions affect function by converging on the brain's electrical components, including neurons, their myelin sheaths, membranes and synapses.
- Synaptic dysfunction or interference with synaptic plasticity can cause brain disorder and neurodivergence.
- Gene variants can impact mechanisms of nerve and synaptic transmission, myelination and synaptic plasticity directly, leading to conditions from epilepsy to intellectual disability.
- Gene variants affecting gene expression, proteostasis or cellular metabolism can indirectly impact brain function, often via complex routes, causing conditions such as Rett syndrome and fragile X syndrome.
- The proximate causes of complex polygenic, multifactorial conditions such as most cases of epilepsy, autism, ADHD, Tourette syndrome and OCD remain relatively poorly understood despite much research.
- Some of these complex conditions might result from a reduction in the stringency of mechanisms that have evolved to limit the inherent variability of developmental mechanisms.

9

Proximate Causes of Conditions Emerging in Adolescence or Adulthood

Some people start to experience neurological or psychiatric symptoms in their teenage years (which we refer to as adolescence), as young adults aged 20–30 years (often referred to as early adulthood) or at older ages. These symptoms range from problems with particular behavioural abilities such as movement, speech and so on, to persistent debilitating types of disordered or divergent perceptions and thoughts. We can, and usually do, think of these sorts of symptoms under two broad headings: those that are a deviation from a typical course of development and those that are degenerative. For example, if a person starts to experience persistent depression or hallucinations in adolescence, we might think of this as a deviation from normal development; if a person starts to experience progressive difficulty with movement or worsening memory loss, we would think of this as degenerative. When we consider the causes of these two groups of symptoms, however, the distinctions are harder to define. Both groups are likely to result from effects on overlapping types of cellular processes, such as cell proliferation, cell death, cell metabolism, **synaptic plasticity** and so on. And, most importantly in the context of a book on brain development, it is likely that both groups are predisposed to by genetic variants that have been affecting the cell biology of the developing brain in as yet undetected ways long before symptoms emerge. I will discuss evidence supporting this view.

In the first part of this chapter, I will discuss conditions that have well-defined sets of genetic variants as their ultimate causes and give examples of how specific environmental agents might impact the developing brain. In the second part, I will discuss multifactorial conditions that we commonly refer to as psychiatric, whose ultimate causes are much less well understood. In the final part, I will discuss the possibility that some conditions that we frequently consider to be entirely degenerative, largely because they are first detected in elderly people, might have their origins much earlier in life.

As with many of the conditions that emerge in childhood that I discussed in the previous chapter, at least some and possibly all of the late-onset conditions I will discuss in this chapter almost certainly have origins as early as prenatal development, although direct evidence is hard to come by. We might infer that this is likely from our knowledge of the causative agents, which are often variants of genes that are expressed in human embryonic brain and function in the embryonic brains of other species. In human embryos, there is usually no obvious way to search prenatally for probably extremely subtle signs of future problems that would be unforeseen at the time. Some notable exceptions to this do, however, support the general contention, as we shall see below. It is also likely, however, that many conditions originate, or are directly contributed to, by developmental changes that occur as late as adolescence or adulthood. This is made more probable by evidence indicating that the functional attributes of human brain neurons develop on an extremely protracted timescale, with some aspects appearing for the first time only during or after late adolescence.

How Defined Genetic Variants Can Cause Disorders After Childhood

As the brain matures into adulthood, the effects of disruptive genetic variants on brain structure and function are likely to accumulate. Many of these effects will be indirect consequences that are many steps removed from the original direct impact of the variant. The original effects of the variant might go undetected for many years, or even decades, and the clinically identifiable symptoms that appear might be a long way downstream of the variant's direct impact. This might happen for many reasons. For example, developing biological systems have evolved the ability to compensate for some disruptions, as occurs when the function of a gene carrying a disruptive variant is taken over by another similar gene.[257] Eventually, however, the developing brain might reach a stage at which a threshold of accumulating uncompensated defects is breached, or compensatory mechanisms no longer work, and symptoms occur. The possible scenarios for how the consequences of any given variant will play out over years and decades in any given individual are extensive, since they will depend heavily on the exact details of a plethora of other interacting biological processes, many of which will be unique to each person. This probably explains why, for the clinical conditions I will describe in this section, similar variants often produce disorders with detectable onsets that can vary widely, in some cases from early childhood to late adulthood. It also makes disentangling the direct and the indirect consequences of a variant very challenging.

Huntington Disease

Huntington disease is an inherited disorder in which neurons gradually die, causing a wide range of symptoms including progressively declining cognitive, behavioural and communication skills and progressively developing involuntary purposeless movements such as **chorea** (Chapter 2). Onset is typically in people over 30 years old. Huntington disease was called Huntington's chorea until relatively recently, but this has changed since the discovery of its genetic cause, which allowed it to be identified independently of symptoms and led to the realization that not everyone with the condition has chorea. This is a good example of how understanding causation can change the way in which we construct diagnostic categories. Although the – mostly unknown – causes of most neurodevelopmental conditions are certain to be much more complicated than Huntington disease, the principle still applies that many of our current commonly used diagnostic categories are likely to be misleading.

The ultimate cause of Huntington disease is expansion of a **trinucleotide repeat** (Chapter 4) that codes for the amino acid glutamine in the *HTT* gene, which causes the protein it produces to gain a toxic function affecting neurons, glia and probably other cell types. The effects of the variant **alleles** are **dominant**: in other words, the production of toxic proteins cannot be overcome by having a normal allele present. The strength of the effect in an individual depends on how many times the trinucleotide is repeated in them or, in other words, how long it is (Figure 9.1). Recent research has suggested that the repeat might continue to expand in length for decades in some people's brain cells, gradually achieving progressively more toxic effects in different cells at different times in each individual, thereby adding to the variability in the phenotypes seen in this condition.[258]

To understand how the trinucleotide expansion has its effects, it would be helpful to know exactly what the HTT protein does in cells and, unfortunately, we do not. What we do know is that HTT is widely expressed in the body, with high levels in the brain, and interacts with many hundreds of other proteins, suggesting its involvement in many cellular processes. Trinucleotide expansion to harmful lengths is thought to produce versions of HTT that somehow trigger, directly or indirectly, pathogenic mechanisms disrupting some or all of **transcription**, **proteostasis**, metabolism and mitochondrial function and causing dopamine imbalance, neuroinflammation and excitotoxicity (Chapter 8, Figure 8.8). These disruptions lead to the death of neurons and other cell types.

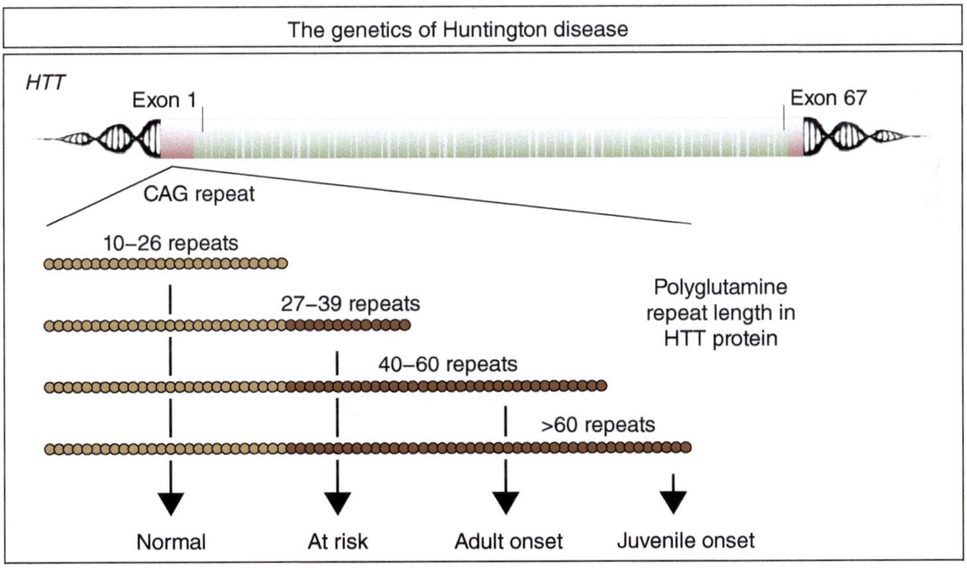

Figure 9.1 The genetic basis of Huntington disease. The human *HTT* gene, located on Chromosome 4, comprises 67 exons. Exon 1 contains a CAG repeat (which translates to a polyglutamine stretch in the protein). If the trinucleotide repeat expands above a certain size, there is increased likelihood that Huntington disease will result. In general, the longer the expansion, the earlier the onset, although other environmental and genetic factors contribute to determine the age of onset and the exact symptoms. Normally, the number of CAG repeats is around 10–26. Individuals with expansions to around 27–39 are at risk for further CAG expansion and may suffer emotional or behavioural symptoms, even though many do not develop Huntington disease. If expansion reaches around 40 repeats or more, Huntington disease results.

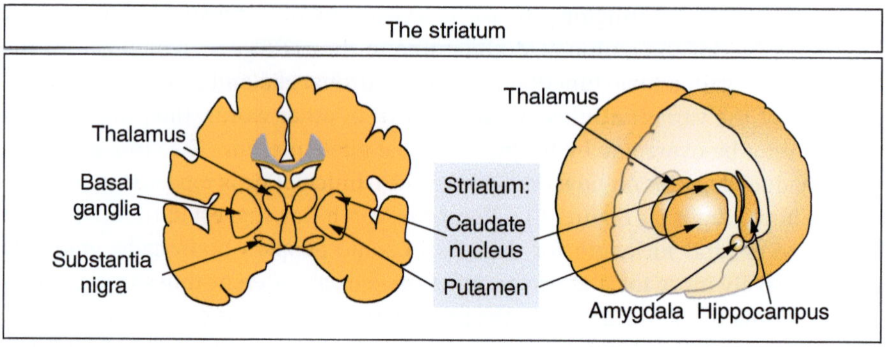

Figure 9.2 The striatum. Diagram shows the location of the striatum and its two major components, the caudate nucleus and the putamen. The diagram is based on Figure 2.1, Chapter 2.

The onset of symptoms is attributed to the death of neurons mainly in regions of the **basal ganglia** called the caudate nucleus and putamen, which together are referred to as the striatum (Figure 9.2). These two structures are parts of the basal ganglia that receive inputs from other regions, mainly **excitatory glutamatergic** ones from the cerebral cortex and **thalamus** and dopaminergic ones from the substantia nigra. The substantia nigra is itself also in the basal ganglia and receives reciprocal **inhibitory** input from the striatum. The striatum is a critical component of the brain's circuitry controlling cognitive decision-making and movement and providing subconscious instinctive or reflexive activity that refines planned movements and behaviours. The gradual attrition of these striatal structures and their functional contributions explains the progressive decline in the cognitive and movement skills of people with Huntington disease. The fact that inhibitors of dopaminergic **neurotransmission** can work as a palliative treatment to alleviate some of the symptoms is probably because it counteracts the effects of the excessive dopaminergic input from the substantia nigra that results from the loss of striatal inhibition.

Huntington disease is an example of a disorder that probably starts prenatally even though it is only diagnosed clinically decades after birth. In common with many genes, *HTT* is expressed during embryonic development in species including humans. Studies in non-human species have shown that the protein is needed for normal early brain development. In humans, Huntington disease is one exception to the general rule that searching for direct evidence for the early origins of adolescent-onset or adult-onset conditions is extremely difficult because, in most cases, we do not know who will end up developing the condition and who will not. Since the ultimate cause of Huntington disease is clearly defined and those at risk of developing the condition later in life can be identified, researchers have been able to look for – and have found – detectable changes in brain structure decades before an affected person shows overt symptoms. The most direct evidence for prenatal changes comes from studies on human fetuses carrying pathogenic variants of *HTT*, whose brains show abnormalities among cerebral cortical **progenitor** cells (Chapter 1) and primary cilia (Figure 7.5 in Chapter 7). We should, therefore, consider Huntington disease as a neurodevelopmental disorder as much as it is a degenerative one, which will surely influence future attempts to understand more clearly its proximate causes.[259]

Amyotrophic Lateral Sclerosis and Frontotemporal Dementia

Amyotrophic lateral sclerosis results from progressive degeneration and death of neurons in the brain and spinal cord that are responsible for voluntary movement and breathing, with secondary muscle wasting. The brain's main functions are not usually affected directly, although depression and other symptoms are a consequence and some affected individuals develop a form of dementia that is known as frontotemporal dementia.

Variants of the *C9ORF72* gene are the ultimate cause of a minority of cases. They account for just under 50% of inherited cases, which are less common than de novo ones arising for other unknown reasons. *C9ORF72* codes for a protein abundant in neurons of the cerebral cortex and spinal cord. It appears to be concentrated at synapses, suggesting a role in synaptic function, but there is evidence for its involvement in many other processes including lysosome function and **autophagy** (Figure 7.17 in Chapter 7). *C9ORF72* contains a hexanucleotide (six nucleotide) sequence that can become repeated extensively (more than about 30 times) to produce pathogenic variants.

Usually, symptoms are first detected between the ages of 55 and 75 years, but researchers who followed individuals with pathogenic variants of *C9ORF72* using brain imaging and psychological tests found changes in the brain much earlier, before the age of 40 years, long before symptoms were detectable. It has been suggesting that, as in Huntington disease, the disorder might originate from disrupted developmental processes at even earlier ages.[260]

Spinocerebellar Ataxia

Spinocerebellar **ataxia** is subdivided into different types depending on the symptoms. They are all called spinocerebellar ataxia followed by a number from 1 to well over 40, with more almost sure to be added. They have different ultimate genetic causes and can have somewhat different symptoms, which are often first detected in adulthood but can appear in adolescence or childhood. Most types include a progressive loss of coordination, balance and muscle tone together with increasing difficulty with movements, including speaking, swallowing, walking and so on, and cognitive problems. The genetic changes cause a degeneration of the structure and function of the cerebellum, hindbrain, upper spinal cord and sometimes other regions too. Some are caused by the effects of nucleotide repeat expansions, others by other types of variants. Types 1, 2, 3 and 6 are the commonest. Types 1, 2 and 3 are caused by trinucleotide repeat expansions in the *ATXN1*, *ATXN2* and *ATXN3* genes respectively. Type 6 is caused by a trinucleotide repeat expansion in the *CACNA1A* gene.

ATXN1 encodes a nuclear protein that is thought to interact with many other proteins including **transcription factors** to help regulate gene expression. *ATXN2* encodes a cytoplasmic protein that is thought to help regulate RNA processing and **translation**. Its interaction with FMRP protein (Chapter 8) has suggested it might have a role in synaptic function and plasticity.[261] *ATXN3* encodes a protein involved in the ubiquitin–proteasome pathway (Chapter 7, Figure 7.17). *CACNA1A* encodes a part of a calcium channel and, as we have seen earlier (Chapters 2 and 8, for example), Ca^{2+} ions are involved in many different cellular functions, including synaptic transmission and plasticity.

The **cerebellar ataxias** are usually described as neurodegenerative rather than as being disorders of development, despite the fact that they can emerge, albeit rarely, within just a few years of birth. As I mentioned for Huntington disease, amyotrophic lateral sclerosis and frontotemporal dementia, it seems likely that the proximate causes of at least some of the symptoms of cerebellar ataxias lie in the way parts of the brain were constructed earlier in development. For example, genetic variants of genes such as *ATXN1*, *ATXN2*, *ATXN3* and *CACNA1A*, all of which are expressed in the developing brain, have the potential to interfere with cellular mechanisms that are essential in many aspects of early brain development, as exemplified in the previous paragraph.

Studies of the cerebellar ataxias have highlighted particularly forcibly a question that I introduced at the start of Chapter 7 and is relevant to many disorders caused by genetic variants: why are the effects of the variants more severe in just some regions of the brain (in this case, the cerebellum, hindbrain and so on) even when the affected genes are expressed at similar levels more widely or ubiquitously? The answers are likely to be complex and to vary from gene to gene. One possibility, illustrated by cerebellar ataxia type 1, is that the effect of the variant depends on what other proteins the cell expresses, which depends largely on the cell type and stage of development. For example, the actions of the *ATXN1* gene, which is expressed ubiquitously, would vary depending what transcription factors the cell expresses (remember that the cocktails of transcription factors expressed by different cells vary greatly, as I discussed in Chapter 7, in the section titled: 'How the Specification of Cell Types Can be Disrupted'). If ATXN1 has particularly important interactions with transcription factors expressed only in the cerebellum and hindbrain, then this is where variants would have their greatest impact. At present, such possibilities remain speculative.

Leukodystrophies

While some forms of **leukodystrophy** are present at birth or appear during childhood (discussed in Chapter 8), some are only detected in adulthood, although brain demyelination might start to occur years before problems develop. Leukodystrophies usually get worse with age, in other words they are progressive.

An example of a leukodystrophy that is usually detected in individuals in their forties or fifties is autosomal dominant leukodystrophy with autonomic disease (caused by variants of *LMNB1*). Its symptoms include problems with movement and its coordination, muscle stiffness or weakness and tremor that worsens during movement (we call this intention tremor because it happens when someone is trying, or intending, to do something). Usually, the ultimate cause is a duplication of the *LMNB1* gene, which makes a **structural protein** lamin B1. This protein is part of a meshwork that maintains nuclear structure and integrity of cells throughout the body. The question posed in the previous section is equally relevant here: how do variants of a widely expressed protein such as this, which has a fundamental part to play in the cell biology of a huge range of cell types, create pathology as specific as demyelination? It appears that **oligodendrocytes**, which create the **myelin** sheath of neurons, are particularly affected, but why this is so remains unclear.

Inherited Cerebral Vasculopathies

In these conditions, the direct effects of variants are on the brain's blood vessels. For example, a condition called cerebral autosomal dominant arteriopathy with subcortical infarcts and leukoencephalopathy (a mouthful that is commonly abbreviated to CADASIL) is caused by variants of the *NOTCH3* gene, which encodes a receptor protein required for the function and survival of cells that make up the walls of the blood vessels (Figure 7.3, Chapter 7). *NOTCH3* variants that interfere with the function of the receptor result in the death by **apoptosis** of these cells. In the brain, the weakening and damage of the blood vessels can cause problems with the function and viability of all brain cells, resulting in symptoms such as strokes and epilepsy at almost any stage of life from childhood to late adulthood.

How Specific Environmental Factors Can Affect the Maturing Brain

After childhood, an individual's horizons, self-directed behaviours and so on expand, so there is an ever-increasing potential for them to encounter environmental factors, such as exposure to specific substances or life experiences (outlined in Chapters 5 and 6), that can have a long-term impact on their brains. Remember, as we saw in Chapter 2, that the brain's circuitry is plastic and is modified by the activity of its components, and this property is retained to some extent throughout life. Environmental factors can, therefore, have long-term effects on the brain at any age.

In this section, I will discuss a few well-studied examples of how *specific* environmental agents can affect the electrical activity of brain cells and, as a result, cause knock-on effects for the subsequent course of brain circuit formation and function through adolescence and adulthood. In the subsequent section, I will touch on the little we know about how variations in less specifically defined external environmental factors – such as interpersonal and societal interactions, socio-economic status and so on (which I discussed in Chapter 5) – can contribute to long-term conditions affecting the brain in adolescence and adulthood.

Some specific environmental toxins cause problems with the brain's ability to function optimally because they directly affect its electrical components. Lead is an exemplar of this. Although it has toxic effects on most organs, the brain is especially sensitive to it. It directly disrupts the brain's electrical components in a number of ways. For example, it affects synapse function and plasticity by substituting for calcium, whose crucial role in these processes I explained in Chapter 2. It interferes with neurotransmitter release, promoting spontaneous release and inhibiting stimulated release. By reducing the activity of an enzyme that converts the excitatory neurotransmitter glutamate to glutamine, it allows glutamate to accumulate to the point where it becomes excitotoxic (Figure 8.8, Chapter 8). It can affect myelin by, for example, decreasing the activity of an enzyme, called CNPase (cyclic nucleotide phosphodiesterase), that is expressed in oligodendrocytes and involved in myelin synthesis. Lead can also affect the brain's activity as a result of its uptake by mitochondria, whose functions it disrupts to create the sorts of cellular problems I outlined in Chapter 8.

Alcohol (ethanol) is an example of a frequently abused substance that can directly affect the ability of the brain's electrical components to function. Central to its direct acute and chronic actions are its effects on glutamatergic neurotransmission (Chapter 2).[262] Current evidence indicates that acute ethanol exposure inhibits **ionotropic NMDA** receptors and chronic exposure leads to a compensatory increase in their function, particularly in some areas of the brain such as the **hippocampus**, which is important for memory, and the striatum (Figure 9.2), which is important for movement. It is possible that this long-term increase happens due to epigenetic modifications that change the expression of NMDA receptor subunits. A change in the ratio between the receptor subunits is likely to affect their involvement in synaptic plasticity, altering the likelihood that long-term potentiation (LTP) rather than long-term depression (LTD) occurs (Figures 2.6 and 2.7 in Chapter 2).

There are many specific environmental toxins that can affect mitochondria, leading to problems with functions of the brain as well as of other organs. Some drugs, including antiviral agents used to treat HIV infection, can have adverse side-effects by inhibiting POLG (Chapter 8). The consequences can induce similar syndromes to some mitochondrial genetic disorders. Other drugs, such as valproic acid, antibiotics, anaesthetics and so on, add to the negative effects of mitochondrial genetic disorders without directly inhibiting POLG or the mitochondrial DNA replication mechanisms. While these drugs do not usually cause problems in individuals who do not have mitochondrial genetic disorders, they can trigger symptoms in those with such genetic conditions. Similarly, stressors including infection and fever, dehydration and anorexia can exacerbate mitochondrial genetic disorders, which brings us on to consider multifactorial conditions in the next section.

How Multifactorial Conditions of Adolescence and Adulthood Might Be Caused

The developing brains of adolescents are undergoing many changes and are particularly vulnerable to environmental variation that can contribute to the emergence of spectra of disorders or types of neurodivergence. Environmental factors include those included in the previous section as well as other specific environmental agents and more broadly defined influences and pressures from social, socio-economic and peer-group sources (discussed in Chapters 5 and 6). The effects of environmental variation on any given adolescent depends on the way that their genetic variants and their earlier environment have combined to construct the detailed cell biology of their brain earlier in their life.

Among the commonest multifactorial conditions to affect adolescents and young adults are those described as psychoses, with schizophrenia and bipolar disorder being prime examples, along with major depression. Despite much research and the identification of both genetic and environmental risk factors that contribute to the development of these conditions, we have little insight into how these factors bring about the symptoms that patients experience.

Schizophrenia and Bipolar Disorder

Schizophrenia is probably one of the most distinct psychiatric conditions since its symptoms generally include perceptual phenomena, such as persistent hallucinations, that most people rarely experience. Despite being relatively clearly definable, figuring out its ultimate causes has proven extremely difficult, as I discussed in Chapter 6. While schizophrenia is usually first diagnosed in adolescence or early adulthood, there is strong evidence that its origins are much earlier in brain development, possibly even prenatally, although direct evidence for this is limited.[263] Disruption of two major neurotransmitter systems, namely the dopaminergic and glutamatergic ones, has been seen as important for schizophrenia's proximate causation because variants of a dopamine receptor (DRD2) and a glutamate receptor subunit (GRIN2A) are both risk factors for the condition, and the receptors themselves are the targets of antipsychotic drugs used to treat schizophrenia. However, what seems clear from research to date is that no single variants on their own cause at least the vast majority of cases of schizophrenia; rather, it is polygenic and multifactorial.[264] It is still entirely plausible that its ultimate causative mechanisms converge onto common cellular processes, such as dopaminergic and glutamatergic neurotransmission and probably others. Any such convergence might be very indirect, since few of the several hundred genomic loci whose variants are risk factors for schizophrenia have an obvious direct role in these processes.[265]

Many studies have tried to identify structural or biochemical correlates of schizophrenia. Some of these might be part of the condition's proximate causation, but it can be difficult to work out whether correlated features are causative or are effects of the condition or of exposure to antipsychotics or illicit drugs (which schizophrenic people are more likely to use). For instance, there is evidence of enlarged **ventricles**, thinned cortical grey matter and a loss of neuronal **dendrites** and synapses in the brains of people with schizophrenia. Biochemical differences associated with schizophrenia include decreased brain levels of proteins that function at synapses, in metabolic – including mitochondrial – processes and in gene expression. Researchers are now trying to distinguish cause from effect using methods such as Mendelian randomization, which I explained in Chapter 5.

Attempts to gain a clearer understanding of the proximate causes of the symptoms of schizophrenia are hampered by the fact that, although schizophrenia seems a clinically relatively discrete condition, its ultimate genetic causes overlap considerably with those of other types of neurodivergence and disorder, mainly major depression, bipolar disorder and autism (Chapter 6). This suggests that a common set of genetic variants might increase the chances of the developing brain acquiring the features of any of these conditions, with the features that emerge depending on other genetic variants or non-genetic factors. Another possibility is that what we currently call schizophrenia, whose characteristics are quite heterogeneous when examined in detail, actually comprises multiple conditions with different causes. This possibility is supported by recent evidence from high-resolution single-cell mRNA sequencing (Chapter 7, Figure 7.15) on post-mortem brain cells from individuals with and without schizophrenia which showed that those with schizophrenia could be separated into several different groups on the basis of their gene expression profiles.[266]

As with schizophrenia, working out the proximate causes of bipolar disorder is still a major challenge.[165] Researchers have identified a number of genetic variants and environmental risk factors that tend to be shared by individuals with schizophrenia and bipolar disorder. The variants associated with these conditions affect genes whose products regulate a range of processes, including **chromatin** modification, synaptic function, glutamate and **GABA** neurotransmission, sodium and calcium signalling, **cytoskeletal** components and cortical neurogenesis.[267–270] Common environmental factors that might increase the risk for both disorders include problems with birth, adverse life events especially during childhood and drug abuse.

Imaging studies have identified similarities in the structural and functional features of the brains of individual with schizophrenia and with bipolar disorder. These include abnormally small brain volumes, reduced cortical thickness and large ventricular volumes.[271] The sizes of these alterations tend to be less in bipolar disorder than in schizophrenia and some of them affect some parts of the brain more than others in condition-specific ways. How such condition-specific alterations bring about condition-specific symptoms is unclear.

Major Depression

Major depressive disorder, which you will also see referred to as clinical depression, is distinguished by its persistence and severity from the episodes of depression that many of us experience in our lives. As with many mental health conditions, it can be difficult to define the borders of the condition. Decisions on how to categorize it can seem somewhat arbitrary given that depression is a continuum. For example, features including several weeks of pervasive low mood, low self-esteem, loss of interest or pleasure and suicidal thoughts requiring clinical intervention are often used, but are unlikely to be universal. Imprecision in our ability to decide unequivocally whether some individuals fall into the category or not adds to the difficulty of identifying causes, reflecting an issue commonly confronted by those researching the causes of mental health problems in general.

Nevertheless, by studying severe cases of depression, scientists have reached some important conclusions on the causes of major depression. It is widely accepted that the genetic variants that a person inherits predisposes them to a greater or lesser degree to major depressive illness. Researchers think that the combined effects of many variants that individually contribute only a small risk are likely to determine a person's likelihood of suffering major depression. Some talk about the load of risk variants an individual carries: those with a bigger load are more likely to be affected. What is also clear, however, is that genetics alone does not explain why some people suffer major depression and others do not. Life events, psychosocial factors and lifestyle choices are likely to be very important in triggering the onset of symptoms.

The question for here is: what biological processes do genetic and environmental factors converge on to trigger the condition? Many hypotheses have been proposed and I will give a few examples of some of the more prominent ones. In no case is it yet clear how genetic and environmental factors act on these proposed mechanisms.

One of the oldest and most influential hypotheses, the biogenic amine hypothesis, suggests that depression is caused by a deficiency or suboptimal functioning of one or more of the brain's monoamine neurotransmitters – for example, noradrenaline (or norepinephrine) and serotonin (or 5-hydroxytryptamine) – at critical synapses.[272] One of the main reasons that this theory gained support is that many antidepressants work by increasing the availability or activity of these monoamines at synapses: examples are monoamine oxidase inhibitors, which increase monoamine availability by blocking their breakdown (Chapter 8). Moreover, the drug reserpine, which depletes brain biogenic amines, can induce symptoms of depression in humans. Clearly, these pieces of evidence could be explained in other ways: the drugs that increase monoamines might be effective because they raise levels above normal rather than by counteracting any deficit in the brain, and just because depression can be induced by a drug does not mean that is how it occurs naturally. More recently, researchers have come to the understanding that the biogenic amine hypothesis is at best only part of the explanation.

Subsequent hypotheses have stressed the likely involvement of altered synaptic plasticity and brain circuitry,[273] disorders of immunological factors (for example, involving **microglia**) and dysregulation of the **hypothalamic**–pituitary–adrenal axis (Figure 9.3). These hypotheses are attractive since they offer possible ways in which environmental influences might trigger depression: synaptic plasticity allows the brain's circuits to be moulded by experience, immunological factors are involved in the brain's inflammatory responses to external agents and the hypothalamic–pituitary–adrenal axis is crucial for the body's stress responses, which are undoubtedly altered in depression.[274] Care is needed in drawing strong conclusions about causes and effects. It is not clear to what extent the alterations in cellular processes trigger depression as opposed to being the result of it. For example, in the case of the hypothalamic–pituitary–adrenal axis, altered signalling – perhaps a reduction in the effectiveness of negative feedback mechanisms, say for subtle genetic reasons – might elevate cortisol and thereby anxiety and stress levels. But equally, a consequence of depression might be to activate this system to unusually high levels, for reasons we do not yet understand. Or, indeed, both might be true and – as in other conditions (I mentioned this earlier when discussing epilepsy in Chapter 8) – a vicious circle might be created. The

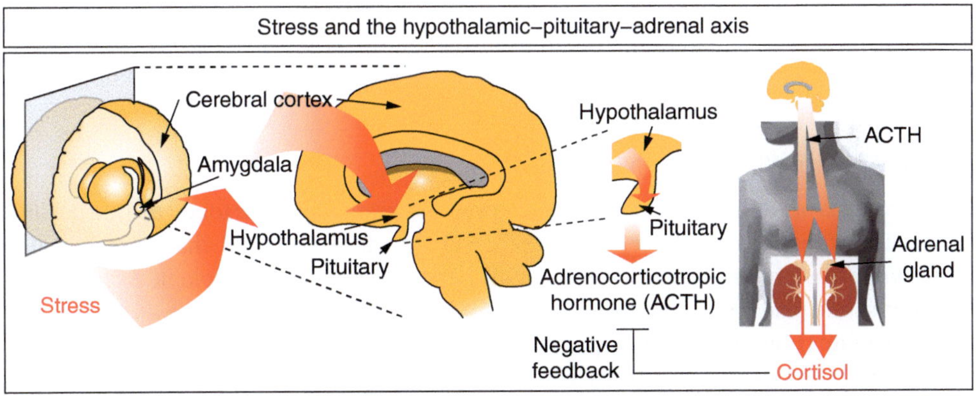

Figure 9.3 The hypothalamic–pituitary–adrenal axis and the effects of stress. Dysregulation of hypothalamic–pituitary–adrenal axis might be involved in the pathogenesis of depression. The body's normal responses to stress, perceived by brain's structures including the cerebral cortex and amygdala, affect the hypothalamus. Hypothalamic neurons release hormones that travel through a specialized blood portal system to affect the activity of the pituitary gland, which secretes adrenocorticotropic hormone (ACTH) into the bloodstream. ACTH stimulates the adrenal cortex to secrete cortisol. Cortisol plays a crucial role in regulating body functions such as metabolism, the immune system, regulation of blood sugar and blood pressure and the reduction of inflammation. Negative feedback by cortisol binding to receptors in the brain helps control levels; some scientists have hypothesized that a reduction in the effectiveness of cortisol receptors in some people might, by reducing negative feedback, allow chronic overactivation of the system. These are still very speculative ideas. Although there is evidence for overactivation of the hypothalamic–pituitary–adrenal axis in people with depression, it remains unclear which is cause and which is effect, or whether a vicious circle is created involving both.

several hypotheses I have included here are not mutually exclusive: they could all have validity, perhaps to a greater or a lesser degree in different individuals. Much more work is needed to get to the bottom of the proximate causes of major depression and the reasons for the variability in the way it presents itself in different people.

How Disrupted Development Might Contribute to Degeneration at Older Ages

In the previous part of this chapter, you might have been surprised to learn that many researchers now consider it likely that psychiatric disorders diagnosed in adolescence or adulthood are predisposed to by neurodevelopmental events that occur much earlier in life, maybe even before birth. If so, you might be even more surprised to hear that some researchers are investigating the possibility that degenerative medical conditions we normally associate with older ages also have origins in the earliest stages of brain development.

These conditions, which include Alzheimer disease and Parkinson disease, are usually described as neurodegenerative, since their features include the progressive loss of neuronal structure and function in the later years of life. Nevertheless, it is widely accepted that causative molecular and cellular changes have begun to affect the brain decades before symptoms become detectable. It has been suggested that, in some cases, these effects might start as early as embryonic development.[2,275] The initial effects might be very subtle, creating a brain with an increased risk of undergoing functionally debilitating degeneration later in life as a result of lifelong stresses from genetic and non-genetic factors. What early developmental differences place some brains at greater risk than others are unknown.

Alzheimer Disease

The symptoms of Alzheimer disease include progressive memory loss followed by a decline in executive functions (Chapter 2) and the development of atypical behaviours. Usually, these symptoms are associated with the brain undergoing several pathological changes: it develops senile plaques and neurofibrillary tangles and its hippocampus and cerebral cortex degenerate (Figure 9.4). These pathological changes very likely cause or contribute to symptoms, although it is not firmly established to what extent they all directly interfere with the ability of the brain's electrical components to function optimally and to what extent some might be a consequence of underlying problems that create pathology while affecting brain function in other ways.

Alzheimer disease is a leading cause of dementia in people over 65 years of age. About 5% of people with Alzheimer disease develop it earlier in life, in which case it is often referred to as early-onset Alzheimer disease. Several heritable genetic variants are risk factors for Alzheimer disease: early-onset Alzheimer disease is associated with variants of the *APP* gene, which encodes β-amyloid precursor protein, and with variants of the *PSEN1* and *PSEN2* genes, which encode presenilin 1 and presenilin 2; a variant of the *APOE* gene (which encodes apolipoprotein E) called *APOE4* is a strong risk factor for late onset Alzheimer disease. The proximate causes of Alzheimer disease likely involve a complex network of interactions between these proteins in their various forms and other types of molecules, especially lipids (Figure 9.4). I will describe these main players and why we

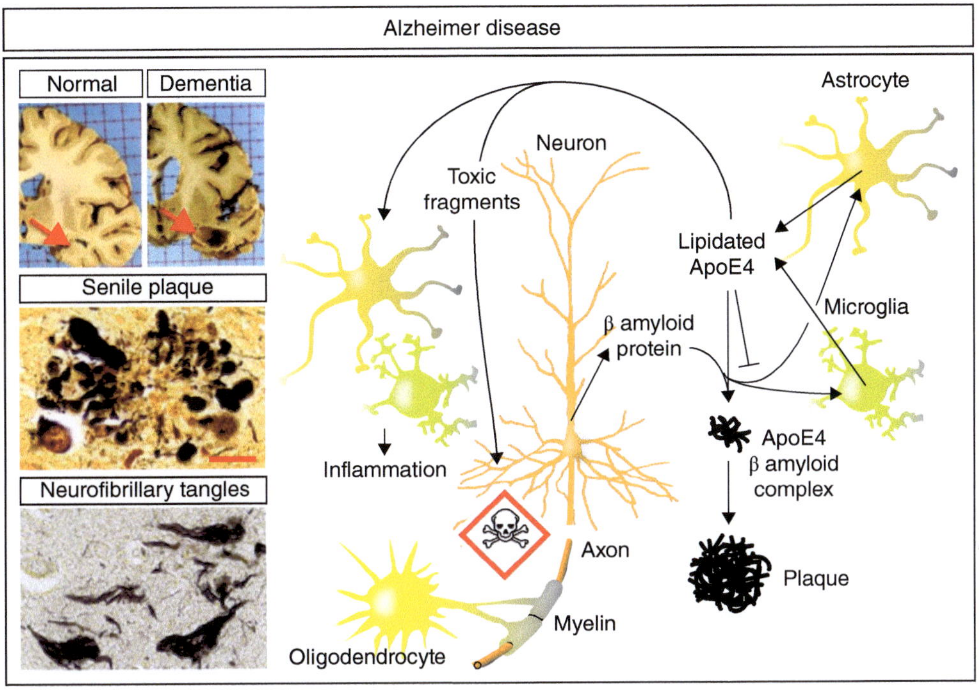

Figure 9.4 Alzheimer disease: pathology and likely causation. Photographs on the left show, at the top, a slice through one side of the brain from an individual who did not have symptoms of Alzheimer disease and one who did. Red arrows point to part of the hippocampus, which shows extreme **atrophy** in the person with Alzheimer disease. Underneath are sections of the brain from an individual with Alzheimer disease, stained with silver and viewed down the microscope. They show an example of a senile plaque and neurofibrillary tangles (which come out looking dark/black with this stain). The red scale bar applies to the lower two pictures and is 0.02 mm, which is about the diameter of a nerve cell body. The diagram on the right illustrates key mechanisms. Starting in the centre: neurons produce β-amyloid, which interacts with lipidated apolipoprotein E (ApoE; lipidated means it has attached itself to a lipid molecule) that is secreted mainly by microglia and astrocytes. If the person carries the ApoE4 variant, interaction leads to the formation of complexes and plaques. To make matters worse, ApoE4 impairs β-amyloid clearance by reducing reuptake by astrocytes and microglia. ApoE4 elicits chronic neuroinflammation and proteolysis of ApoE4 creates toxic protein fragments, both of which promote neurodegeneration. *Reproduced from Niedowicz et al.*[276] */ with permission of Bentham Science Publishers.*

think problems in their functions might start to impact the brain long before an individual develops detectable symptoms of Alzheimer disease.

First, β-amyloid protein: it is produced mainly by neurons, as a result of cleavage of the larger protein, β-amyloid precursor protein. Both proteins are involved in cell–cell signalling: β-amyloid protein is a physiological intercellular **signalling molecule**; β-amyloid precursor protein is a transmembrane protein whose extracellular **domain** is a receptor for signalling molecules (such as WNT family members, which are important regulators of early brain development; Chapter 7, Figure 7.3) and whose intracellular domain activates **second messengers** to regulate transcription. Prenatal brain cells express *APP* and there is evidence that variants affecting β-amyloid precursor protein function significantly shorten the time human cerebral cortical progenitors take to divide (which might be due to interference with intercellular signalling; this remains speculative). Disturbing the choreography of prenatal cerebral cortical neurogenesis, during which precisely timed cell–cell interactions are critical for specifying the fates of newly created neurons (as discussed in Chapter 1), is likely to impact almost all aspects of subsequent brain development. Genetic variants affecting *APP* expression or function might put the brain at risk of Alzheimer disease later in life for other reasons. For example, they might increase the amount of β-amyloid protein in the developing brain or cause the production of a longer and stickier form of it. People with Down syndrome are at high risk of developing Alzheimer disease and, in part, this might be because *APP* is one of the triplicated genes on chromosome 21. This triplication might result in the accumulation of β-amyloid precursor protein and β-amyloid protein, lowering resilience to other genetic or environmental factors that promote degenerative changes later in life. Clearly, the triplication of other genes on chromosome 21 is sure to be important as well.[277]

Second, *PSEN1* and *PSEN2*: these genes encode proteins that are part of the enzyme γ-secretase, a membrane protease that cleaves many other membrane proteins including amyloid precursor protein and, notably in the context of the present discussion, notch receptors, which are involved in short-range signalling regulating many aspects of brain development (Chapter 7, Figure 7.3).[278] There is evidence that variants with impaired *PSEN1* function have defects of early cerebral cortical neurogenesis.

Finally, apolipoprotein E: this is made both in the brain and in the rest of the body, mainly in the liver, and the two pools are kept separate by the blood–brain barrier. Outside the brain, it is a lipid transporter that helps distribute phospholipids and cholesterol throughout the body. In the brain, it is made and secreted primarily by **astrocytes** and microglia. It has been implicated in processes that include synapse formation, glutamatergic neurotransmission, synaptic plasticity, neuronal dendritic development and tissue repair. The *APOE4* variant can impair all of these processes, many of which start early in brain development. Later in life, apolipoprotein E4 can start to combine with β-amyloid protein outside brain cells to form clumps, which are the plaques characteristic of Alzheimer disease (Figure 9.4). It appears, therefore, that the *APOE4* variants produce a protein that imparts an increased risk of Alzheimer disease that starts in the early stages of brain neurogenesis and continues to accumulate throughout life.

Overall, it is likely that all the major genetic risk variants for Alzheimer disease have effects on brain development, which might predispose to the development of the condition later in life for reasons that remain unclear for now.[279–281] In early onset Alzheimer disease, the causative variants are likely to directly trigger the production of β-amyloid plaques, leading to the onset of detectable symptoms. How late-onset Alzheimer disease is triggered is less clear. Various external and internal environmental factors and types of early life experiences have been suggested to contribute.

Parkinson Disease

Parkinson disease affects about 1% of the over 60s. Its symptoms include problems with movement such as tremor at rest, unusually slow movement (bradykinesia) and rigidity, progressive cognitive decline and behavioural and neuropsychiatric changes. Its proximate causes involve a loss of neurons, especially dopaminergic neurons in the substantia nigra, and in most cases a gain of structures called Lewy bodies in neurons in the substantia nigra, cerebral cortex and elsewhere (Figure 9.5). Lewy bodies are thought to be toxic to cells because they damage mitochondria and probably create other problems too. They are formed by misfolded α-synuclein combined with other proteins. The normal functions of α-synuclein are not very well understood, but they probably include actions on processes such as vesicle transport, which is involved in neurotransmitter release from **presynaptic** terminals (particularly dopaminergic ones) and the mechanisms of DNA repair. Its incorrect folding would compromise these functions, disrupting dopaminergic neurons and leading to an accumulation of damaged DNA increasing the likelihood of cell death.

As with Alzheimer disease, a small percentage of Parkinson disease cases are caused by variants of single genes. They often, but not invariably, show symptoms at an earlier age, very occasionally even in childhood or adolescence. Dominant variants are of genes that

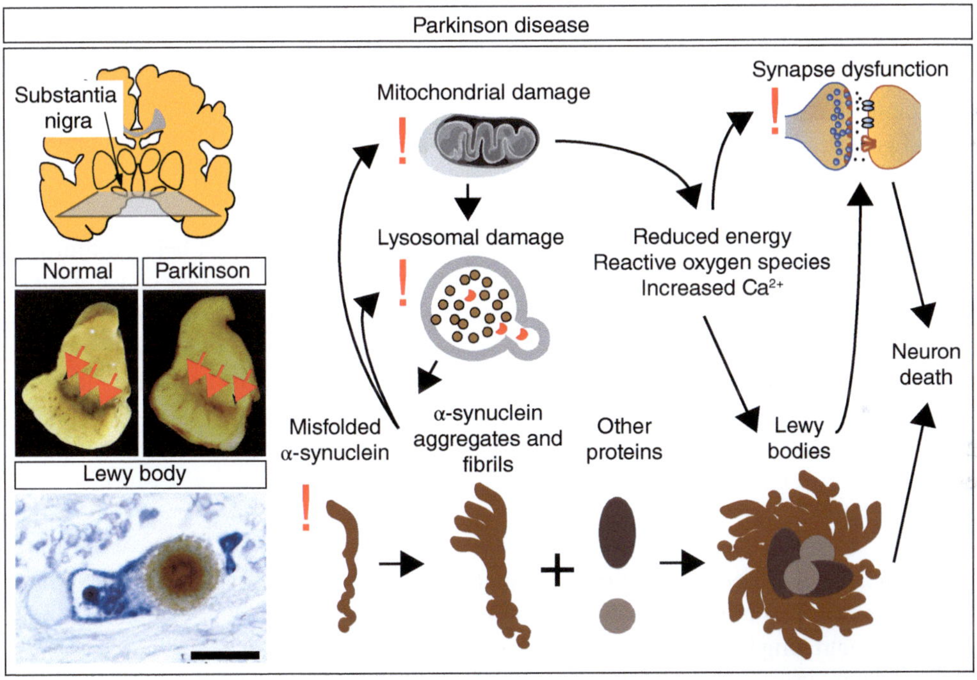

Figure 9.5 **Parkinson disease: pathology and likely causation.** On the left are sections through the substantia nigra (on one side of the brain). Pigmentation due to neuromelanin within the dopaminergic neurons is noticeably reduced in the brain from a person with Parkinson disease (red arrows). This is due to the depletion of this population of neurons. Lewy bodies, which are composed of α-synuclein and other proteins, form within neurons and impair their functions (the scale bar applies to this photograph and is 0.02 mm, which is about the diameter of a nerve cell body). The diagram illustrates how misfolded α-synuclein aggregates forming fibrils and Lewy bodies with a range of other proteins; this occurs in the gut and cerebral cortex as well as in the substantia nigra, raising the prospect that disruption elsewhere is likely to contribute in some way to the condition. A vicious circle is created in which mitochondrial and lysosomal damage due to genetic variants and environmental factors enhances α-synuclein abnormalities which further damage the mitochondria and lysosomes. The effects of mitochondrial damage are felt throughout the cell, reducing energy production (in the form of adenosine triphosphate, or ATP: Figure 8.7, Chapter 8), releasing reactive oxygen species (Chapter 4) and increasing cytoplasmic Ca^{2+} levels. These changes promote Lewy body formation and harm synaptic function, leading eventually to neuronal death. *Reproduced with permission from Dauer et al.*[282] */ with permission of Elsevier.*

include *SNCA*, which encodes α-synuclein, and *LRRK2*, which encodes leucine-rich repeat kinase 2, a cytoplasmic protein of uncertain function that might be involved in mitochondrial regulation and vesicle transport. Recessive variants are of genes that include *PRKN*, which encodes parkin, a protein involved in the ubiquitin–proteasome system and possibly other processes; *PINK1*, which encodes PTEN-induced kinase 1; *PARK7*, which encodes Parkinsonism-associated deglycase. Both parkin and PTEN-induced kinase 1 are thought to help control mitochondrial quality in neurons and Parkinsonism-associated deglycase appears to help protect cells against oxidative stress (which involves the production of harmful **reactive oxygen species**, Chapter 4). Variants affecting some of these genes, and over 200 others, have been identified by **genome wide association studies** (GWAS) as risk factors for the bulk of Parkinson disease cases.[283] These cases are often referred to as idiopathic because their exact causes are unknown. They are multifactorial, with both genetic and environmental risk factors. There is evidence that the latter include chemicals such as pesticides that inhibit oxidative phosphorylation, which occurs in mitochondria and is often dysfunctional in Parkinson disease, as well as other detrimental factors such as severe and/or repeated head trauma. One so-called 'multiple hit model' suggests that exposure of genetically vulnerable individuals to multiple harmful environmental factors from gestation through to old age leads to the progressively accelerating emergence of the symptoms of Parkinson disease.

Overall, much of what we know points to mitochondrial dysfunction being a common proximate cause onto which the many ultimate causes of Parkinson disease converge. As I have explained earlier in this and other chapters, mitochondrial dysfunction is likely to disturb many aspects of neurodevelopment. Factors that cause the symptoms of Parkinson disease by impairing mitochondrial function probably make their first causative marks on the brain early in its development.[284,285]

Summary

- Neurological and psychiatric symptoms that emerge in adolescence or adulthood might be caused at least in part by genetic variants and environmental factors affecting brain development long before symptoms appear.
- Huntington disease is a good example of a neurological condition with a defined genetic cause that is diagnosed relatively late in life but probably begins much earlier during brain development.
- From adolescence on, there is an ever-increasing potential for individuals to encounter environmental factors, from specific substances to life experiences, that can have long-term impacts on subsequent brain development.
- Most cases of relatively common psychiatric disorders, like schizophrenia, bipolar disorder and major depression, are caused by combinations of individually low-impact genetic variants and environmental triggers affecting the brain as it undergoes significant fine structural and functional change.
- The origins of degenerative conditions like Alzheimer disease and Parkinson disease are likely to be found in predisposing events that occur early in brain development, maybe even prenatally.

Afterword: Looking Forward

Every one of our brains has been unique since it first began to develop during early embryonic life. From the outset, its combination of inherited genetic variants and new errors carried in its DNA, its fine structural details and the moment-to-moment dynamics of its biochemical reactions would have been unique. As it progressed through prenatal life, childhood and adolescence to adulthood, the differences between it and all other brains – past, present and future – would have been amplified immeasurably by variation in events occurring both inside and around it. In other words, diversity embraces all of our brains at every stage of their existence.

To be clear, in this book I distinguish between this universal neurodiversity and neurodivergence. All of our brains exist within structural or functional ranges – or spectra – of whatever features we care to look at: it might be their physical dimensions and internal organisation, or the way they perceive, think or act. Some brains whose attributes place them towards the extremes of such spectra might be described as neurodivergent or disordered, but they are all encompassed within our collective neurodiversity.

It is not necessarily disadvantageous for someone to have a brain with attributes located towards the extremes of spectra. Indeed, neurodivergent minds can be of great benefit in many kinds of human activities, for example those that rely for their success on particular intellectual, cognitive, perceptual, creative or performative skills, or on combinations of different viewpoints, insights, experiences and so on. Unfortunately, however, some neurodivergent and most disordered brains create various types and degrees of hardship. What kinds of fixes or mitigations might become possible for disorders that are unquestionably harmful to an individual's chances of thriving or even surviving? How might we help those living with conditions that are disabling? How best can we help a person with a potentially problematic neurodivergent condition if and when they run into difficulties? Can we support efforts in these directions by learning more about the mechanisms that create our brains and the ultimate and proximate causes of disabling brain conditions? Before trying to address these questions, I will briefly summarize some key points arising throughout the book that are particularly relevant here.

Our neurodiversity results from variation in our genomes, our brains' external and internal environment and the random noise inherent in complex biological processes. Anything that affects these sources of variation will affect the diversity of our brains.

Regarding our genetics, some variants act alone to harm brain development in ways that are severe and easily detectable. Other variants, that individually might cause small undetectable differences in brain development, combine with each other and with non-genetic factors to influence how the brain that they contribute to develops. Identifying these small-effect variants and how they combine is difficult, requiring large studies of the actions of many or all genes in large collections of people.

Like genetic effects, environmental ones can range from the obvious to the undetectable. They might impact an individual negatively or positively: examples of the former include substances, such as alcohol, that can cause congenital defects of the brain; examples of the latter include vaccines that can protect against the risk of brain disorders. Both the external environment and the environment inside our own bodies that surrounds and nurtures the development of our brains are immensely variable in their detail. Before birth, that variability is constrained by the range of physiological possibilities compatible with successful gestation. After birth, the environmental variability experienced by an individual increases hugely and progressively, with many aspects created or influenced by humans, all the way from ourselves, our mothers and immediate family members to anyone and everyone else on the planet. We hold sway over many aspects of our environment, from local to global, and our environment reciprocates by having a big impact on how our brains develop.

How Can We Affect the Diversity of Human Brain Development?

We can influence the diversity of brain development, deliberately or inadvertently, by affecting our external environment, the internal environment surrounding our brains and/or our genomes. We can think about our power to influence at three main levels: that of the entire world population; that of groups of various types and sizes; that of individuals.

At a global level, we have become Earth's dominant large mammal by working collectively to create social structures and technologies that either buffer us against our environment or change aspects of it. These activities have mitigated selection pressures to a variable extent that depends on factors such as geographical location and socio-economic status. For example, advances in medical sciences, increases in the distribution of welfare and so on have allowed many individuals with potentially life-threatening conditions to survive, in some cases to reproductive age. This has an overall impact on the diversity of brains across the entire human population because it affects the numbers of people able to live with potentially harmful brain conditions and because genetic variants that might contribute to brain diversity are likely to be retained at increased frequency in the human gene pool. The overall magnitude of the impact this has had – and is likely to have in the future – is hard to work out. Brain diversity can also be affected by the ways in which we have changed our global environment. For example, there is evidence that factors such as pollution and global warming can have a negative impact at all stages of brain development,[286] with the extent and nature of their impact varying with geographical location and/or with the susceptibility of different individuals to the changes we have caused.

At the level of groups (which might, for example, be based on ethnicity, religious beliefs, geographical location or nationality), the diversity of brain development can be influenced in many ways, such as how breeding partners are selected. Influences over breeding might either be through the exertion of top-down control or by agreed practices within a population. Regarding the first of these, mass eugenic programmes aimed at ensuring the preferential breeding of humans possessing whatever phenotypes those in power deem desirable (for example, high scores on intelligence tests[37]) are no longer acceptable to most of us (and possibly never were). Regarding the second, selective breeding based on attributes such as ethnicity, appearance or religious beliefs has been, and still is, practiced at least to some extent in many, maybe most, communities around the world. Selective breeding is likely to affect the frequencies of variants that influence brain development. One well-known example is that the proportions of people carrying recessive potentially harmful variants capable of causing certain neurodevelopmental disorders is particularly high in those of Ashkenazi Jewish descent.[287] Another is that the culturally ingrained practice of breeding with close relatives among some groups of Bedouin Arabs is associated with an unusually high rate of intellectual disability.[288] The diversity of brain development can also be affected by the nature of the support, education, health and social care systems a group creates or is denied. For example, the degree to which groups can adapt their systems to both meet the needs and nurture the development of those with neurodevelopmental conditions will affect brain diversity within those groups.

At the level of individuals, we can try to exert control over how our own brains or those of others develop by affecting any of a wide range of aspects of our local environment, using chemical, physical or interpersonal interventions. Parents try to do this by offering their children opportunities for learning or discouraging activities that might be harmful for their development. Doctors try to do it by prescribing or administering medications of various sorts or using other procedures designed to affect brain development and bring long-term benefits. These procedures might, for example, involve surgery (such as the insertion of tubes, or shunts, to lower pressure on developing brain tissues in **hydrocephalus**), psychotherapy (particularly for psychiatric conditions) or the administration of chemical substances to developing brain cells. These substances can be designed to work in a broad range of ways: some therapeutic compounds might directly affect the biochemical processes of cells; others might allow cells to make beneficial substances themselves. The second of these approaches is starting to yield some very promising results. For example, successful treatment of a fetus developing with spinal muscular atrophy, a severe neurological condition caused by loss of function of the *SMN1* gene (which produces SMN protein) was achieved by administering a drug called risdiplam to boost production of functional SMN protein from another related gene, *SMN2*.[289] Another example, which looks promising in cultured human cells but has not yet been tried in patients, is the possibility of rescuing cells that cannot produce a protein called UBE3A – because the gene they inherited from the mother contains a loss-of-function variant – by administering a drug that can activate the normally silenced gene inherited from the father (*UBE3A* is an example of an imprinted gene whose paternal copy is silenced, see Chapter 4).[290] Interventions such as these might bring tangible improvements in the course of brain development in some individuals with severe genetic conditions. In many other cases, however, they might bring only short-term or palliative relief for as long as, or not much longer than, the manipulations are applied. Can we do more? Can we reduce the numbers of individuals suffering debilitating genetic conditions? Can we permanently reverse the progressive development of genetic conditions by treating the underlying causes in the genome itself?

Approaches to reducing the numbers of individuals with debilitating genetic conditions include screening embryos generated by *in vitro* fertilization for harmful genetic variants, a procedure known as preimplantation genetic testing. This method is used to identify embryos carrying variants that cause specific monogenic conditions or harmful chromosomal abnormalities. The method can help couples avoid passing on a genetic condition to their children by selecting embryos that lack harmful variants for implantation into the uterus. What about polygenic and multifactorial conditions? Methods have been developed to screen embryos for the risk of conditions such as autism, schizophrenia and bipolar disorder. As you will appreciate from all I have said in this book, the best that can be done is to assign risk scores (explained in Chapter 4) to embryos based on the genetic variants they contain. This shrouds any decisions made about individual embryos in uncertainty. The extent to which the potential benefits of this approach outweigh the possible harms to individuals or society is still being discussed.[291]

Most recently, we have learned how we might alter our genomes directly using methods often referred to as gene editing (changing an existing gene), gene silencing (deactivating an existing gene) or gene addition (adding a new gene): these are all examples of what is referred to as gene therapy. Using these methods, we hope to make specific genomic changes that should bring long-term benefits without repeated application of a therapeutic agent. How much of a global impact these types of gene therapy are likely to have in the foreseeable future is dubious, given its technical complexity and cost. Even in wealthy societies where technologically advanced methods are available, whether any given genetic variant is likely to be treatable in this way will probably depend on when and in what way the variant affects development. For example, if the variant has a major impact very early in embryogenesis, then changing it later in life might bring limited benefits. Nevertheless, some conditions might be amenable to treatment in this way.

At the moment, much research in this area is focused on refining the delivery of nucleic acid sequences (for example, using non-pathogenic viral vectors[292]) and developing gene editing technologies. Conditions such as aromatic L-amino acid decarboxylase deficiency, Rett syndrome, fragile X syndrome, Angelman syndrome, Huntington's disease and Tay–Sachs disease are among those that researchers are currently focused on.[293–298] Many challenges remain, however; they revolve around the issues of how to ensure the long-term specificity, efficacy and safety of the therapies. For example, CRISPR/Cas technology, which has revolutionized genome engineering (and is described in our previous book *Building Brains; An Introduction to Neural Development*[10]) can cause off-target genomic alterations, raising safety concerns that will need to be addressed in the future.[299,300]

Discovering More

Throughout this book, I have highlighted the major challenges that we face as we try to understand how the complexity and diversity of our brains develop. At the top of the pile of difficult-to-answer questions are those that ask how the particular structural and functional brain phenotypes *unique to any single individual* were brought into being. Despite many overall similarities, our brains think, react, learn, memorize and so on differently, giving each of us a unique personality. In the midst of all this variability, it is little wonder that trying to work out the developmental causes of any one person's way of thinking, reacting, learning, memorizing and so on is so hard. And yet, in our everyday lives we try to do it all the time: how often, for example, have we heard people confidently ascribe a specific individual's abilities and behaviour to some particular cause, maybe the nature of their childhood, their education, their habits or other similar things. Our views might have validity in some cases and at some level, but at best they are likely to be incomplete and often they are likely to be misleading or wrong. For example, we might hypothesize that a particular person who displays symptoms of a psychiatric disorder does so at least in part because that person had a traumatic childhood. There might be truth in this or there might not; we have no way of knowing for sure. We can extrapolate from our general understanding of factors that contribute to the disorder in question across the population, but this does not guarantee that we are correct in the case of this individual.

The reason for this problem is because scientific methods do not deal easily with unique events, and each individual's brain is the product of a unique sequence of internal molecular and cellular events influenced by the variants they start out with, those they acquire and their internal and external environment as each day of their life unfolds. The scientific method is based on a search for patterns, unifying principles and laws of nature. Scientists rely for their knowledge on attempts to *repeatedly reproduce* observations or experiments. The more reproducible a finding is, the more likely it is to be true. In order to look for patterns and unifying principles in humans, we need to group people with similar phenotypes together so that we can look for patterns of variation in their genomes and/or their environments that might point to the ultimate causes of those phenotypes. Likewise, we need to group people to search for consistent differences in developmental processes that might explain how the different ultimate causes affect phenotypes. As we have seen in this book, grouping people together based on their phenotypes is often fraught with problems.

Some of the problems stem from the sheer complexity of those aspects of a living brain's phenotype that we can potentially measure. The aspects we might measure in most detail are mainly functional, since even modern scanning methods give us limited ability to analyse the details of the brain's fine structure and circuitry in a living person. We should never lose sight of the truly mindboggling magnitude of the brain's functional complexity, being as it is the product of the activity of many billions of neurons with many trillions of cellular connections between them. Categorizing complex and variable phenomena is difficult at the best of times, no matter what the phenomena are. Historically, attempts to develop an understanding of brain complexity, diversity, divergence and disorder have been led by clinical scientists. They have defined broad diagnostic categories based on similarities and differences in peoples' brains and the way they function, in other words, on variation in phenotypes. This is entirely understandable since, as I have described throughout this book, many conditions were defined before we had any understanding of genetics or had gained the ability to identify the effects of many external environmental factors. Subsequent research seeking explanations for brain phenotypes has been dominated by the framework these diagnostic categories provide, using them to try to discover what genetic and/or environmental factors are shared by people within each group. This has worked quite well for understanding relatively clearly definable conditions, for example ones that are monogenic or due to chromosomal abnormalities, but less well for understanding neurodivergent or psychiatric conditions, which can be much harder to delineate precisely and unambiguously. In these conditions at least, we can anticipate without too much difficulty – since it is starting to happen already[301] – that there will be a shift to a different approach in which people are classified according to the genetic variants they carry and/or the environmental factors that impact them rather than the phenotypes they display. Within this framework, we might then ask: what traits are manifest by individuals who share similar combination of variants or environmental exposures or experience? Approaching things from the side of the causes rather than the effects might make it easier to allow for the fact that similar ultimate causes can contribute to many different sets of phenotypes and that many different factors can cause similar phenotypes. Asking what a particular combination of factors can do to brain development is, conceptually, a relatively straightforward way of thinking.

While this approach may be relatively straightforward conceptually, in practice it is anything but. As I pointed out above, the scientific method requires us to group individuals together. If we start from the side of causation, we could use data obtained by sequencing peoples' genomes to group them based on similarities in the combinations of genetic variants they carry. That alone involves handling and analysing a lot of data on each person, but current computing technologies should allow for this and our genomes are finite entities. Far more complex, however, would be to include information on each person's environmental exposures and experiences, which are, for practical purposes, infinitely variable. We could, in theory, get vast amounts of information on aspects of individuals' exposures and experiences by monitoring them continuously in longitudinal studies. This could be seen as an extension of the monitoring of individuals by commercial companies who, for example, analyse things like our mobile phone usage all the time. At some point, however, the limits of current computational power would be exceeded. It seems likely that the step-change in computing capacity that will be needed to comprehensively analyse massive environmental and phenotypic variability will have to wait for the advent of quantum computers, maybe 10 years or more into the future.

As we continue to build our knowledge of the ultimate causes of our diversity, we also need to learn more about how these causes have their effects. This challenge is also huge. The cellular and molecular routes from ultimate causes to phenotypes are numerous, subject to innate random variability (or noise) and are inherently probabilistic in nature. This means that, even if we were somehow able to apply the exact same genetic and environmental factors to two identical individuals (to be clear, this would be impossible in reality), there would still be only a certain *probability* that particular cellular and molecular events would follow. This means that, no matter what approach we take to discovering the steps between ultimate causes and phenotypes, we need to understand and contend with the nature of variability in developmental mechanisms. Understanding the mechanisms that have evolved to constrain this variability, and how genetic variants might jeopardize their effectiveness, is likely to be an important avenue of research.

Progress in understanding how ultimate causes have their effects is being accelerated in several ways. One of these involves the application of microfluidic, molecular biological and computational techniques to isolate large numbers of single cells from a tissue and study their genomes (the DNA and its state), transcriptomes (the RNA), proteomes (the proteins) and metabolomes (the metabolites) (Chapter 7, Figure 7.16). We refer to this as single cell multi-omics. It allows us to work out what is going on in different cells: which parts of the **chromatin** are accessible, which parts are inaccessible, which genes are being expressed and at what levels, which **splice** variants are being made, what proteins are being produced and what physiological processes are occurring in each cell. It allows us to identify different cell types and different cell states, such as proliferation, quiescence or migration. In further developments, it is rapidly becoming possible to use multi-omics methods to analyse the molecular state of individual cells in their natural location in tissues. We call this spatial multi-omics. The approach relies on keeping samples of tissue from donors intact before slicing them thinly and then extracting cells from specific identified locations for multi-omic analysis. In this way, we can tell what the molecular profiles of cells are in their specific locations within a tissue and how they relate to the profiles of their neighbours. The approach goes beyond what is possible in more conventional methods, in which the cells of a tissue must be separated from each other prior to multi-omic analysis. It allows scientists to gain more powerful insights into how cells interact to form tissues.[302]

Single-cell multi-omic and spatial multi-omic data can be derived from any set of cells taken, for example, from post-mortem embryonic, fetal or postnatal brains, from biopsy material taken at surgery or from cultured human brain cells. Material can come from individuals who might or might not have an identified clinical condition. There is enormous scope for combining such single-cell multi-omic data with data on potentially harmful genetic variants from **genome wide association studies** (Chapter 4).[303–305] Using this approach, we can learn about the cell types and cell states in which the variants have their effects. By studying single cells from donated brains of individuals who lived with identified conditions, we can learn more about the cellular and molecular changes underpinning those conditions.[306,307]

The rapid expansion of culture methods for growing human brain-like structures in the laboratory (for example, brain organoids: Chapter 1) is providing more than a potential source of material in studies of how human brain cell types develop. It is also offering a means of testing hypotheses on how developmental mechanisms might be disrupted by genetic variants or environmental agents. The fact that these techniques involve human cells developing in cellular contexts that model those in developing human brains is likely to increase the relevance of the discoveries they enable us to make. Culture methods do, however, come with many caveats. They inevitably either introduce features that are absent or remove elements that are present in the developing brain in a human being. We have, of course, learned a great deal that is relevant to humans from research on other species, but it is also the case that many animal models fail, or are unable in principle, to mimic human brain conditions.[308] It is questionable whether it is even possible to create meaningful animal models of complex human psychiatric or neurodivergent conditions that rely for their diagnosis on attributes unique to human brains, such as communication using understandable languages capable of conveying abstract thoughts, nuanced emotions and a wide range of complex ideas relevant to human life experiences. At the end of the day, we will continue to need multiple different approaches to increase our understanding of the routes from ultimate causes to phenotypes.[309]

And finally, what should we do with our increasing knowledge of how our brains develop, how their diversity is created, how some people come to be diagnosed as neurodivergent and how others develop with clear disorders? Probably most of us immediately jump to the possibility of using it to improve treatments for clinical disorders. Some might consider other applications: for example, some scientists are actively pursuing the possibility that we can learn how to make more powerful computers inspired by the mechanisms that generate a human brain (this is often referred to as neuromorphic computing[310]). In addition, I think one of the most impactful things we can do is to inform as many people as possible about the principles governing how our brains develop and how they vary. We often struggle to accommodate and interact effectively with those of us whose behaviours, perceptions and/or cognitive abilities lie towards the extremes of diversity spectra. More understanding might help us all to appreciate that many of the assumptions we often make about the reasons for a given person's personality or behaviours or abilities are most likely wrong or at best naive. It might improve the integration of many of those with neurodevelopmental disorders and types of neurodivergence into our social, educational and support environments in more effective ways, to the benefit of all. New therapies will not help with this; a better collective understanding of neurodiversity and its origins might.

References

1 Waters F, Collerton D, Ffytche DH, et al. Visual hallucinations in the psychosis spectrum and comparative information from neurodegenerative disorders and eye disease. *Schizophr Bull*. Jul 2014;40(Suppl 4):S233–S245. doi:10.1093/schbul/sbu036

2 Shabani K, Hassan BA. The brain on time: links between development and neurodegeneration. *Development*. May 15 2023;150(10). doi:10.1242/dev.200397

3 Shapson-Coe A, Januszewski M, Berger DR, et al. A petavoxel fragment of human cerebral cortex reconstructed at nanoscale resolution. *Science*. May 10 2024;384(6696):eadk4858. doi:10.1126/science.adk4858

4 Rugg-Gunn PJ, Moris N, Tam PPL. Technical challenges of studying early human development. *Development*. Jun 1 2023;150(11). doi:10.1242/dev.201797

5 Kelley KW, Pasca SP. Human brain organogenesis: toward a cellular understanding of development and disease. *Cell*. Jan 6 2022;185(1):42–61. doi:10.1016/j.cell.2021.10.003

6 Xue X, Kim YS, Ponce-Arias AI, et al. A patterned human neural tube model using microfluidic gradients. *Nature*. Apr 2024;628(8007):391–399. doi:10.1038/s41586-024-07204-7

7 Shahbazi MN, Pasque V. Early human development and stem cell-based human embryo models. *Cell Stem Cell*. Oct 3 2024;31(10):1398–1418. doi:10.1016/j.stem.2024.09.002

8 Hendriks D, Pagliaro A, Andreatta F, et al. Human fetal brain self-organizes into long-term expanding organoids. *Cell*. Feb 1 2024;187(3):712–732.e38. doi:10.1016/j.cell.2023.12.012

9 Turner DA, Martinez AA. Three-dimensional stem cell models of mammalian gastrulation. *BioEssays*. Dec 2024;46(12):e2400123. doi:10.1002/bies.202400123

10 Price DJ, Jarman AP, Mason JO, Kind PC. *Building Brains: An Introduction to Neural Development*. John Wiley & Sons, Inc.; 2018.

11 Moris N, Anlas K, van den Brink SC, et al. An in vitro model of early anteroposterior organization during human development. *Nature*. 2020;582:410–415.

12 Hamazaki N, Yang W, Kubo CA, et al. Retinoic acid induces human gastruloids with posterior embryo-like structures. *Nat Cell Biol*. 2024;26:1790–1803.

13 Zhai J, Xiao Z, Wang Y, Wang H. Human embryonic development: from peri-implantation to gastrulation. *Trends Cell Biol*. Jan 2022;32(1):18–29. doi:10.1016/j.tcb.2021.07.008

14 Ghimire S, Mantziou V, Moris N, Martinez AA. Human gastrulation: the embryo and its models. *Dev Biol*. Jun 2021;474:100–108. doi:10.1016/j.ydbio.2021.01.006

15 Casingal CR, Descant KD, Anton ES. Coordinating cerebral cortical construction and connectivity: unifying influence of radial progenitors. *Neuron*. Apr 6 2022;110(7):1100–1115. doi:10.1016/j.neuron.2022.01.034

16 Nascimento MA, Biagiotti S, Herranz-Perez V, et al. Protracted neuronal recruitment in the temporal lobes of young children. *Nature*. Feb 2024;626(8001):1056–1065. doi:10.1038/s41586-023-06981-x

17 de Faria O, Pivonkova H, Varga B, et al. Periods of synchronized myelin changes shape brain function and plasticity. *Nat Neurosci*. 2021;24:1508–1521.

18 Huttenlocher PR, Dabholkar AS. Regional differences in synaptogenesis in human cerebral cortex. *J Comp Neurol*. 1997;387:167–178.

19 Khakh BS. Astrocyte-neuron interactions in the striatum: insights on identity, form, and function. *Trends Neurosci*. Sep 2019;42(9):617–630. doi:10.1016/j.tins.2019.06.003

20 Thion MS, Ginhoux F, Garel S. Microglia and early brain development: an intimate journey. *Science*. Oct 12 2018;362(6411):185–189. doi:10.1126/science.aat0474

21 Wälchli T, Wacker A, Frei K, et al. Wiring the vascular network with neural cues: a CNS perspective. *Neuron*. 2015;87:271–296.

22 Petersen SE, Seitzman BA, Nelson SM, Wig GS, Gordon EM. Principles of cortical areas and their implications for neuroimaging. *Neuron*. Sep 4 2024;112(17):2837–2853. doi:10.1016/j.neuron.2024.05.008

23 Sydnor VJ, Larsen B, Bassett DS, et al. Neurodevelopment of the association cortices: patterns, mechanisms, and implications for psychopathology. *Neuron*. Sep 15 2021;109(18):2820–2846. doi:10.1016/j.neuron.2021.06.016

24 Miller EK. The prefrontal cortex and cognitive control. *Nat Rev Neurosci*. Oct 2000;1(1):59–65. doi:10.1038/35036228

25 Squire LR. The legacy of patient H.M. for neuroscience. *Neuron*. Jan 15 2009;61(1):6–9. doi:10.1016/j.neuron.2008.12.023

26 Lisman J, Buzsaki G, Eichenbaum H, Nadel L, Ranganath C, Redish AD. Publisher correction: viewpoints: how the hippocampus contributes to memory, navigation and cognition. *Nat Neurosci*. Jul 2018;21(7):1018. doi:10.1038/s41593-017-0034-8

27 Manoli A, Van Overwalle F, Grosse Wiesmann C, Valk SL. Functional recruitment and connectivity of the cerebellum is associated with the emergence of theory of mind in early childhood. *Nat Commun*. Jun 6 2025;16(1):5273. doi:10.1038/s41467-025-60523-9

28 Van Overwalle F. Social and emotional learning in the cerebellum. *Nat Rev Neurosci*. Dec 2024;25(12):776–791. doi:10.1038/s41583-024-00871-5

29 Yuan Q, Li H, Du B, et al. The cerebellum and cognition: further evidence for its role in language control. *Cereb Cortex*. Dec 15 2022;33(1):35–49. doi:10.1093/cercor/bhac051

30 Carey MR. The cerebellum. *Curr Biol*. Jan 8 2024;34(1):R7–R11. doi:10.1016/j.cub.2023.11.048

31 Rocha GS, Freire MAM, Britto AM, et al. Basal ganglia for beginners: the basic concepts you need to know and their role in movement control. *Front Syst Neurosci*. 2023;17:1242929. doi:10.3389/fnsys.2023.1242929

32 Hebb DO. *The Organization of Behavior: A Neuropsychological Theory*. New York: Wiley; 1949.

33 Nicoll RA. A brief history of long-term potentiation. *Neuron*. Jan 18 2017;93(2):281–290. doi:10.1016/j.neuron.2016.12.015

34 Osso LA, Hughes EG. Dynamics of mature myelin. *Nat Neurosci*. Aug 2024;27(8):1449–1461. doi:10.1038/s41593-024-01642-2

35 Hodebourg R, Scofield MD, Kalivas PW, Kuhn BN. Nonneuronal contributions to synaptic function. *Neuron*. Aug 6 2025;113(15):2399–2415. doi:10.1016/j.neuron.2025.04.004

36 Turk-Browne NB, Aslin RN. Infant neuroscience: how to measure brain activity in the youngest minds. *Trends Neurosci.* May 2024;47(5):338–354. doi:10.1016/j.tins.2024.02.003

37 Gould SJ. *The Mismeasure of Man.* Norton; 1981.

38 Baranger J, Demene C, Frerot A, et al. Bedside functional monitoring of the dynamic brain connectivity in human neonates. *Nat Commun.* 2021;12:1080.

39 Lankinen K, Ahveninen J, Jas M, Raij T, Ahlfors SP. Neuronal modeling of cross-sensory visual evoked magnetoencephalography responses in the auditory cortex. *J Neurosci.* 2024;44:e1119232024.

40 Hill RM, Boto E, Rea M, et al. Multi-channel whole-head OPM-MEG: Helmet design and a comparison with a conventional system. *NeuroImage.* 2020;219:116995.

41 Soltanlou M, Artemenko C, Ehlis AC, et al. Reduction but no shift in brain activation after arithmetic learning in children: a simultaneous fNIRS-EEG study. *Sci Rep.* 2018;8:1707.

42 Lotze M, Domin M, Langner S, Platz T. Functional MRI in radiology—a personal review. *Healthcare (Basel).* 2022;10:1646.

43 Tymofiyeva O, Gaschler R. Training-induced neural plasticity in youth: a systematic review of structural and functional MRI studies. *Front Hum Neurosci.* 2021;14:497245.

44 Takahashi M, Akamatsu G, Iwao Y, et al. Small nuclei identification with a hemispherical brain PET EJNMMI. *Physics.* 2022;9:69.

45 Verger A, Grimaldi S, Ribeiro M-J, Frismand S, Guedj E, et al. Single photon emission computed tomography/positron emission tomography molecular imaging for Parkinsonism: a fast-developing field. *Ann Neurol.* 2021;90:711–719.

46 Zhou F, Becker B. Understanding human brain function in real-world environments. *PLoS Biol.* Jun 2025;23(6):e3003210. doi:10.1371/journal.pbio.3003210

47 Shim M, Hwang HJ, Kim DW, Lee SH, Im CH. Machine-learning-based diagnosis of schizophrenia using combined sensor-level and source-level EEG features. *Schizophr Res.* Oct 2016;176(2–3):314–319. doi:10.1016/j.schres.2016.05.007

48 Hansen L, Bernstorff M, Enevoldsen K, et al. Predicting diagnostic progression to schizophrenia or bipolar disorder via machine learning. *JAMA Psychiatry.* May 1 2025;82(5):459–469. doi:10.1001/jamapsychiatry.2024.4702

49 Wong AK, Sealfon RSG, Theesfeld CL, Troyanskaya OG. Decoding disease: from genomes to networks to phenotypes. *Nat Rev Genet.* Dec 2021;22(12):774–790. doi:10.1038/s41576-021-00389-x

50 Kohane IS. Finding a new balance between a genetics-first or phenotype-first approach to the study of disease. *Neuron.* Jul 21 2021;109(14):2216–2219. doi:10.1016/j.neuron.2021.07.001

51 Zaretskaya N. When sensory input meets spontaneous brain activity. *Trends Neurosci.* Oct 2024;47(10):749–750. doi:10.1016/j.tins.2024.08.010

52 Lotto RB, Purves D. The effects of color on brightness. *Nat Neurosci.* 1999;2:1010–1014.

53 Simons DJ, Chabris CF. Gorillas in our midst: sustained inattentional blindness for dynamic events. *Perception.* 1999;28(9):1059–1074. doi:10.1068/p281059

54 Sengpiel F. The critical period. *Curr Biol.* Sep 4 2007;17(17):R742–R743. doi:10.1016/j.cub.2007.06.017

55 Kujala T, Partanen E, Virtala P, Winkler I. Prerequisites of language acquisition in the newborn brain. *Trends Neurosci.* Sep 2023;46(9):726–737. doi:10.1016/j.tins.2023.05.011

56 Blakemore SJ. Development of the social brain in adolescence. *J R Soc Med.* Mar 2012;105(3):111–116. doi:10.1258/jrsm.2011.110221

57 Fuhrmann D, Knoll LJ, Blakemore SJ. Adolescence as a sensitive period of brain development. *Trends Cognit Sci.* Oct 2015;19(10):558–566. doi:10.1016/j.tics.2015.07.008

58 Lee Langman E, Hertzberg BS, Boyd BK, Gupta RT. Dichorionic, diamnionic twin pregnancy discordant for anencephaly: report of two cases and literature review. *Radiol Case Rep.* 2015;8(2):843. doi:10.2484/rcr.v8i2.843. eCollection 2013.

59 Hahn JS, Plawner LL. Evaluation and management of children with holoprosencephaly. *Pediatr Neurol.* 2004;31:79–88.

60 Koenig M, Dobyns WB, Di Donato N. Lissencephaly: update on diagnostics and clinical management. *Eur J Paediatr Neurol.* 2021;35:147–152.

61 Hino-Fukuyo N, Togashi N, Takahashi R, et al. Neuroepidemiology of porencephaly, schizencephaly, and hydranencephaly in Miyagi Prefecture, Japan. *Pediatr Neurol.* 2016;54:39–42.e1.

62 Iida S, Seino H, Nagahata F, et al. Cerebral ventriculomegaly in myotonic dystrophy type 1: normal pressure hydrocephalus?like appearances on magnetic resonance imaging. *BMC Neuroscience.* 2021;22:62.

63 Gonçalves F, G, Caschera L, Teixeira SR, et al. Intracranial calcifications in childhood: Part 2. *Pediatr Radiol.* 2020;50:1448–1475.

64 Poretti A, Boltshauser E. Terminology in morphological anomalies of the cerebellum does matter. *Cerebellum Ataxias.* 2015;2:8. doi:10.1186/s40673-015-0027-x eCollection 2015

65 Leventer RJ, Guerrini R, Dobyns WB. Malformations of cortical development and epilepsy. *Dialogues Clin Neurosci.* 2008;10:47–62.

66 Kosmeri C, Dermitzaki N, Xydis VG, Drougia A. A rare case of a neonate with agenesis of the corpus callosum and severe laryngomalacia. *Hippokratia.* 2022;26(3):118–120.

67 Galardi MM, Gaudioso C, Ahmadi S, et al. Differential diagnosis of pediatric multiple sclerosis. *Children.* 2019;6:75.

68 Gharaylou MA, Sahraian MH, et al. Widespread disruptions of white matter in familial multiple sclerosis: DTI and NODDI study. *Front Neurol.* 2021;12:678245.

69 WHO. *International Classification of Diseases.* 11th ed. The World Health Organization; 2019.

70 APA. *Diagnostic and Statistical Manual of Mental Disorders.* 5th ed. American Psychiatric Association Publishing; 2022.

71 Morris-Rosendahl DJ, Crocq MA. Neurodevelopmental disorders-the history and future of a diagnostic concept. *Dialogues Clin Neurosci.* Mar 2020;22(1):65–72. doi:10.31887/DCNS.2020.22.1/macrocq

72 Ryan AM. Electrographic monitoring for seizure detection in the neonatal unit: current status and future direction. *Pediatr Res.* 2024;96:896–904.

73 Matsubara K, Hattori T, Narumi S. Achievement of developmental milestones recorded in real time: a mobile app-based study. *J Pediatr.* 2022;245:201–207.e9.

74 Plomin R, von Stumm S. The new genetics of intelligence. *Nat Rev Genet.* Mar 2018;19(3):148–159. doi:10.1038/nrg.2017.104

75 Deary IJ, Cox SR, Hill WD. Genetic variation, brain, and intelligence differences. *Mol Psychiatry.* Jan 2022;27(1):335–353. doi:10.1038/s41380-021-01027-y

76 Saulnier CA, Klaiman C, McQueen E. Adaptive behavior profiles in autism spectrum disorder. *Curr Psychiatry Rep.* Dec 2022;24(12):749–756. doi:10.1007/s11920-022-01381-w

77 Manohari SM, Raman V, Ashok MV. Use of Vineland adaptive behavior scales - II in children with autism - an Indian experience. *J Indian Assoc Child Adolesc Ment Health.* 2013;9:5–12. doi:10.1177/0973134220130102

78 Haslam J. *Illustrations of Madness.* G. Hayden; 1810.

79 Capgras JR-LJ. L'illusion des "sosies" dans un délire systématisé chronique. *Bulletin de la Société clinique de médecine mentale.* 1923;11:6–16.

80 Bottini R, Nava E, De Cuntis I, Benetti S, Collignon O. Synesthesia in a congenitally blind individual. *Neuropsychologia.* Jun 6 2022;170:108226. doi:10.1016/j.neuropsychologia.2022.108226

81 Luo S, Valencia CA, Zhang J, et al. Biparental inheritance of mitochondrial DNA in humans. *PNAS.* Dec 18 2018;115(51):13039–13044. doi:10.1073/pnas.1810946115

82 Wallace DC. Mitochondrial genetic medicine. *Nat Genet.* Dec 2018;50(12):1642–1649. doi:10.1038/s41588-018-0264-z

83 Bizzotto S, Dou Y, Ganz J, et al. Landmarks of human embryonic development inscribed in somatic mutations. *Science.* Mar 19 2021;371(6535):1249–1253. doi:10.1126/science.abe1544

84 Naxerova K. Mutation fingerprints encode cellular histories. *Nature.* Sep 2021;597(7876):334–336. doi:10.1038/d41586-021-02269-0

85 Chatterjee N, Walker GC. Mechanisms of DNA damage, repair, and mutagenesis. *Environ Mol Mutagen.* Jun 2017;58(5):235–263. doi:10.1002/em.22087

86 Nickoloff JA, Sharma N, Taylor L, Allen SJ, Hromas R. The safe path at the fork: ensuring replication-associated DNA double-strand breaks are repaired by homologous recombination. *Front Genet.* 2021;12:748033. doi:10.3389/fgene.2021.748033

87 Bizzotto S, Walsh CA. Genetic mosaicism in the human brain: from lineage tracing to neuropsychiatric disorders. *Nat Rev Neurosci.* May 2022;23(5):275–286. doi:10.1038/s41583-022-00572-x

88 Waldvogel SM, Posey JE, Goodell MA. Human embryonic genetic mosaicism and its effects on development and disease. *Nat Rev Genet.* Oct 2024;25(10):698–714. doi:10.1038/s41576-024-00715-z

89 Jonsson H, Magnusdottir E, Eggertsson HP, et al. Differences between germline genomes of monozygotic twins. *Nat Genet.* Jan 2021;53(1):27–34. doi:10.1038/s41588-020-00755-1

90 Stewart JB, Chinnery PF. Extreme heterogeneity of human mitochondrial DNA from organelles to populations. *Nat Rev Genet.* 2021;22(2):106–118. doi:10.1038/s41576-020-00284-x

91 Silva-Pinheiro P, Minczuk M. The potential of mitochondrial genome engineering. *Nat Rev Genet.* Apr 2022;23(4):199–214. doi:10.1038/s41576-021-00432-x

92 Gorman GS, Chinnery PF, S DM, et al. Mitochondrial diseases. *Nat Rev Dis Primers.* Oct 20 2016;2:16080. doi:10.1038/nrdp.2016.80

93 Acuna-Hidalgo R, Veltman JA, Hoischen A. New insights into the generation and role of de novo mutations in health and disease. *Genome Biol.* Nov 28 2016;17(1):241. doi:10.1186/s13059-016-1110-1

94 Wood KA, Goriely A. The impact of paternal age on new mutations and disease in the next generation. *Fertil Steril.* Dec 2022;118(6):1001–1012. doi:10.1016/j.fertnstert.2022.10.017

95 Zarrei M, MacDonald JR, Merico D, Scherer SW. A copy number variation map of the human genome. *Nat Rev Genet.* Mar 2015;16(3):172–183. doi:10.1038/nrg3871

96 Iafrate AJ, Feuk L, Rivera MN, et al. Detection of large-scale variation in the human genome. *Nat Genet.* Sep 2004;36(9):949–951. doi:10.1038/ng1416

97 Sebat J, Lakshmi B, Troge J, et al. Large-scale copy number polymorphism in the human genome. *Science.* Jul 23 2004;305(5683):525–528. doi:10.1126/science.1098918

98 Hayward A, Gilbert C. Transposable elements. *Curr Biol.* Sep 12 2022;32(17):R904–R909. doi:10.1016/j.cub.2022.07.044

99 Lawson HA, Liang Y, Wang T. Transposable elements in mammalian chromatin organization. *Nat Rev Genet.* Oct 2023;24(10):712–723. doi:10.1038/s41576-023-00609-6

100 Salzberg SL. Open questions: How many genes do we have? *BMC Biol.* Aug 20 2018; 16(1):94. doi:10.1186/s12915-018-0564-x

101 Palazzo AF, Koonin EV. Functional long non-coding RNAs evolve from junk transcripts. *Cell.* Nov 25 2020;183(5):1151–1161. doi:10.1016/j.cell.2020.09.047

102 Petric Howe M, Patani R. Nonsense-mediated mRNA decay in neuronal physiology and neurodegeneration. *Trends Neurosci.* Oct 2023;46(10):879–892. doi:10.1016/j.tins.2023.07.001

103 Girskis KM, Stergachis AB, DeGennaro EM, et al. Rewiring of human neurodevelopmental gene regulatory programs by human accelerated regions. *Neuron.* Oct 20 2021;109(20):3239–3251.e7. doi:10.1016/j.neuron.2021.08.005

104 Doan RN, Bae BI, Cubelos B, et al. Mutations in human accelerated regions disrupt cognition and social behavior. *Cell.* Oct 6 2016;167(2):341–354.e12. doi:10.1016/j.cell.2016.08.071

105 Ruiz-Jimenez JM, Santpere G. The impact of human accelerated regions on neuronal development. *Trends Genet.* Jun 2025;41(6):459–461. doi:10.1016/j.tig.2025.03.005

106 Cheng J, Novati G, Pan J, et al. Accurate proteome-wide missense variant effect prediction with AlphaMissense. *Science.* Sep 22 2023;381(6664):eadg7492. doi:10.1126/science.adg7492

107 Zlotorynski E. Nucleosomes as blueprints of genome architecture. *Nat Rev Mol Cell Biol.* Jul 2025;26(7):500. doi:10.1038/s41580-025-00866-9

108 Garcia-Cazorla A, Oyarzabal A, Saudubray JM, Martinelli D, Dionisi-Vici C. Genetic disorders of cellular trafficking. *Trends Genet.* Jul 2022;38(7):724–751. doi:10.1016/j.tig.2022.02.012

109 Zschocke J, Byers PH, Wilkie AOM. Mendelian inheritance revisited: dominance and recessiveness in medical genetics. *Nat Rev Genet.* Jul 2023;24(7):442–463. doi:10.1038/s41576-023-00574-0

110 Loda A, Collombet S, Heard E. Gene regulation in time and space during X-chromosome inactivation. *Nat Rev Mol Cell Biol.* Apr 2022;23(4):231–249. doi:10.1038/s41580-021-00438-7

111 Isles AR. The contribution of imprinted genes to neurodevelopmental and neuropsychiatric disorders. *Transl Psychiatry.* May 21 2022;12(1):210. doi:10.1038/s41398-022-01972-4

112 Fabo T, Khavari P. Functional characterization of human genomic variation linked to polygenic diseases. *Trends Genet.* Jun 2023;39(6):462–490. doi:10.1016/j.tig.2023.02.014

113 Mitchell KJ. *Innate: How the Wiring of Our Brains Shapes Who We Are.* Princeton University Press; 2018. 293 pages

114 Friedman NP, Banich MT, Keller MC. Twin studies to GWAS: there and back again. *Trends Cognit Sci.* Oct 2021;25(10):855–869. doi:10.1016/j.tics.2021.06.007

115 Assary E, Coleman JRI, Hemani G, et al. Genetics of monozygotic twins reveals the impact of environmental sensitivity on psychiatric and neurodevelopmental phenotypes. *Nat Hum Behav.* Jun 10 2025;9:1683–1696. doi:10.1038/s41562-025-02193-7

116 Tam V, Patel N, Turcotte M, Bosse Y, Pare G, Meyre D. Benefits and limitations of genome-wide association studies. *Nat Rev Genet.* Aug 2019;20(8):467–484. doi:10.1038/s41576-019-0127-1

117 Qi T, Song L, Guo Y, Chen C, Yang J. From genetic associations to genes: methods, applications, and challenges. *Trends Genet.* Aug 2024;40(8):642–667. doi:10.1016/j.tig.2024.04.008

118 Kachuri L, Chatterjee N, Hirbo J, et al. Principles and methods for transferring polygenic risk scores across global populations. *Nat Rev Genet.* Jan 2024;25(1):8–25. doi:10.1038/s41576-023-00637-2

119 Clark H, Coll-Seck AM, Banerjee A, et al. A future for the world's children? A WHO–UNICEF–Lancet Commission. *The Lancet.* 2020;395:605–658.

120 Sanderson E, Glymour MM, Holmes MV, et al. Mendelian randomization. *Nat Rev Methods Primers.* Feb 10 2022:2. doi:10.1038/s43586-021-00092-5

121 Fitzgerald M. What do we really know about newborn infant pain? *Exp Physiol.* Dec 2015;100(12):1451–1457. doi:10.1113/EP085134

122 Rupawala M, Bucsea O, Laudiano-Dray MP, et al. A developmental shift in habituation to pain in human neonates. *Curr Biol.* Apr 24 2023;33(8):1397–1406.e5. doi:10.1016/j.cub.2023.02.071

123 Fenn-Moltu S, Fitzgibbon SP, Ciarrusta J, et al. Development of neonatal brain functional centrality and alterations associated with preterm birth. *Cereb Cortex.* Apr 25 2023;33(9):5585–5596. doi:10.1093/cercor/bhac444

124 Hee Chung E, Chou J, Brown KA. Neurodevelopmental outcomes of preterm infants: a recent literature review. *Transl Pediatr.* Feb 2020;9(Suppl 1):S3–S8. doi:10.21037/tp.2019.09.10

125 Kratimenos P, Sanidas G, Simonti G, Byrd C, Gallo V. The shifting landscape of the preterm brain. *Neuron.* Jul 9 2025;113(13):2042–2064. doi:10.1016/j.neuron.2025.03.024

126 Thalhammer M, Seidlitz J, Neubauer A, et al. Heterogeneous, temporally consistent, and plastic brain development after preterm birth. *Nat Commun.* Sep 12 2025;16(1):8269. doi:10.1038/s41467-025-63967-1

127 Ene D, Der G, Fletcher-Watson S, et al. Associations of socioeconomic deprivation and preterm birth with speech, language, and communication concerns among children aged 27 to 30 months. *JAMA Netw Open.* 2019;2:e1911027.

128 Abdellaoui A, Dolan CV, Verweij KJH, Nivard MG. Gene-environment correlations across geographic regions affect genome-wide association studies. *Nat Genet.* Sep 2022;54(9):1345–1354. doi:10.1038/s41588-022-01158-0

129 Mathan J, Maximino-Pinheiro M, He Q, et al. Effects of parental socioeconomic status on offspring's fetal neurodevelopment. *Cereb Cortex.* Nov 5 2024;34(11). doi:10.1093/cercor/bhae443

130 Maitre L, Bustamante M, Hernandez-Ferrer C, et al. Multi-omics signatures of the human early life exposome. *Nat Commun*. Nov 21 2022;13:7024. doi:10.1038/s41467-022-34422-2 ISSN:1

131 Kurbatfinski S, Dosani A, Dewey DM, Letourneau N. Proposed physiological mechanisms underlying the association between adverse childhood experiences and mental health conditions: a narrative review. *Children (Basel)*. Sep 12 2024;11(9). doi:10.3390/children11091112

132 Benjamin TA, Bennett MM, Davis KE. Environmental and interpersonal factors impact the developing brain. *J Neurosci*. Jul 31 2024;44(31). doi:10.1523/JNEUROSCI.0361-24.2024

133 Blakemore SJ. Adolescence and mental health. *Lancet*. May 18 2019;393(10185):2030–2031. doi:10.1016/S0140-6736(19)31013-X

134 Miller J, Mills KL, Vuorre M, Orben A, Przybylski AK. Impact of digital screen media activity on functional brain organization in late childhood: evidence from the ABCD study. *Cortex*. Dec 2023;169:290–308. doi:10.1016/j.cortex.2023.09.009

135 Orben A, Blakemore SJ. How social media affects teen mental health: a missing link. *Nature*. Feb 2023;614(7948):410–412. doi:10.1038/d41586-023-00402-9

136 Lengersdorff LL, Wagner IC, Mittmann G, et al. Neuroimaging and behavioral evidence that violent video games exert no negative effect on human empathy for pain and emotional reactivity to violence. *Elife*. Nov 17 2023;12. doi:10.7554/eLife.84951

137 Wang J, Wang Y, Ou Q, Yang S, Jing J, Fang J. Computer gaming alters resting-state brain networks, enhancing cognitive and fluid intelligence in players: evidence from brain imaging-derived phenotypes-wide Mendelian randomization. *Cereb Cortex*. Mar 1 2024;34(3). doi:10.1093/cercor/bhae061

138 Cryan JF, O'Riordan KJ, Cowan CSM, et al. The microbiota-gut-brain axis. *Physiol Rev*. Oct 1 2019;99(4):1877–2013. doi:10.1152/physrev.00018.2018

139 Ahrens AP, Hyotylainen T, Petrone JR, et al. Infant microbes and metabolites point to childhood neurodevelopmental disorders. *Cell*. Apr 11 2024;187(8):1853–1873.e15. doi:10.1016/j.cell.2024.02.035

140 Coley EJL, Hsiao EY. Malnutrition and the microbiome as modifiers of early neurodevelopment. *Trends Neurosci*. Sep 2021;44(9):753–764. doi:10.1016/j.tins.2021.06.004

141 Lepeule J. Child brains respond to climate change. *Nat Clim Change*. Jul 2024;14(7):679–680. doi:10.1038/s41558-024-02029-8

142 Hussein IR, Magbooli A, Huwait E, et al. Genome wide array-CGH and qPCR analysis for the identification of genome defects in Williams' syndrome patients in Saudi Arabia. *Mol Cytogenet*. 2016;9:65.

143 Bates GP. History of genetic disease: the molecular genetics of Huntington disease – a history. *Nat Rev Genet*. Oct 2005;6(10):766–773. doi:10.1038/nrg1686

144 Rood JE, Regev A. The legacy of the Human Genome Project. *Science*. Sep 24 2021;373(6562):1442–1443. doi:10.1126/science.abl5403

145 Deciphering Developmental Disorders S. Prevalence and architecture of de novo mutations in developmental disorders. *Nature*. Feb 23 2017;542(7642):433–438. doi:10.1038/nature21062

146 Garrity M, Kavus H, Rojas-Vasquez M, et al. Neurodevelopmental phenotypes in individuals with pathogenic variants in CHAMP1. *Cold Spring Harb Mol Case Stud.* Aug 2021;7(4):a006092. doi:10.1101/mcs.a006092

147 Sambo D, Goldman D. Genetic influences on fetal alcohol spectrum disorder. *Genes (Basel).* Jan 12 2023;14(1). doi:10.3390/genes14010195

148 Little WJ. The classic: hospital for the cure of deformities: course of lectures on the deformities of the human frame. 1843. *Clin Orthop Relat Res.* May 2012;470(5):1252–1256. doi:10.1007/s11999-012-2302-y

149 Sadowska M, Sarecka-Hujar B, Kopyta I. Cerebral palsy: current opinions on definition, epidemiology, risk factors, classification and treatment options. *Neuropsychiatr Dis Treat.* 2020;16:1505–1518. doi:10.2147/NDT.S235165

150 Fehlings DL, Zarrei M, Engchuan W, et al. Comprehensive whole-genome sequence analyses provide insights into the genomic architecture of cerebral palsy. *Nat Genet.* Apr 2024;56(4):585–594. doi:10.1038/s41588-024-01686-x

151 Pearson H. Autism is on the rise: what's really behind the increase? *Nature.* Aug 2025;644(8078):860–863. doi:10.1038/d41586-025-02636-1

152 Hours C, Recasens C, Baleyte JM. ASD and ADHD comorbidity: what are we talking about? *Front Psychiatry.* 2022;13:837424. doi:10.3389/fpsyt.2022.837424

153 Rolland T, Cliquet F, Anney RJL, et al. Phenotypic effects of genetic variants associated with autism. *Nat Med.* Jul 2023;29(7):1671–1680. doi:10.1038/s41591-023-02408-2

154 Willsey HR, Willsey AJ, Wang B, State MW. Genomics, convergent neuroscience and progress in understanding autism spectrum disorder. *Nat Rev Neurosci.* Jun 2022;23(6):323–341. doi:10.1038/s41583-022-00576-7

155 Wisniowiecka-Kowalnik B, Nowakowska BA. Genetics and epigenetics of autism spectrum disorder-current evidence in the field. *J Appl Genet.* Feb 2019;60(1):37–47. doi:10.1007/s13353-018-00480-w

156 Vorstman JAS, Scherer SW. Contemplating syndromic autism. *Genet Med.* Oct 2023;25(10):100919. doi:10.1016/j.gim.2023.100919

157 Patowary A, Zhang P, Jops C, et al. Developmental isoform diversity in the human neocortex informs neuropsychiatric risk mechanisms. *Science.* May 24 2024;384(6698):eadh7688. doi:10.1126/science.adh7688

158 Geschwind DH. Autism: many genes, common pathways? *Cell.* Oct 31 2008;135(3): 391–395. doi:10.1016/j.cell.2008.10.016

159 Roman-Urrestarazu A, van Kessel R, Allison C, Matthews FE, Brayne C, Baron-Cohen S. Association of race/ethnicity and social disadvantage with autism prevalence in 7 million school children in England. *JAMA Pediatr.* Jun 1 2021;175(6):e210054. doi:10.1001/jamapediatrics.2021.0054

160 Vuong HE, Hsiao EY. Emerging roles for the gut microbiome in autism spectrum disorder. *Biol Psychiatry.* Mar 1 2017;81(5):411–423. doi:10.1016/j.biopsych.2016.08.024

161 Deer B. *The Doctor Who Fooled the World : Science, Deception, and the War on Vaccines.* vol. 1. Johns Hopkins University Press; 2020. online resource

162 Mattheisen M, Grove J, Als TD, et al. Identification of shared and differentiating genetic architecture for autism spectrum disorder, attention-deficit hyperactivity disorder and case subgroups. *Nat Genet.* Oct 2022;54(10):1470–1478. doi:10.1038/s41588-022-01171-3

163 Lin BD, Pries LK, Sarac HS, et al. Nongenetic factors associated with psychotic experiences among UK biobank participants: exposome-wide analysis and Mendelian randomization analysis. *JAMA Psychiatry*. Sep 1 2022;79(9):857–868. doi:10.1001/jamapsychiatry.2022.1655

164 Borcuk C, Parihar M, Sportelli L, et al. Network-wide risk convergence in gene co-expression identifies reproducible genetic hubs of schizophrenia risk. *Neuron*. Nov 6 2024;112(21):3551–3566.e6. doi:10.1016/j.neuron.2024.08.005

165 O'Connell KS, Koromina M, van der Veen T, et al. Genomics yields biological and phenotypic insights into bipolar disorder. *Nature*. Mar 2025;639(8056):968–975. doi:10.1038/s41586-024-08468-9

166 Lee S, McAfee JC, Lee J, et al. Massively parallel reporter assay investigates shared genetic variants of eight psychiatric disorders. *Cell*. Mar 6 2025;188(5):1409–1424.e21. doi:10.1016/j.cell.2024.12.022

167 Gordovez FJA, McMahon FJ. The genetics of bipolar disorder. *Mol Psychiatry*. Mar 2020;25(3):544–559. doi:10.1038/s41380-019-0634-7

168 Wu Y, Cao H, Baranova A, et al. Multi-trait analysis for genome-wide association study of five psychiatric disorders. *Transl Psychiatry*. Jun 30 2020;10(1):209. doi:10.1038/s41398-020-00902-6

169 An JY, Kim Y. Genomic patterns in the schizophrenia brain. *Science*. Oct 11 2024; 386(6718):146–148. doi:10.1126/science.ads6781

170 Romero C, Werme J, Jansen PR, et al. Exploring the genetic overlap between twelve psychiatric disorders. *Nat Genet*. Dec 2022;54(12):1795–1802. doi:10.1038/s41588-022-01245-2

171 Cross-Disorder Group of the Psychiatric Genomics Consortium. Genomic relationships, novel loci, and pleiotropic mechanisms across eight psychiatric disorders. *Cell*. Dec 12 2019;179(7):1469–1482.e11. doi:10.1016/j.cell.2019.11.020

172 Xu J, Huckins LM. Psychiatric disorders: teasing apart genetic similarities and differences. *Trends Genet*. Jul 4 2025;41:640–642. doi:10.1016/j.tig.2025.06.012

173 Lin WD, Tsai FJ, Chou IC. Current understanding of the genetics of tourette syndrome. *Biom J*. Apr 2022;45(2):271–279. doi:10.1016/j.bj.2022.01.008

174 Meijsen J, Hu K, Wei D, et al. Shared genetic architecture between eating disorders, mental health conditions, and cardiometabolic diseases: a comprehensive population-wide study across two countries. *Nat Commun*. Jul 4 2025;16(1):6193. doi:10.1038/s41467-025-61496-5

175 Grotzinger AD, Werme J, Peyrot WJ, et al. Mapping the genetic landscape across 14 psychiatric disorders. *Nature*. 2026;649(8096):406–415. doi:10.1038/s41586-025-09820-3

176 Klingler E, Francis F, Jabaudon D, Cappello S. Mapping the molecular and cellular complexity of cortical malformations. *Science*. Jan 22 2021;371(6527). doi:10.1126/science.aba4517

177 Kim S, Lee J, Koh IG, et al. An integrative single-cell atlas for exploring the cellular and temporal specificity of genes related to neurological disorders during human brain development. *Exp Mol Med*. Oct 2024;56(10):2271–2282. doi:10.1038/s12276-024-01328-6

178 Di Liberto V, Mudo G, Belluardo N. Crosstalk between receptor tyrosine kinases (RTKs) and G protein-coupled receptors (GPCR) in the brain: focus on heteroreceptor complexes and related functional neurotrophic effects. *Neuropharmacology*. Jul 1 2019;152:67–77. doi:10.1016/j.neuropharm.2018.11.018

179 Mill P, Christensen ST, Pedersen LB. Primary cilia as dynamic and diverse signalling hubs in development and disease. *Nat Rev Genet*. Jul 2023;24(7):421–441. doi:10.1038/s41576-023-00587-9

180 Jurisch-Yaksi N, Wachten D, Gopalakrishnan J. The neuronal cilium – a highly diverse and dynamic organelle involved in sensory detection and neuromodulation. *Trends Neurosci*. May 2024;47(5):383–394. doi:10.1016/j.tins.2024.03.004

181 Hilgendorf KI, Myers BR, Reiter JF. Emerging mechanistic understanding of cilia function in cellular signalling. *Nat Rev Mol Cell Biol*. Jul 2024;25(7):555–573. doi:10.1038/s41580-023-00698-5

182 Wang L, Guo Q, Acharya S, et al. Primary cilia signaling in astrocytes mediates development and regional-specific functional specification. *Nat Neurosci*. Sep 2024;27(9):1708–1720. doi:10.1038/s41593-024-01726-z

183 Matlik K, Govek EE, Hatten ME. Histone bivalency in CNS development. *Genes Dev*. Apr 1 2025;39(7–8):428–444. doi:10.1101/gad.352306.124

184 Macrae TA, Fothergill-Robinson J, Ramalho-Santos M. Regulation, functions and transmission of bivalent chromatin during mammalian development. *Nat Rev Mol Cell Biol*. Jan 2023;24(1):6–26. doi:10.1038/s41580-022-00518-2

185 Ernst C, Jefri M. Epigenetic priming in neurodevelopmental disorders. *Trends Mol Med*. Dec 2021;27(12):1106–1114. doi:10.1016/j.molmed.2021.09.005

186 Qu L, Wang L, Ji H, et al. Toxic mechanism and biological detoxification of fumonisins. *Toxins (Basel)*. Mar 1 2022;14(3). doi:10.3390/toxins14030182

187 Yoshino Y, Fang Z, Chiba N. The centrosome: a critical hub for cell cycle regulation. *Trends Cell Biol*. Jun 2024;34(6):437–439. doi:10.1016/j.tcb.2024.04.003

188 Colas P. Cyclin-dependent kinases and rare developmental disorders. *Orphanet J Rare Dis*. Aug 6 2020;15(1):203. doi:10.1186/s13023-020-01472-y

189 Derderian C, Canales GI, Reiter JF. Seriously cilia: a tiny organelle illuminates evolution, disease, and intercellular communication. *Dev Cell*. Aug 7 2023;58(15):1333–1349. doi:10.1016/j.devcel.2023.06.013

190 Deretic J, Odabasi E, Firat-Karalar EN. The multifaceted roles of microtubule-associated proteins in the primary cilium and ciliopathies. *J Cell Sci*. Dec 1 2023;136(23). doi:10.1242/jcs.261148

191 Jewett CE, McCurdy BL, O'Toole ET, et al. Trisomy 21 induces pericentrosomal crowding delaying primary ciliogenesis and mouse cerebellar development. *Elife*. Jan 19 2023;12. doi:10.7554/eLife.78202

192 Krontira AC, Cruceanu C, Dony L, et al. Human cortical neurogenesis is altered via glucocorticoid-mediated regulation of ZBTB16 expression. *Neuron*. May 1 2024;112(9): 1426–1443.e11. doi:10.1016/j.neuron.2024.02.005

193 Singh K, Lau CK, Manigrasso G, Gama JB, Gassmann R, Carter AP. Molecular mechanism of dynein-dynactin complex assembly by LIS1. *Science*. Mar 29 2024;383(6690):eadk8544. doi:10.1126/science.adk8544

194 Sebastien M, Paquette AL, Prowse ENP, Hendricks AG, Brouhard GJ. Doublecortin restricts neuronal branching by regulating tubulin polyglutamylation. *Nat Commun*. Feb 18 2025;16(1):1749. doi:10.1038/s41467-025-56951-2

195 Friocourt G, Kanatani S, Tabata H, et al. Cell-autonomous roles of ARX in cell proliferation and neuronal migration during corticogenesis. *J Neurosci*. May 28 2008;28(22):5794–5805. doi:10.1523/JNEUROSCI.1067-08.2008

196 Birey F, Li MY, Gordon A, et al. Dissecting the molecular basis of human interneuron migration in forebrain assembloids from Timothy syndrome. *Cell Stem Cell*. Feb 3 2022;29(2):248–264.e7. doi:10.1016/j.stem.2021.11.011

197 Weninger A, Arlotta P. A family portrait of human brain cells. *Science*. Oct 13 2023;382(6667):168–169. doi:10.1126/science.adk4857

198 Siletti K, Hodge R, Mossi Albiach A, et al. Transcriptomic diversity of cell types across the adult human brain. *Science*. Oct 13 2023;382(6667):eadd7046. doi:10.1126/science.add7046

199 Braun E, Danan-Gotthold M, Borm LE, et al. Comprehensive cell atlas of the first-trimester developing human brain. *Science*. Oct 13 2023;382(6667):eadf1226. doi:10.1126/science.adf1226

200 Tyser RCV, Mahammadov E, Nakanoh S, Vallier L, Scialdone A, Srinivas S. Single-cell transcriptomic characterization of a gastrulating human embryo. *Nature*. Dec 2021; 600(7888):285–289. doi:10.1038/s41586-021-04158-y

201 Berg J, Sorensen SA, Ting JT, et al. Human neocortical expansion involves glutamatergic neuron diversification. *Nature*. Oct 2021;598(7879):151–158. doi:10.1038/s41586-021-03813-8

202 Zeng B, Liu Z, Lu Y, et al. The single-cell and spatial transcriptional landscape of human gastrulation and early brain development. *Cell Stem Cell*. Jun 1 2023;30(6):851–866.e7. doi:10.1016/j.stem.2023.04.016

203 Clarence T, Bendl J, Cao X, et al. Multiomic single-cell profiling identifies critical regulators of postnatal brain. *Nat Genet*. Mar 2025;57(3):591–603. doi:10.1038/s41588-025-02083-8

204 Chan WK, Negro D, Munro VM, Marshall H, et al. Loss of PAX6 alters the excitatory/inhibitory neuronal ratio in human cerebral organoids. *bioRxiv*. 2023. doi:10.1101/2023.07.31.551262

205 Fleck JS, Jansen SMJ, Wollny D, et al. Inferring and perturbing cell fate regulomes in human brain organoids. *Nature*. 2023;621:365–372.

206 Mauriac SA, Hien YE, Bird JE, et al. Defective Gpsm2/Galpha(i3) signalling disrupts stereocilia development and growth cone actin dynamics in Chudley-McCullough syndrome. *Nat Commun*. Apr 7 2017;8:14907. doi:10.1038/ncomms14907

207 Engle EC. Human genetic disorders of axon guidance. *Cold Spring Harbor Perspect Biol*. Mar 2010;2(3):a001784. doi:10.1101/cshperspect.a001784

208 Guo J, Otis JM, Suciu SK, et al. Primary cilia signaling promotes axonal tract development and is disrupted in Joubert syndrome-related disorders models. *Dev Cell*. Dec 16 2019;51(6):759–774.e5. doi:10.1016/j.devcel.2019.11.005

209 Rastogi M, Bartolucci M, Nanni M, et al. Integrative multi-omic analysis reveals conserved cell-projection deficits in human Down syndrome brains. *Neuron*. Aug 7 2024;112(15):2503–2523.e10. doi:10.1016/j.neuron.2024.05.002

210 Mukherjee K, LaConte LEW, Srivastava S. The non-linear path from gene dysfunction to genetic disease: lessons from the MICPCH mouse model. *Cells*. Mar 28 2022;11(7). doi:10.3390/cells11071131

211 Washbourne P. Synapse assembly and neurodevelopmental disorders. *Neuropsychopharmacology*. Jan 2015;40(1):4–15. doi:10.1038/npp.2014.163

212 Clapier CR, Iwasa J, Cairns BR, Peterson CL. Mechanisms of action and regulation of ATP-dependent chromatin-remodelling complexes. *Nat Rev Mol Cell Biol.* Jul 2017;18(7):407–422. doi:10.1038/nrm.2017.26

213 Danovi S. RNU4-2 variants cause neurodevelopmental disorders. *Nat Genet.* Aug 2024;56(8):1541. doi:10.1038/s41588-024-01882-9

214 Greene D, Thys C, Berry IR, et al. Mutations in the U4 snRNA gene RNU4-2 cause one of the most prevalent monogenic neurodevelopmental disorders. *Nat Med.* Aug 2024;30(8):2165–2169. doi:10.1038/s41591-024-03085-5

215 Burns VF, Radford EJ. ReNU syndrome – a newly discovered prevalent neurodevelopmental disorder. *Trends Genet.* Nov 2024;40(11):914–916. doi:10.1016/j.tig.2024.09.005

216 Fiorenzani C, Mossa A, De Rubeis S. DEAD/DEAH-box RNA helicases shape the risk of neurodevelopmental disorders. *Trends Genet.* May 2025;41(5):437–449. doi:10.1016/j.tig.2024.12.006

217 Vasquez GE, Medinas DB, Urra H, Hetz C. Emerging roles of endoplasmic reticulum proteostasis in brain development. *Cells Dev.* Jun 2022;170:203781. doi:10.1016/j.cdev.2022.203781

218 Stanley P. Genetics of glycosylation in mammalian development and disease. *Nat Rev Genet.* Oct 2024;25(10):715–729. doi:10.1038/s41576-024-00725-x

219 Wisnovsky S, Bertozzi CR. Reading the glyco-code: new approaches to studying protein-carbohydrate interactions. *Curr Opin Struct Biol.* Aug 2022;75:102395. doi:10.1016/j.sbi.2022.102395

220 He Z, Li Y, Zhao X, Li B. Dravet syndrome: advances in etiology, clinical presentation, and treatment. *Epilepsy Res.* Dec 2022;188:107041. doi:10.1016/j.eplepsyres.2022.107041

221 Lo J, Forst AL, Warth R, Zdebik AA. EAST/SeSAME syndrome and beyond: the spectrum of Kir4.1- and Kir5.1-associated channelopathies. *Front Physiol.* 2022;13:852674. doi:10.3389/fphys.2022.852674

222 Garner CC, Wetmore DZ. Synaptic pathology of Down syndrome. *Adv Exp Med Biol.* 2012;970:451–468. doi:10.1007/978-3-7091-0932-8_20

223 Russo ML, Sousa AMM, Bhattacharyya A. Consequences of trisomy 21 for brain development in Down syndrome. *Nat Rev Neurosci.* Nov 2024;25(11):740–755. doi:10.1038/s41583-024-00866-2

224 Cases O, Vitalis T, Seif I, De Maeyer E, Sotelo C, Gaspar P. Lack of barrels in the somatosensory cortex of monoamine oxidase A-deficient mice: role of a serotonin excess during the critical period. *Neuron.* Feb 1996;16(2):297–307. doi:10.1016/s0896-6273(00)80048-3

225 Kaizuka T, Takumi T. Postsynaptic density proteins and their involvement in neurodevelopmental disorders. *J Biochem.* Jun 1 2018;163(6):447–455. doi:10.1093/jb/mvy022

226 Kennedy MB. The protein biochemistry of the postsynaptic density in glutamatergic synapses mediates learning in neural networks. *Biochemistry.* Jul 10 2018;57(27):4005–4009. doi:10.1021/acs.biochem.8b00496

227 Vermaercke B, Iwata R, Wierda K, et al. SYNGAP1 deficiency disrupts synaptic neoteny in xenotransplanted human cortical neurons in vivo. *Neuron.* Sep 25 2024;112(18):3058–3068.e8. doi:10.1016/j.neuron.2024.07.007

228 Libe-Philippot B, Iwata R, Recupero AJ, et al. Synaptic neoteny of human cortical neurons requires species-specific balancing of SRGAP2-SYNGAP1 cross-inhibition. *Neuron*. Nov 6 2024;112(21):3602–3617.e9. doi:10.1016/j.neuron.2024.08.021

229 Araki Y, Rajkovich KE, Gerber EE, et al. SynGAP regulates synaptic plasticity and cognition independently of its catalytic activity. *Science*. Mar 2024;383(6686):eadk1291. doi:10.1126/science.adk1291

230 Birtele M, Del Dosso A, Xu T, et al. Non-synaptic function of the autism spectrum disorder-associated gene SYNGAP1 in cortical neurogenesis. *Nat Neurosci*. Dec 2023;26(12):2090–2103. doi:10.1038/s41593-023-01477-3

231 Liu Y, Whitfield TW, Bell GW, et al. Exploring the complexity of MECP2 function in Rett syndrome. *Nat Rev Neurosci*. Jul 2025;26(7):379–398. doi:10.1038/s41583-025-00926-1

232 Liu Y, Flamier A, Bell GW, et al. MECP2 directly interacts with RNA polymerase II to modulate transcription in human neurons. *Neuron*. Jun 19 2024;112(12):1943–1958.e10. doi:10.1016/j.neuron.2024.04.007

233 Tillotson R, Bird A. The molecular basis of MeCP2 function in the brain. *J Mol Biol*. Mar 13 2020;432(6):1602–1623. doi:10.1016/j.jmb.2019.10.004

234 Guy J, Gan J, Selfridge J, Cobb S, Bird A. Reversal of neurological defects in a mouse model of Rett syndrome. *Science*. Feb 23 2007;315(5815):1143–1147. doi:10.1126/science.1138389

235 Mellios N, Feldman DA, Sheridan SD, et al. MeCP2-regulated miRNAs control early human neurogenesis through differential effects on ERK and AKT signaling. *Mol Psychiatry*. Apr 2018;23(4):1051–1065. doi:10.1038/mp.2017.86

236 Darnell JC, Van Driesche SJ, Zhang C, et al. FMRP stalls ribosomal translocation on mRNAs linked to synaptic function and autism. *Cell*. Jul 22 2011;146(2):247–261. doi:10.1016/j.cell.2011.06.013

237 Ramesh V, Tsoukala E, Kougianou I, et al. The fragile X messenger ribonucleoprotein 1 regulates the morphology and maturation of human and rat oligodendrocytes. *Glia*. Jun 2025;73(6):1203–1220. doi:10.1002/glia.24680

238 Bear MF, Huber KM, Warren ST. The mGluR theory of fragile X mental retardation. *Trends Neurosci*. Jul 2004;27(7):370–377. doi:10.1016/j.tins.2004.04.009

239 Louros SR, Seo SS, Maio B, et al. Excessive proteostasis contributes to pathology in fragile X syndrome. *Neuron*. Feb 15 2023;111(4):508–525.e7. doi:10.1016/j.neuron.2022.11.012

240 Auerbach BD, Osterweil EK, Bear MF. Mutations causing syndromic autism define an axis of synaptic pathophysiology. *Nature*. Nov 23 2011;480(7375):63–68. doi:10.1038/nature10658

241 Shen M, Sirois CL, Guo Y, et al. Species-specific FMRP regulation of RACK1 is critical for prenatal cortical development. *Neuron*. Dec 20 2023;111(24):3988–4005.e11. doi:10.1016/j.neuron.2023.09.014

242 Hagemann TL. Alexander disease: models, mechanisms, and medicine. *Curr Opin Neurobiol*. Feb 2022;72:140–147. doi:10.1016/j.conb.2021.10.002

243 Gan Y, Wei Z, Liu C, Li G, Feng Y, Deng Y. Solute carrier transporter disease and developmental and epileptic encephalopathy. *Front Neurol*. 2022;13:1013903. doi:10.3389/fneur.2022.1013903

244 Suomalainen A, Nunnari J. Mitochondria at the crossroads of health and disease. *Cell*. May 23 2024;187(11):2601–2627. doi:10.1016/j.cell.2024.04.037

245 Rahman S, Copeland WC. POLG-related disorders and their neurological manifestations. *Nat Rev Neurol*. 2019;15(1):40–52. doi:10.1038/s41582-018-0101-0

246 Plotegher N, Filadi R, Pizzo P, Duchen MR. Excitotoxicity revisited: mitochondria on the verge of a nervous breakdown. *Trends Neurosci*. May 2021;44(5):342–351. doi:10.1016/j.tins.2021.01.001

247 Cartes-Saavedra B, Ghosh A, Hajnoczky G. The roles of mitochondria in global and local intracellular calcium signalling. *Nat Rev Mol Cell Biol*. Jun 2025;26(6):456–475. doi:10.1038/s41580-024-00820-1

248 Jiang D, Li T, Guo C, Tang TS, Liu H. Small molecule modulators of chromatin remodeling: from neurodevelopment to neurodegeneration. *Cell Biosci*. Jan 16 2023;13(1):10. doi:10.1186/s13578-023-00953-4

249 Murtaza N, Cheng AA, Brown CO, et al. Neuron-specific protein network mapping of autism risk genes identifies shared biological mechanisms and disease-relevant pathologies. *Cell Rep*. Nov 22 2022;41(8):111678. doi:10.1016/j.celrep.2022.111678

250 Contractor A, Ethell IM, Portera-Cailliau C. Cortical interneurons in autism. *Nat Neurosci*. Dec 2021;24(12):1648–1659. doi:10.1038/s41593-021-00967-6

251 Tang X, Jaenisch R, Sur M. The role of GABAergic signalling in neurodevelopmental disorders. *Nat Rev Neurosci*. May 2021;22(5):290–307. doi:10.1038/s41583-021-00443-x

252 Galvez-Contreras AY, Zarate-Lopez D, Torres-Chavez AL, Gonzalez-Perez O. Role of oligodendrocytes and myelin in the pathophysiology of autism spectrum disorder. *Brain Sci*. Dec 8 2020;10(12). doi:10.3390/brainsci10120951

253 Gandal MJ, Haney JR, Wamsley B, et al. Broad transcriptomic dysregulation occurs across the cerebral cortex in ASD. *Nature*. Nov 2022;611(7936):532–539. doi:10.1038/s41586-022-05377-7

254 Strom NI, Burton CL, Iyegbe C, et al. Genome-Wide Association Study of Obsessive-Compulsive Symptoms including 33,943 individuals from the general population. *Mol Psychiatry*. Sep 2024;29(9):2714–2723. doi:10.1038/s41380-024-02489-6

255 Robbins TW, Banca P, Belin D. From compulsivity to compulsion: the neural basis of compulsive disorders. *Nat Rev Neurosci*. May 2024;25(5):313–333. doi:10.1038/s41583-024-00807-z

256 Perera MPN, Gotsis ES, Bailey NW, Fitzgibbon BM, Fitzgerald PB. Exploring functional connectivity in large-scale brain networks in obsessive-compulsive disorder: a systematic review of EEG and fMRI studies. *Cereb Cortex*. Aug 1 2024;34(8). doi:10.1093/cercor/bhae327

257 Paryani F, Kwon JS, Ng CW, et al. Multi-omic analysis of Huntington's disease reveals a compensatory astrocyte state. *Nat Commun*. Aug 8 2024;15(1):6742. doi:10.1038/s41467-024-50626-0

258 Handsaker RE, Kashin S, Reed NM, et al. Long somatic DNA-repeat expansion drives neurodegeneration in Huntington's disease. *Cell*. Feb 6 2025;188(3):623–639 e19. doi:10.1016/j.cell.2024.11.038

259 Barnat M, Capizzi M, Aparicio E, et al. Huntington's disease alters human neurodevelopment. *Science*. Aug 14 2020;369(6505):787–793. doi:10.1126/science.aax3338

260 Bede P, Siah WF, McKenna MC, Shing LH, S. Consideration of C9orf72-associated ALS-FTD as a neurodevel-opmental disorder: insights from neuroimaging. *J Neurol Neurosurg Psychiatry*. Nov 2020;91(11):1138. doi:10.1136/jnnp-2020-324416

261 Laffita-Mesa JM, Paucar M, Svenningsson P. Ataxin-2 gene: a powerful modulator of neurological disorders. *Curr Opin Neurol.* Aug 1 2021;34(4):578–588.

262 Naassila M, Pierrefiche O. GluN2B subunit of the NMDA receptor: the keystone of the effects of alcohol during neurodevelopment. *Neurochem Res.* Jan 2019;44(1):78–88. doi:10.1007/s11064-017-2462-y

263 Owen MJ, O'Donovan MC, Thapar A, Craddock N. Neurodevelopmental hypothesis of schizophrenia. *Br J Psychiatry.* Mar 2011;198(3):173–175. doi:10.1192/bjp.bp.110.084384

264 Duncan LE, Li T, Salem M, et al. Mapping the cellular etiology of schizophrenia and complex brain phenotypes. *Nat Neurosci.* Feb 2025;28(2):248–258. doi:10.1038/s41593-024-01834-w

265 Sullivan PF, Yao S, Hjerling-Leffler J. Schizophrenia genomics: genetic complexity and functional insights. *Nat Rev Neurosci.* Sep 2024;25(9):611–624. doi:10.1038/s41583-024-00837-7

266 Ruzicka WB, Mohammadi S, Fullard JF. Single-cell multi-cohort dissection of the schizophrenia transcriptome. *Science.* May 24 2024;384:eadg5136. doi:10.1126/science.adg5136

267 Girdhar K, Hoffman GE, Bendl J, et al. Chromatin domain alterations linked to 3D genome organization in a large cohort of schizophrenia and bipolar disorder brains. *Nat Neurosci.* Apr 2022;25(4):474–483. doi:10.1038/s41593-022-01032-6

268 Trubetskoy V, Pardinas AF, Qi T, et al. Mapping genomic loci implicates genes and synaptic biology in schizophrenia. *Nature.* Apr 2022;604(7906):502–508. doi:10.1038/s41586-022-04434-5

269 Singh T, Poterba T, Curtis D, et al. Rare coding variants in ten genes confer substantial risk for schizophrenia. *Nature.* Apr 2022;604(7906):509–516. doi:10.1038/s41586-022-04556-w

270 Grotzinger AD, Mallard TT, Akingbuwa WA, et al. Genetic architecture of 11 major psychiatric disorders at biobehavioral, functional genomic and molecular genetic levels of analysis. *Nat Genet.* May 2022;54(5):548–559. doi:10.1038/s41588-022-01057-4

271 de Zwarte SMC, Brouwer RM, Agartz I, et al. The association between familial risk and brain abnormalities is disease specific: an ENIGMA-relatives study of schizophrenia and bipolar disorder. *Biol Psychiatry.* Oct 1 2019;86(7):545–556. doi:10.1016/j.biopsych.2019.03.985

272 Ligneul R, Mainen ZF. Serotonin. *Curr Biol.* Dec 4 2023;33(23):R1216–R1221. doi:10.1016/j.cub.2023.09.068

273 Lynch CJ, Elbau IG, Ng T. Frontostriatal salience network expansion in individuals in depression. *Nature.* Sep 2024;633(8030):624–633. doi:10.1038/s41586-024-07805-2

274 Jesulola E, Micalos P, Baguley IJ. Understanding the pathophysiology of depression: From monoamines to the neurogenesis hypothesis model – are we there yet? *Behav Brain Res.* Apr 2 2018;341:79–90. doi:10.1016/j.bbr.2017.12.025

275 Catlin JP, Schaner Tooley CE. Exploring potential developmental origins of common neurodegenerative disorders. *Biochem Soc Trans.* Jun 26 2024;52(3):1035–1044. doi:10.1042/BST20230422

276 Niedowicz DM, Nelson PT, Paul Murphy M. Alzheimer's disease: pathological mechanisms and recent insights. *Curr Neuropharmacol.* 2011;9:674–684.

277 Martini AC, Gross TJ, Head E, Mapstone M. Beyond amyloid: immune, cerebrovascular, and metabolic contributions to Alzheimer disease in people with Down syndrome. *Neuron.* Jul 6 2022;110(13):2063–2079. doi:10.1016/j.neuron.2022.04.001

278 Hurley EM, Mozolewski P, Dobrowolski R, Hsieh J. Familial Alzheimer's disease-associated PSEN1 mutations affect neurodevelopment through increased Notch signaling. *Stem Cell Rep.* Jul 11 2023;18(7):1516–1533. doi:10.1016/j.stemcr.2023.05.018

279 Meyer-Acosta KK, Diaz-Guerra E, Varma P, et al. APOE4 impacts cortical neurodevelopment and alters network formation in human brain organoids. *Stem Cell Rep.* Jul 8 2025;20(7):102537. doi:10.1016/j.stemcr.2025.102537

280 Arendt T, Stieler J, Ueberham U. Is sporadic Alzheimer's disease a developmental disorder? *J Neurochem.* Nov 2017;143(4):396–408. doi:10.1111/jnc.14036

281 Shabani K, Pigeon J, Benaissa Touil Zariouh M, et al. The temporal balance between self-renewal and differentiation of human neural stem cells requires the amyloid precursor protein. *Sci Adv.* Jun 16 2023;9(24):eadd5002. doi:10.1126/sciadv.add5002

282 Dauer W, Przedborski S. Parkinson's disease: mechanisms and models. *Neuron.* 2003;39:889–909.

283 Funayama M, Nishioka K, Li Y, Hattori N. Molecular genetics of Parkinson's disease: contributions and global trends. *J Hum Genet.* Mar 2023;68(3):125–130. doi:10.1038/s10038-022-01058-5

284 Barlow BK, Cory-Slechta DA, Richfield EK, Thiruchelvam M. The gestational environment and Parkinson's disease: evidence for neurodevelopmental origins of a neurodegenerative disorder. *Reprod Toxicol.* Apr-May 2007;23(3):457–470. doi:10.1016/j.reprotox.2007.01.007

285 Schwamborn JC. Is Parkinson's disease a neurodevelopmental disorder and will brain organoids help us to understand it? *Stem Cells Dev.* Jul 15 2018;27(14):968–975. doi:10.1089/scd.2017.0289

286 Kidd SA, Gong J, Massazza A, Bezgrebelna M, Zhang YL, Hajat S. Climate change and its implications for developing brains – In utero to youth: a scoping review. *J Clim Change Health.* Sep–Oct 2023;13:ARTN 100258. doi:10.1016/j.joclim.2023.100258

287 Gross SJ, Pletcher BA, Monaghan KG. Carrier screening in individuals of Ashkenazi Jewish descent. *Genet Med.* Jan 2008;10(1):54–56. doi:10.1097/GIM.0b013e31815f247c

288 Manor-Binyamini I, Abu-Ajaj O. Coping of siblings of children with developmental disabilities in the Bedouin community. *Res Dev Disabil.* May-Jun 2012;33(3):825–831. doi:10.1016/j.ridd.2011.11.016

289 Finkel RS, Hughes SH, Parker J, et al. Risdiplam for prenatal therapy of spinal muscular atrophy. *N Engl J Med.* Mar 13 2025;392(11):1138–1140. doi:10.1056/NEJMc2300802

290 Vihma H, Li K, Welton-Arndt A, et al. Ube3a unsilencer for the potential treatment of Angelman syndrome. *Nat Commun.* Jul 8 2024;15:5558. doi:10.1038/s41467-024-49788-8

291 Capalbo A, de Wert G, Mertes H, et al. Screening embryos for polygenic disease risk: a review of epidemiological, clinical, and ethical considerations. *Hum Reprod Update.* Oct 1 2024;30(5):529–557. doi:10.1093/humupd/dmae012

292 Raguram A, Banskota S, Liu DR. Therapeutic in vivo delivery of gene editing agents. *Cell.* Jul 21 2022;185(15):2806–2827. doi:10.1016/j.cell.2022.03.045

293 Flotte TR, Cataltepe O, Puri A, et al. AAV gene therapy for Tay-Sachs disease. *Nat Med.* Feb 2022;28(2):251–259. doi:10.1038/s41591-021-01664-4

294 Wang N, Lv L, Huang X, et al. Gene editing in monogenic autism spectrum disorder: animal models and gene therapies. *Front Mol Neurosci.* 2022;15:1043018. doi:10.3389/fnmol.2022.1043018

295 Panayotis N, Ehinger Y, Felix MS, Roux JC. State-of-the-art therapies for Rett syndrome. *Dev Med Child Neurol*. Feb 2023;65(2):162–170. doi:10.1111/dmcn.15383

296 Mullard A. FDA approves L-amino acid decarboxylase deficiency gene therapy. *Nat Rev Drug Discov*. Jan 2025;24(1):6. doi:10.1038/d41573-024-00184-3

297 Pearson TS, Gupta N, San Sebastian W, et al. Gene therapy for aromatic L-amino acid decarboxylase deficiency by MR-guided direct delivery of AAV2-AADC to midbrain dopaminergic neurons. *Nat Commun*. Jul 12 2021;12:4251. doi:10.1038/s41467-021-24524-8

298 Duan W, Urani E, Mattson MP. The potential of gene editing for Huntington's disease. *Trends Neurosci*. May 2023;46(5):365–376. doi:10.1016/j.tins.2023.02.005

299 Ledford H. Why CRISPR babies are still too risky - embryo studies highlight challenges. *Nature*. Mar 2023;615(7953):568–569. doi:10.1038/d41586-023-00756-0

300 Aussel C, Cathomen T, Fuster-Garcia C. The hidden risks of CRISPR/Cas: structural variations and genome integrity. *Nat Commun*. Aug 5 2025;16(1):7208. doi:10.1038/s41467-025-62606-z

301 Warrier V, Zhang X, Reed P, et al. Genetic correlates of phenotypic heterogeneity in autism. *Nat Genet*. Sep 2022;54(9):1293–1304. doi:10.1038/s41588-022-01072-5

302 Vandereyken K, Sifrim A, Thienpont B, Voet T. Methods and applications for single-cell and spatial multi-omics. *Nat Rev Genet*. Aug 2023;24(8):494–515. doi:10.1038/s41576-023-00580-2

303 Chung C, Yang X, Bae T, et al. Comprehensive multi-omic profiling of somatic mutations in malformations of cortical development. *Nat Genet*. Feb 2023;55(2):209–220. doi:10.1038/s41588-022-01276-9

304 Dorr A. Single-cell RNA-seq relates GWAS variants to disease risk. *Nat Biotechnol*. Nov 2022;40(11):1574. doi:10.1038/s41587-022-01570-1

305 Sumida TS, Hafler DA. Population genetics meets single-cell sequencing. *Science*. Apr 8 2022;376(6589):134–135. doi:10.1126/science.abq0426

306 Risgaard RD, Arachchilage KH, Knaack SA, et al. Molecular and cellular processes disrupted in the early postnatal Down syndrome prefrontal cortex. *bioRxiv*. Jul 4 2025. doi:10.1101/2025.06.30.662385

307 Vuong CK, Weber A, Seong P, et al. A single cell multi-omic analysis identifies molecular and gene-regulatory mechanisms dysregulated in the developing Down syndrome neocortex. *bioRxiv*. Jul 4 2025. doi:10.1101/2025.06.30.662136

308 Zhao X, Bhattacharyya A. Human models are needed for studying human neurodevelopmental disorders. *Am J Hum Genet*. Dec 6 2018;103(6):829–857. doi:10.1016/j.ajhg.2018.10.009

309 Sun J, Noss S, Banerjee D, Das M, Girirajan S. Strategies for dissecting the complexity of neurodevelopmental disorders. *Trends Genet*. Feb 2024;40(2):187–202. doi:10.1016/j.tig.2023.10.009

310 Kudithipudi D, Schuman C, Vineyard CM, et al. Neuromorphic computing at scale. *Nature*. Jan 2025;637(8047):801–812. doi:10.1038/s41586-024-08253-8

Index

b

c

h

t

v